Thyristor Control

F. F. MAZDA, M.PHIL., D.F.H., M.I.E.E.

SENIOR ENGINEER,
ITT COMPONENTS GROUP,
EUROPE

LONDON
NEWNES-BUTTERWORTHS

THE BUTTERWORTH GROUP

ENGLAND

Butterworth & Co (Publishers) Ltd
London: 88 Kingsway, WC2B 6AB

AUSTRALIA

Butterworths Pty Ltd
Sydney: 586 Pacific Highway, NSW 2067
Melbourne: 343 Little Collins Street, 3000
Brisbane: 240 Queen Street, 4000

CANADA

Butterworth & Co (Canada) Ltd
Toronto: 14 Curity Avenue, 374

NEW ZEALAND

Butterworths of New Zealand Ltd
Wellington: 26–28 Waring Taylor Street, 1

SOUTH AFRICA

Butterworth & Co (South Africa) (Pty) Ltd
Durban: 152–154 Gale Street

First published in 1973 by
Newnes-Butterworths, an imprint
of the Butterworth Group

ISBN 0 408 00091 0

Filmset by Keyspools Ltd, Golborne, Lancs

Printed and bound in England by C. Tinling & Co. Ltd, Prescot and London

Preface

The growth of power electronics during the last decade has been phenomenal. Although originally intended as a tool for electrical engineers, these devices are rapidly gaining acceptance in all branches of engineering. This has created an urgent problem in education, since presently available text books on thyristors tend to be either very basic or extremely specialised. Even those books which try to cover as many topics as possible tend to give scanty coverage of the 'newer' applications such as choppers, inverters and cycloconverters.

This book is intended to be a detailed treatise on all aspects of thyristor circuits. Starting with relative basics all the standard types of thyristor configurations are described. As such it is hoped that the book will find acceptance in Universities and Colleges with a Power Engineering curriculum, as well as serving as a reference manual for the industrial design engineer.

The most elementary and yet the most important chapter of the book is the first. This gives an overview of all standard thyristor circuits, and explains how they are based on the elementary switch concept of the thyristor. Surprisingly, even engineers who have worked for many years in one particular aspect of thyristor control are totally ignorant about other applications of the device.

Chapter 2 introduces the reader to the characteristics of the thyristor, and its correct use and protection against abnormal circuit conditions. The most basic application of a thyristor, that of switching, is described in Chapter 3. This is followed by a.c. phase control in the next chapter.

Chapter 5 covers the application of thyristors in choppers, and presents a classification system for forced-commutated circuits. It is hoped that this will assist the design engineer in the choice of a suitable circuit for any particular application. An appendix to this chapter describes an electrical vehicle controller, in which chopper systems are now extensively used.

Naturally commutated circuits, used as rectifiers, are described in Chapter 6, and this is followed in Chapter 7 by a very neglected subject, that of cycloconverter frequency changers. The next chapter has been devoted to the design of forced-commutated circuits since I have found from personal experience that, although considerable information exists on the design of phase-control circuits, chopper and inverter design is very poorly documented. All the design curves given here have been produced with the aid of an IBM 360 computer.

Forced-commutated inverter circuits are described in Chapter 9, and the classification system of Chapter 5 is extended to cover inverters. In this chapter, as indeed throughout the book, care has been taken to ensure that the reader appreciates how thyristor circuits have developed. This has been done to guide the designer when he is faced with a bewildering choice between different systems, and also to encourage him not to accept standard solutions but to design a circuit which is tailor made for his application. To this end Chapter 10 illustrates how the circuits of Chapters 5 and 9 can be modified to enhance some particular aspect of their performance.

Without doubt the most popular application for thyristors is in electrical-machine control. Chapter 11 introduces the basic principles of electrical machines, for the benefit of the non-electrical reader, and then goes on to describe some typical machine control applications. This chapter does not introduce any substantially new circuit techniques, since these have all been covered in previous chapters. Instead the operational principles of the circuits are illustrated by descriptions of their uses in different machine applications. The book concludes with a chapter on other power devices such as triacs, power transistors and sidacs, as well as some low-power semiconductors which have been used to control these power elements.

The book is based on a series of twelve articles which appeared in the journal *Electronic Components*. I am very grateful to the magazine for allowing me to use this material freely, and for assisting the publishers by supplying the original drawings from the series for use in this book. I am specially grateful to the editor of *Electronic Components*, Roy Atterbury, for invaluable assistance during both the preparation of the articles and in their adaptation to book form. I can unhesitatingly say that without his help this book would never have been published in its present form.

The information presented in this book has been acquired over many years, and I am grateful to the many, many people who, either personally or through their writing, have contributed to my education. In this respect I am specially indebted to two of my colleagues, John Storr-Best of ITT Components Group Europe, and

David Bradley, formerly of Bush Electrical Engineering Co. and presently with Thorn Automation, under both of whom I had the privilege of working for many years. My thanks are also due to John Storr-Best for permission to draw freely from his own prolific writings. I am also grateful to my personal, unpaid, overworked publicity agent, Tony Devereux. Apart from his professional occupation as Press Officer for ITT Components Group Europe, he has found time for the rather demanding task of placing my articles with the technical press.

It gives me great pleasure to acknowledge the following journals for permission to use material from various articles of mine: *Electrical Review*, for the contents of Chapter 12. *Electrical Times*, for the section on electronic commutators in Chapter 11. *Electronic Engineering*, and *Control and Instrumentation* for various sections through the book, and of course *Electronic Components*.

And of course no list is complete without an acknowledgement to the author's wife. My gratitude to my wife Roshan cannot be fully expressed in this short sentence. Apart from bringing up two demanding young children she has found time to do all the typing for the articles and the book, and offered some invaluable advice ('I wish you would improve your handwriting!'). Thank you for everything, darling, and specially for turning a blind eye every time I retired to the study with a handful of files under one arm and a can of midnight oil under the other!

F.F.M.

Contents

1

A résumé of thyristor circuits

Since its discovery in 1957 the thyristor has established itself as a semiconductor device second only to the transistor in popularity for use in switching. Although intended originally as a power device it is being used to an increasing extent in all fields of engineering.

Much has already been written about thyristors. Most of this has been of a highly specialised nature and tended to concentrate on particular aspects of thyristor operation. This can confuse the non-specialist, who is led to the mistaken conclusion that thyristors have a limited application.

The thyristor is basically a unidirectional switch, and can be considered as nothing but a switch. This must be kept firmly in mind. However, its switching property can be used to produce circuits with distinctive features. These standard configurations will be examined briefly with a view to establishing the principles of thyristor control.

THE THYRISTOR AS A SWITCH

The basic operation of a thyristor can best be explained by reference to its electromechanical analogy, as in *Figure 1.1.* A thyristor has three terminals, anode, cathode and gate. Diode D1 in series with the anode makes it unidirectional in operation. RL is a relay with normally open contact RL1. Diode D2 feeds into one end of the relay. When main switch SW is closed no load current can flow from d.c. supply V_B since the thyristor is de-energised. Suppose the gate terminal is now made positive to the cathode. Current will flow through RL and, provided this is of sufficient value, the relay will be energised and RL1 closed. Load current now flows from V_B via D1 and RL. If the gate supply is removed, the thyristor will not turn off, since it is maintained in the energised state by the passage of load current through RL (provided this is of a sufficiently large value). The

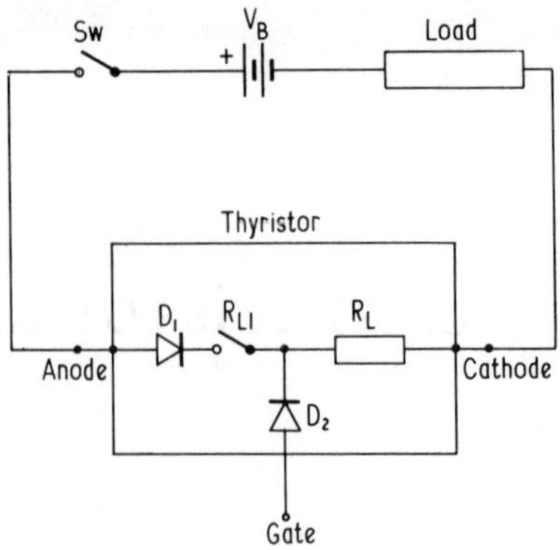

Figure 1.1. Thyristor electromechanical analogy

only way that the thyristor can be returned to its off state is to open SW for a time in excess of that required to de-energise the relay coil, after which RL1 opens and SW can be reclosed.

Diode D2 makes this device relatively unaffected by negative gate voltages. If this diode was not present and the thyristor was conducting, a negative gate potential would divert some of the current from relay RL, enabling it to turn off, even though SW remained closed. Such a device is called a gate turn-off switch. It will not be

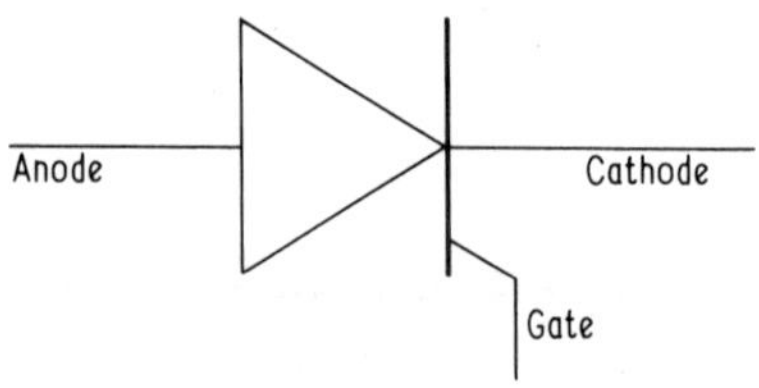

Figure 1.2. Thyristor symbol

considered here since it is normally only available as a low power device having a low turn-off gain. It is covered briefly in Chapter 12.

Figure 1.2 shows the symbol for a thyristor.

A.C. LINE CONTROL

Perhaps the most basic thyristor circuit is that used to control a.c. power flowing to a load. Input and output are both a.c., and since the thyristor is a unidirectional device it is necessary to use an antiparallel pair as in *Figure 1.3*. The operation of this circuit can be explained by reference to the waveforms of *Figure 1.4*.

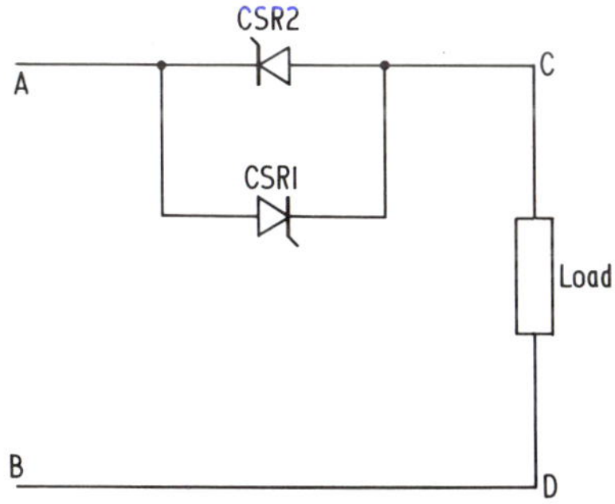

Figure 1.3. A.C. line control

The input voltage V_{AB} is shown in *Figure 1.4(a)*. Line A is positive to B over the first half cycle, and negative during the second half. Output voltage waveform for a resistive load is shown in *Figure 1.4(b)*. At t_0 the supply voltage goes positive. However, thyristors CSR1 and CSR2 are non-conducting so that the voltage across the load is zero. At t_0 thyristor CSR1 is turned on. It now presents a very low impedance so that the potential at C is almost the same as that at A and the load voltage jumps to practically the a.c. input value.

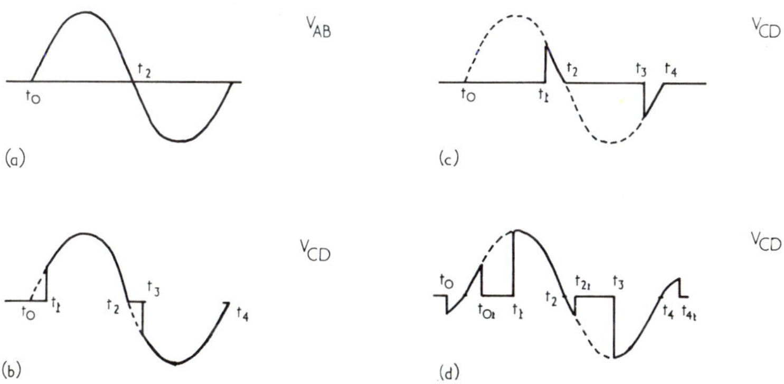

Figure 1.4. A.C. line control circuit waveforms

When the line voltage reverses at time t_2, thyristor CSR1 is reverse-biased. Its current decays and it turns off. Once again the output voltage is zero until thyristor CSR2 is fired at t_3, when it rises to the negative line potential. Clearly the thyristor firing instants t_1 and t_3 are variable. *Figure 1.4(c)* shows a longer delay from the zero a.c. voltage points than *Figure 1.4(b)* and illustrates the principle of variable mean a.c. voltage control.

If the load is inductive the thyristors will not turn off directly the supply reverses. The waveform for inductive load is shown in *Figure 1.4(d)*. Thyristor CSR1 is fired at t_1. It conducts until time t_2, since it is forward-biased. Load current flows from the supply. After time t_2 the input voltage reverses. However since the energy stored in the inductive load cannot be dissipated instantaneously, it forces CSR1 to remain conducting while current flows against a positive potential. This is known as regeneration, energy being transferred from load to supply. At t_{21} load current decays to a sufficiently low value and CSR1 turns off. The load voltage will remain at zero till the next thyristor is fired at t_3.

D.C. LINE CONTROL

In the previous section the thyristor was used to regulate the amount of energy between an a.c. input line and an a.c. load. The circuit illustrated in *Figure 1.5* controls a d.c. load operating from a d.c. supply.

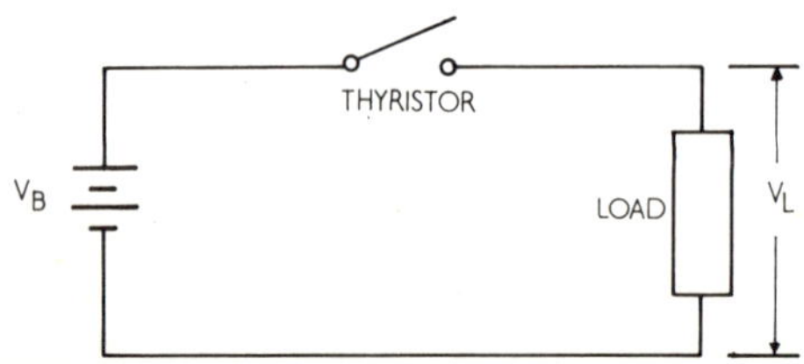

Figure 1.5. D.C. line control

The thyristor is here shown as a simple switch, which, after all, is what it is. Such a circuit is frequently referred to as a chopper. *Figure 1.6* shows the output voltage waveforms.

When the thyristor is closed, the load voltage is equal to the supply V_B (voltage drop across the device being neglected). With the thyristor open the voltage is zero (assuming negligible leakage current). In *Figure 1.6(a)* the thyristor is closed at t_1, opened at t_2, closed at t_3, opened at t_4, etc. If t_c and t_0 are the closed and open times during one cycle, the mean load voltage is given by:

$$V_{L(M)} = V_B \frac{t_c}{t_c + t_0} \tag{1.1}$$

This equation suggests a method of controlling the output voltage. For instance in *Figure 1.6(b)* although the switch is closed at the instances t_1 and t_3 as before, it is not opened till t_2' and t_4', such that the ratio of the closed time to total periodic time has changed, even though the operating cycle time is constant. Such a system is known as variable mark-space control. Alternatively t_0 can be changed while t_c is kept fixed as in *Figure 1.6(c)*. Clearly the period has now been changed from *Figure 1.6(a)*. This is known as variable frequency control.

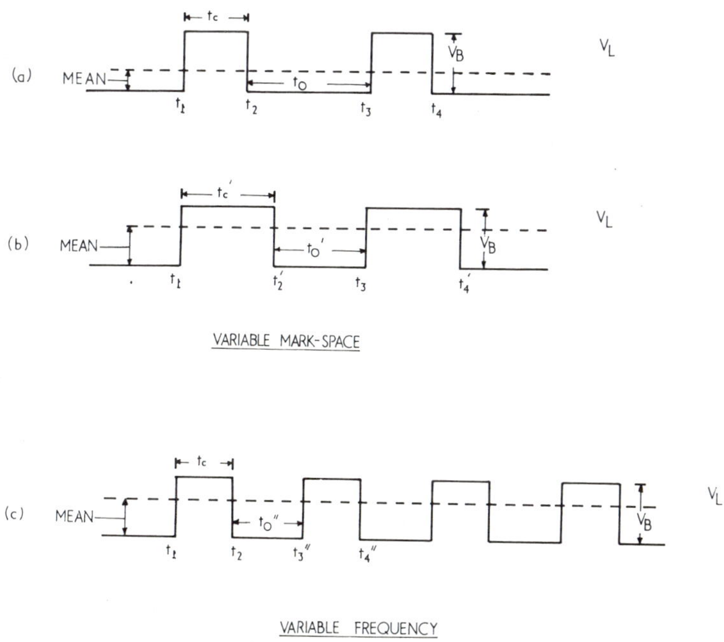

Figure 1.6. Chopper control modes

No matter what method is used equation (1.1) shows that the minimum load voltage is zero, when $t_c = 0$, and is equal to V_B when $t_0 = 0$, *i.e.* the switch is always closed. Such a system is called a step-down chopper, since it always controls voltages below that of the supply.

Figure 1.7 illustrates a step-up chopper. With the thyristor closed, current flows in inductor L storing energy. When the thyristor is open this energy transfers to the load, boosting its voltage above the

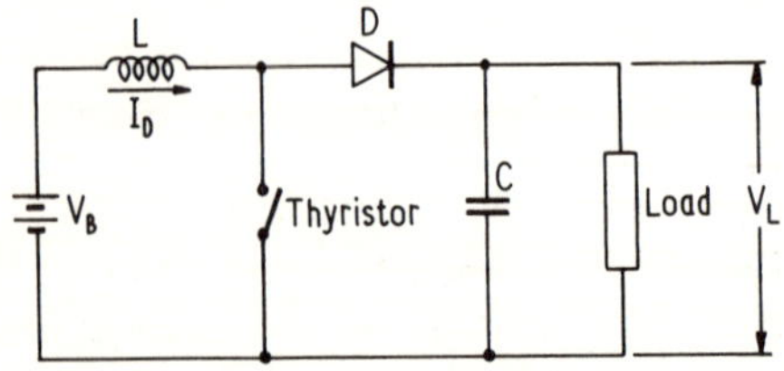

Figure 1.7. Step-up chopper

supply. To analyse the circuit, L is assumed large enough to maintain the current I_D substantially constant, and the load voltage is assumed to be smooth. If the thyristor is closed for time t_c and open for t_0 then equating the energy transfers in the choke during these periods:

$$V_B I_D t_c = (V_L - V_B) I_D t_0$$
$$\therefore V_L = V_B \frac{t_c + t_0}{t_0} \tag{1.2}$$

Equations (1.1) and (1.2) are plotted in *Figure 1.8* which illustrates the two different chopper types.

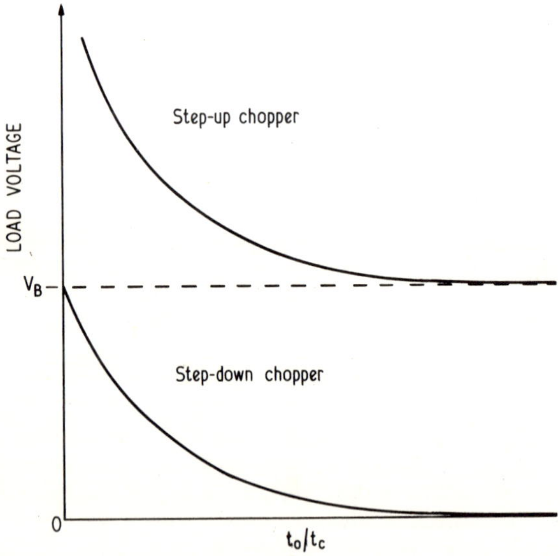

Figure 1.8. Chopper characteristics

The thyristors in *Figures 1.5* and *1.7* have been shown as switches for a reason. Since they operate from a d.c. supply it is no longer sufficient to turn them on and expect conduction to cease naturally when the supply polarity reverses, as in a.c. line control. They must be forcibly turned off, and such a system is referred to as forced commutation to differentiate it from a.c. line commutation (also called natural commutation).

There are several methods of forced commutating of thyristors. *Figure 1.9* shows one possible circuit and *Figure 1.10* its capacitor

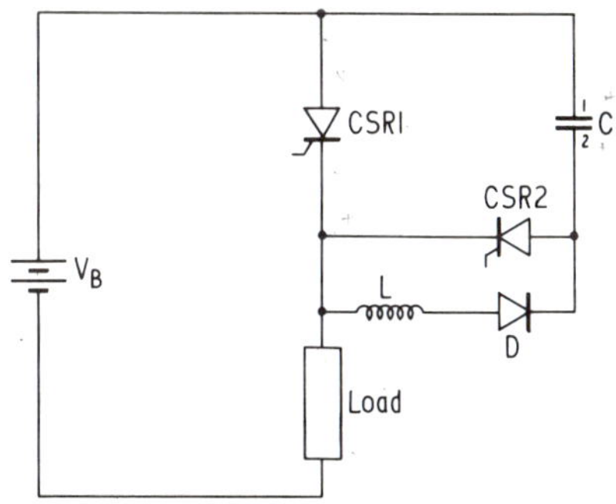

Figure 1.9. Typical chopper circuit

voltage waveform. At time t_0 capacitor C is charged to the supply voltage with plate 1 positive due to a previous cycle (CSR2 is fired on first switch-on to ensure this). At t_1 the main thyristor CSR1 is turned on to commence the load cycle. This also discharges C

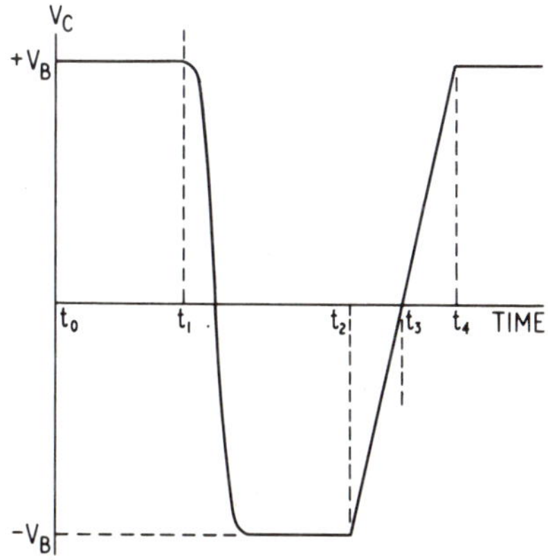

Figure 1.10. Capacitor voltage waveform for Figure 1.9

through L and D such that it resonates, recharging to V_B with plate 2 positive (resonant losses being ignored). To turn thyristor CSR1 off at any time t_2, CSR2 is fired. C discharges through the load and recharges again with plate 1 positive. Provided the time for which CSR1 is reverse-biased, (t_3 to t_2) in *Figure 1.10*, is sufficiently large, this thyristor will remain in the de-energised state, until refired by a gate signal. Commutation is then said to have been successful.

To differentiate forced commutated thyristors from those which are naturally turned off they can be drawn with the addition of an arrow as in *Figure 1.11*. This arrow then represents all auxiliary thyristors, diodes, chokes and capacitors used in the forced turn-off process.

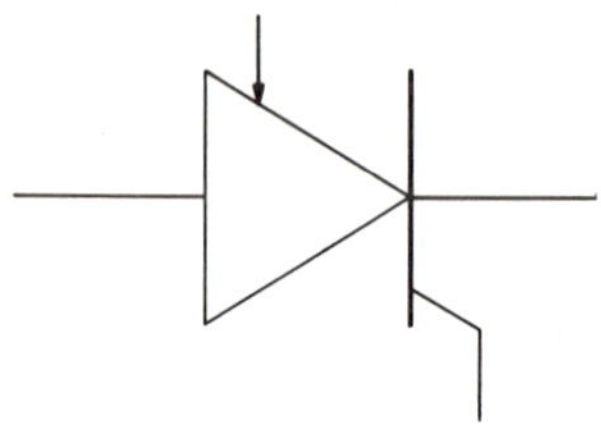

Figure 1.11. Forced-commutated thyristor symbol

It should be mentioned that the loads in the chopper circuits shown have been assumed to be resistive. If they were inductive it would be necessary to add a free-wheeling diode across them to carry the inductive current, and so prevent voltage transients during switch off, which could destroy the semiconductors.

CONTROLLED RECTIFICATION

Owing to its unidirectional conduction properties, a thyristor is ideally suited for use as a controlled rectifier. One such circuit which regulates the power from an a.c. source to a d.c. load is shown in *Figure 1.12(a)*. The load could be resistive, or inductive with a free-wheeling diode.

Figure 1.12(b) shows the a.c. line voltage and *Figure 1.12(c)* the load voltage. At time t_1 CSR1 and CSR4 are fired so that the load voltage is almost equal to that of the supply. When the polarity of the input reverses at t_2 these thyristors turn off. The load voltage is zero till at t_3 thyristors CSR2 and CSR3 are fired. Even though the input line polarity has reversed, end C of the load is again connected to the most positive side, so that the load voltage is maintained unidirectional. *Figure 1.12(d)* illustrates how the mean value of the d.c. load voltage can be changed by altering the firing point of the thyristors during the half-cycle.

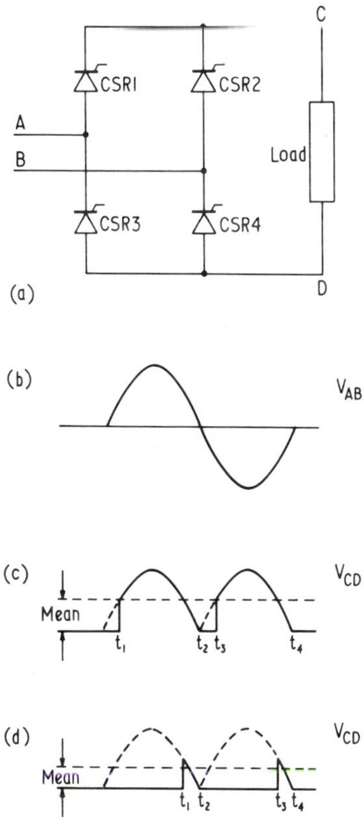

Figure 1.12. Controlled rectification

If the load is inductive, and is not by-passed by a free-wheeling diode, then the main thyristors will not turn off exactly at the zero input voltage points. As seen in the case of a.c. line control, the energy stored in the load during the rectification period forces the main thyristors to remain in conduction, as current is fed back from the load to a negative supply. This is regeneration, or inversion as it is better known here, since current flows from a d.c. load to an a.c. source. Clearly the maximum inversion period is 90°E since only as much energy can be taken out of the load as is put in. However, if the load includes an independent d.c. source, as in *Figure 1.13*, inversion over a full 180°E is possible.

When the firing points of the thyristors in *Figure 1.13* are delayed, the output voltage of the bridge is not zero but follows the negative contours of the line voltage, since the devices are kept conducting by current provided by E_B. In *Figure 1.13*(*c*) up to time t_2 thyristors CSR2 and CSR3 were conducting and supplying energy from the a.c. source to the load. If now the firing of CSR1 and CSR4 is delayed until t_3, then from t_2 to t_3 thyristors CSR2 and CSR3 are kept conducting by current flowing from E_B, through CSR3, against the negative line voltage, through CSR2 and back to E_B. This continues until at t_3 thyristors CSR1 and CSR4 are fired. Point C now rises to the potential of line B and D falls to that of A, so that CSR2 and CSR3 are reverse-biased and turn off.

In *Figure 1.13*(*c*) the mean output voltage is positive so that the a.c. lines give out more energy than they receive from the d.c. source.

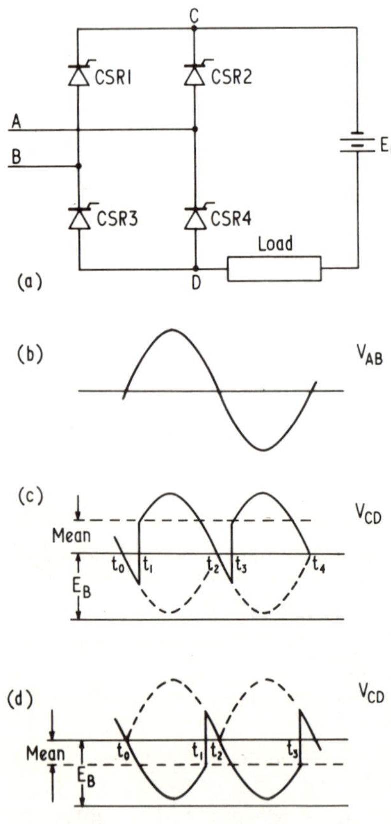

Figure 1.13. Controlled inversion

Cutting the pulses further back, beyond 90°E, as in *Figure 1.13*(*d*), the mean voltage is made negative. There is now a net transfer of power from the d.c. source to the a.c. lines, *i.e.* an overall inversion.

FREQUENCY CONVERSION

To change from an input at one frequency to another at a different frequency, one can either go directly from a.c. input to a.c. output, or pass first through an intermediate d.c. stage. The former is called a cycloconverter and the latter an inverter. *Figure 1.14* shows a typical cycloconverter with its waveforms, assuming a resistive load.

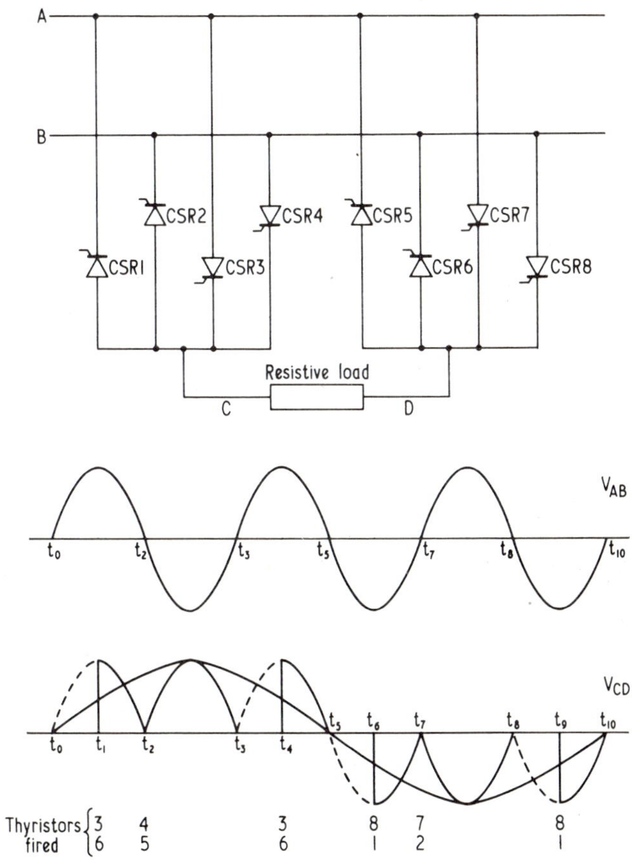

Figure 1.14. Cycloconverter

To obtain a frequency reduction between input and output of 3:1 the following firing sequence can be observed. At t_0 when line A is positive, CSR3 and CSR6 are fired. At t_2 thyristors CSR4 and CSR5 and at t_3 thyristors CSR3 and CSR6. These four thyristors are known as the positive group since they make the load voltage positive irrespective of the input polarity. Closer examination of *Figure 1.14* will show that they form a bridge, similar to that of *Figure 1.12*. Similarly CSR1, CSR8, CSR2 and CSR7 form a second bridge which can cause the load voltage to be negative, independently of the supply polarity. They are fired over times t_5, t_7, t_8. The firing points of the thyristors in *Figure 1.14* are shown delayed, the mean voltage being reduced with frequency. This is often necessary when the load is iron-cored so that its flux is maintained at the most optimum value.

An inverter operates from a d.c. input. If the supply is a.c. it must first be rectified and smoothed. *Figure 1.15* shows one form of inverter circuit. Note that the thyristors are all forced-commutated by external circuitry which is not shown. The load is again assumed to be resistive. At time t_0 thyristors CSR1 and CSR2 are turned on. The load voltage is positive and equal to that of the supply, if the loss across devices be assumed negligible. At t_3 thyristors CSR1 and CSR2 are turned off and CSR3 and CSR4 fired. The load voltage swings negative. The half cycle is repeated at t_6 by turning off 3 and 4 and refiring 1 and 2 to give the waveforms shown in *Figure 1.15(b)*. Voltage control of the load, as frequency changes, can be obtained by using the mark-space technique described in *Figure 1.6*. At t_0 all four thyristors are maintained non-conducting. At t_1 thyristors CSR1 and CSR2 are turned on. At t_2 all thyristors are again off until CSR3 and CSR4 are fired at t_4. The output voltage waveform is shown in *Figure 1.15(c)*. It must be emphasised that this description only applies for a resistive load, which has been adopted to explain the principles of operation.

Although the circuit of *Figure 1.13*, when operating in the mode shown in *Figure 1.13(d)*, is referred to as an inverter, it is important to differentiate it from the system given in *Figure 1.15*. The thyristors in *Figure 1.15* are forced-commutated by external circuitry which is not shown, whereas in *Figure 1.13* commutation occurs naturally owing to the presence of an a.c. input source. Therefore these latter inverters are often called a.c. line commutated. Although they can vary the magnitude of the inverted power, they cannot alter its frequency, which is fixed by that of the a.c. lines.

Comparison of *Figures 1.14* and *1.15* will show three important differences between an inverter and a cycloconverter:

(i) The cycloconverter thyristors are all naturally commutated whereas an inverter uses auxiliary circuitry for forced commutation.

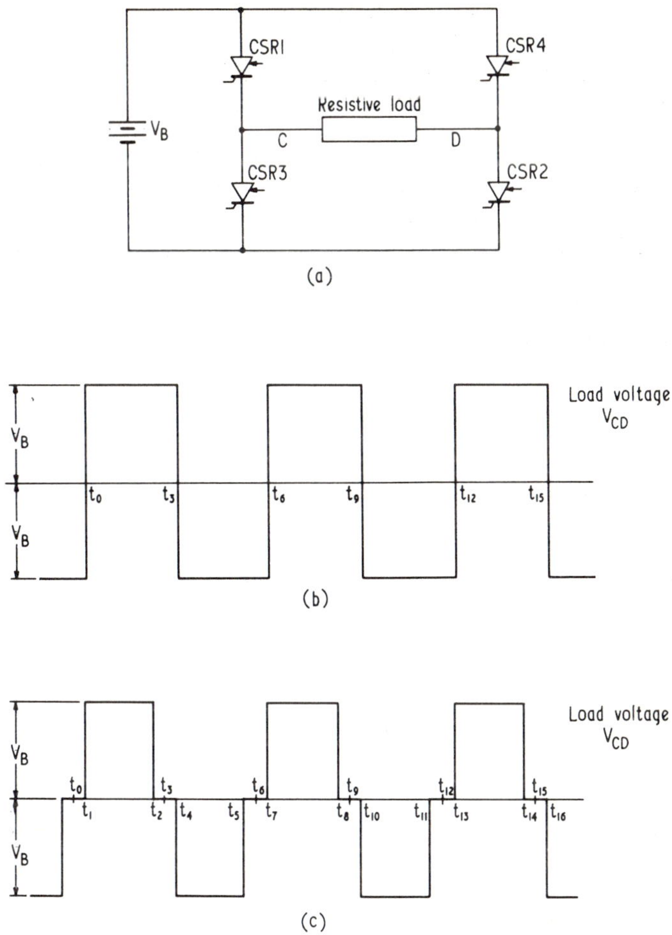

Figure 1.15. Inverter

(ii) An inverter operates from a d.c. source whereas a cycloconverter follows the envelope of the a.c. input.

(iii) The inverter can give an output range of frequencies which is infinitely variable. The naturally commutated cycloconverter, on the other hand, can only change the frequency in discrete steps of the input, and can only operate up to a maximum frequency equal to that of the input.

Frequency changing systems will be discussed in detail in Chapter 7.

CONCLUSIONS

In this introductory chapter thyristor circuits have been considered as variable voltage converters of a.c. to a.c., d.c. to d.c., a.c. to d.c. and d.c. to a.c. They have also been shown to be capable of frequency changing. However, it must be emphasised that the thyristor itself is a switch, albeit a very powerful and efficient semiconductor switch, and nothing more. The various circuits described here will be considered again in later chapters in further detail. It is, however, important at this early stage to have a good overall picture of all the basic thyristor circuit principles and so avoid being confused by excessive detail.

2

Using the thyristor

The thyristor was introduced in Chapter 1 by reference to its electromechanical analogy. This analogy is extremely useful in explaining many of the device characteristics, and will be developed further in this chapter. The performance of most modern devices is given in the form of data sheets, and it is essential that these be understood in order to obtain the optimum output from the thyristor. Reference will be made here to a typical low-power thyristor data sheet in order to illustrate its characteristics and ratings.

Although extensive use is made of the thyristor analogy it must not be forgotten that the device is in practice a semiconductor. Its manufacture and physical basis are introduced briefly later in the chapter. This is then followed by discussions on practical aspects of using the thyristor, for example cooling, protection and series/parallel operation.

AN ELECTROMECHANICAL ANALOGY

A thyristor is a semiconductor. Its characteristics can be explained very conveniently by reference to an electromechanical analogy, as in *Figure 2.1*. This is essentially the same as that introduced in Chapter 1 but with the addition of capacitor *C* across the relay contacts. Operation of the circuit has been described before, the effect of *C* being ignored for the moment.

When the gate is made positive with respect to the cathode the relay is energised. RL1 closes and the thyristor stays on even though its gate drive is disconnected. To turn the thyristor off, the load current must be interrupted, either by opening SW or by reversing the voltage across RL by an external source, for a time long enough to allow the relay to de-energise. Diode D1 makes the thyristor unidirectional in operation, and D2 prevents negative gate signals from turning it off.

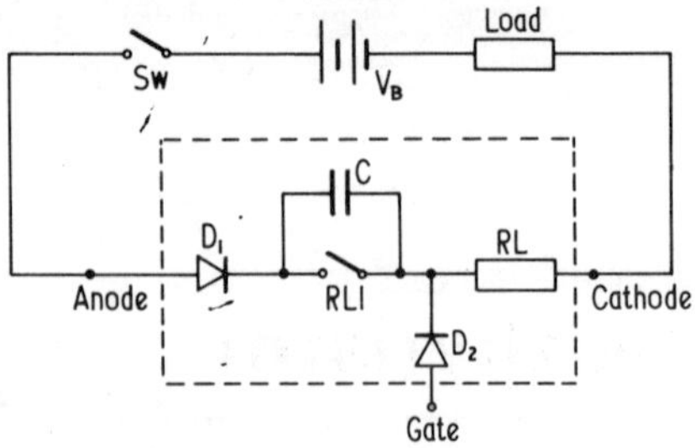

Figure 2.1. Thyristor electromechanical analogy

There are two classes of information given by data sheets. The first is device rating, which must not be exceeded or damage to the thyristor could result. The second is its electrical characteristics which define how the thyristor performs under certain conditions. These groups will be examined with reference to the model given in *Figure 2.1.*

THYRISTOR RATINGS

Figure 2.2 reproduces two pages of the CRS3 (ITT Semiconductors) data sheets which refer to its ratings. There are basically five main parameters, voltage, current, frequency, power and temperature.

VOLTAGE RATING

In the forward direction the thyristor in *Figure 2.1* can withstand a voltage equal to the flashover rating of contact RL1. If this is exceeded, current will flow in relay coil RL and the thyristor is turned on. However, no damage should occur to the device. This is called the 'continuous forward blocking voltage'. The peak forward blocking voltage represents the breakdown voltage of capacitor *C*. Voltages greater than this could destroy the thyristor.

In the reverse direction the crest working voltage is the peak of the sinusoidal voltage which may be safely applied. If it is exceeded diode D1 could be destroyed, but for short-duration pulses it is possible to go higher—up to the repetitive peak reverse voltage. Provided this maximum value is not exceeded the pulse energy is insufficient to flashover across contact RL1. On a non-repetitive basis an even higher voltage can be used. It stresses the contact insulation resistance and should not be applied frequently.

THYRISTORS
Silicon Controlled Rectifiers
CRS3 Series

DESCRIPTION

The CRS3 Reverse Blocking Triode Thyristors are silicon all-diffused *p-n-p-n* devices with gate turn-on control. They conform to the IEC A13 outline (VASCA SO–35, JEDEC TO64 equivalent).

VOLTAGE RATINGS

Maximum rated values up to a maximum stud temperature of 125°C.

Code	*Crest working reverse voltage*	*Repetitive peak reverse voltage*	*Non-repetitive peak reverse voltage*	*Continuous reverse voltage*	*Peak forward blocking voltage**†	*Continuous forward blocking voltage*†
CRS3/05AF	50	55	60	50	60	50
CRS3/10AF	100	110	120	100	120	100
CRS3/20AF	200	220	240	200	240	200
CRS3/30AF	300	330	360	300	360	300
CRS3/35AF	350	385	420	350	420	350
CRS3/40AF	400	440	480	400	480	400
CRS3/50AF	500	550	600	500	600	500
CRS3/60AF	600	660	720	600	720	600

* Maximum rate of rise 15V/μs   † Gate open circuit.

Peak forward gate voltage 15Ω minimum source resistance (anode positive with respect to cathode)	10V (see graph 1)
Peak forward gate voltage (anode negative with respect to cathode)	0·25V
Peak reverse gate voltage 15Ω minimum source resistance	10V

CURRENT RATINGS

Maximum values at maximum stud temperature of 100°C	
Rated mean forward current (single-phase, half-wave circuit with resistive load and 180° conduction angle)	3A
For conduction angles below 180°	see graph 2
Rated peak repetitive forward current (at any conduction angle)	15A
Rated overload forward current	see graph 4
Rated surge (non-repetitive) forward current half sine wave 50Hz	
from rated load	40A (peak)
from no load	55A (peak)
Rated continuous forward current (90°C stud temperature)	4·5A
I^2t for fusing ($t \leqslant 0{\cdot}01$s)	10A^2 s
Rated peak forward gate current (anode positive with respect to cathode)	1·0A

Figure 2.2. Thyristor ratings (continued overleaf)

FREQUENCY RATING

Maximum operating frequency without derating	400Hz

For frequencies between 400 and 2 500Hz derate maximum stud temperature by $\delta T_s = 2VI\,(f-400)+10^{-6}$ °C

Where V = instantaneous voltage immediately prior to turn on (V)
I = instantaneous current immediately following turn on (A)
f = frequency (Hz)

POWER RATINGS

Maximum values at maximum stud temperature of 100°C

Rated total device power dissipation		see graphs 2 and 4
Rated mean gate power	see graph 1	0·5W
Rated peak gate power	see graph 1	2·0W

TEMPERATURE RATINGS

Working junction temperature	125°C
Maximum stud temperature	see graph 2
Storage temperature range	−65°C to +150°C

Figure 2.2. Thyristor ratings (continued)

Forward blocking voltages are specified at a certain rate of rise of voltage dv/dt. This causes current to flow through capacitor C, of a value given by: $I = C \times dv/dt$. If the current is sufficiently large the thyristor will turn on even though its peak forward blocking voltage has not been exceeded.

The gate ratings are determined by diode D2. The peak forward gate voltage is principally dependent on gate current and can be increased if the source impedance is also increased. In the reverse direction the breakdown of D2 limits the voltage to the value specified. The problem of gate drive, when the thyristor is reverse biased, will be examined later.

CURRENT RATING

The current rating of the thyristor is determined by the thermal dissipation within the device (relay coil and D1). The larger the heat conductive area to which the thyristor is connected, the greater the current it can carry without getting too hot. The limiting value in the case of the CRS3 device is 3A for 180°, *i.e.* half wave conduction. The lower the conduction angle the less the mean current. This is shown in *Figure 2.3*. However, the r.m.s. (or d.c.) current in all cases is nearly the same, except at very low conduction angles when peak currents can cause local hot points. To illustrate this, consider the case of 180° conduction shown in *Figure 2.4*.

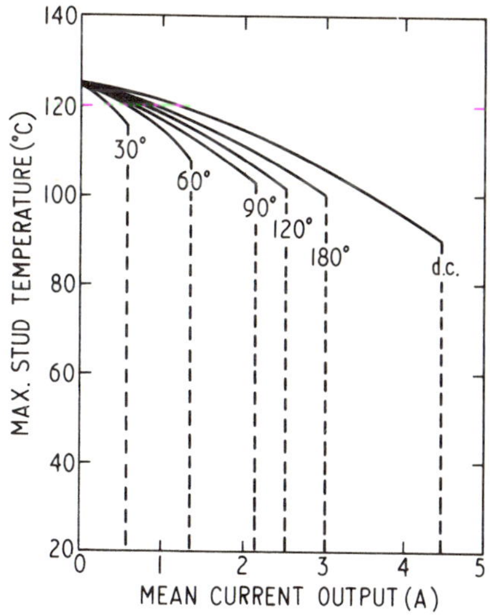

Figure 2.3. Thyristor current curves

The half cycle mean is now 3A. Therefore

$$3 = \frac{1}{2\pi}\int_0^{\pi} I_M \sin\theta \, d\theta$$

$$= \frac{I_M}{\pi}$$

or $$I_M = 3\pi \qquad (2.1)$$

Effective d.c. (r.m.s.) current

$$= \left[\frac{1}{2\pi}\int_0^{\pi} I_M^2 \sin^2\theta \, d\theta\right]^{1/2}$$

$$= \frac{I_M}{2} \qquad (2.2)$$

Substituting equation (2.1) in equation (2.2) gives:

$$I_{DC} = 4{\cdot}5\text{A} \qquad (2.3)$$

For 30° conduction the firing point is delayed by 150° as shown in *Figure 2.4*.

Therefore

$$(I_{RMS})_{30^\circ} = \left[\frac{1}{2\pi}\int_{5\pi/6}^{\pi} I_M^2 \sin^2\theta \, d\theta\right]^{1/2}$$
$$= 0{\cdot}087 I_M$$
$$= 0{\cdot}8\text{A}$$

from equation (2.1).

As seen from *Figure 2.3* these results are approximately correct. Furthermore, when the temperature increases the current-carrying capability of the thyristor also falls off.

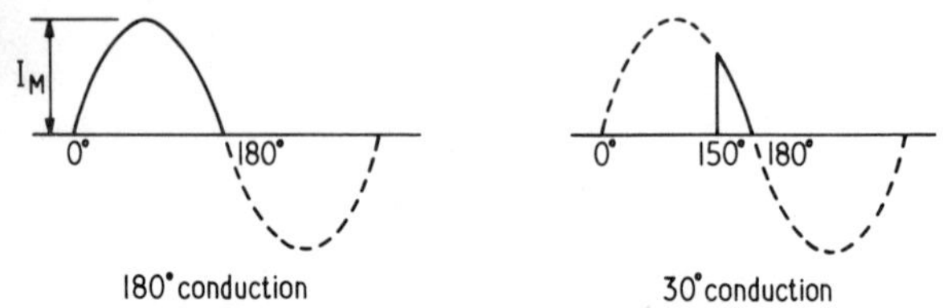

Figure 2.4. Variation in conduction angle

The shorter the time for which the current flows the greater its possible overload value. *Figure 2.5* shows that for only one cycle (one half cycle to be strictly correct since the thyristor blocks over the negative half cycle) a peak of 55A can flow. If rated load current of 3A mean, 180° conduction, was flowing before overload occurred, the thyristor is already hot. This give a lower surge current rating, hence the two curves in *Figure 2.5*.

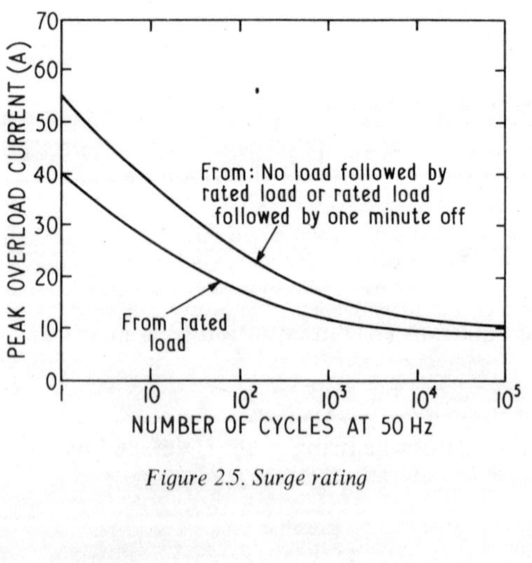

Figure 2.5. Surge rating

Sometimes the *di/dt* rating of a thyristor is given. This is the fastest rate at which the load current can rise as the thyristor turns on. It represents the rate at which the contact resistance of RL1 (*Figure 2.1*) changes as the relay is energised.

Gate current is limited by the dissipation in diode D2 to 1·0A. I^2t ratings are considered later.

FREQUENCY RATING

As with all devices the thyristor dissipates heat when switching on and off. It must be derated at high frequencies, which tend to reduce the life of contact RL1 shown in *Figure 2.1.* Data sheets give derating figures.

POWER AND TEMPERATURE RATING

These will be considered later when cooling of the thyristor is considered.

THYRISTOR CHARACTERISTICS

Characteristics describe operation of the thyristor under various conditions. They are shown in *Figure 2.6.*

CHARACTERISTICS	
Maximum on-voltage at 4·5 amperes continuous forward current (at 25°C stud temperature)	1·3V see graph 6
Maximum holding current gate open circuit. For typical variation with temperature	25mA see graph 5
Maximum forward leakage current at rated continuous forward blocking voltage and maximum stud temperature (gate open circuit)	1mA
Maximum reverse current at rated crest working reverse voltage and maximum rating (gate open circuit)	1mA (peak)
Minimum gate trigger current to turn on all devices (6V anode positive with respect to cathode)	20mA see graph 1
Minimum gate trigger voltage to turn on all devices (6V anode positive with respect to cathode)	3V
Maximum gate non-triggering current at rated peak forward blocking voltage and maximum stud temperature	0·75mA
Maximum gate non-triggering voltage at rated peak forward blocking voltage and maximum stud temperature	0·25V
Typical turn-on time. (See graphs 7 and 8)	2μs
Typical circuit commutated turn-off time	17μs
Total power loss	see graph 3
Maximum thermal resistance (junction to stud)	6°C/W

Figure 2.6. Thyristor characteristics

When the thyristor conducts current a voltage drop occurs across it. This is due to current flowing in *RL* and D1. The voltage varies in a nonlinear manner with current, as for a diode, and is usually shown by a curve (graph 6 of the data sheets which is not reproduced here). The maximum value at 4·5A is 1·3V.

Holding current defines the minimum relay current below which it will no longer remain energised. For a CRS3 device if the load current falls below 25mA it will cease conduction. Holding current decreases with device temperature (graph 5 shows this but is not reproduced here).

Leakage current of the thyristor when forward biased, with RL1 open, is the current flowing through *C*. In the reverse direction it is the leakage through *C* and D1. For the CRS3 device, D1 has a low reverse impedance so that in both cases the current is determined by *C*, *i.e.*, 1mA at peak voltage.

Graph 1 of the data sheets gives the gate characteristics of the thyristor and is shown in *Figure 2.7*. The maximum gate non-triggering voltage and current (*i.e.* values which will not energise relay RL1) are shown. To turn on all devices at 25°C current and voltage must exceed 20mA and 3·0V respectively. This graph is very useful in determining the gate drive impedance. For instance, suppose the thyristor is driven from a source of V volts through a resistance of $R\Omega$. Peak gate dissipation would occur when the gate characteristic is such that it has a voltage drop equal to $V/2$. The dissipation is then $V^2/4R$. For a duty cycle of $x\%$:

$$\frac{x}{100} = \frac{V_{GM}}{V^2/4R}$$

where V_{GM} is the mean gate power.

or

$$R = \frac{xV^2}{400V_{GM}} \qquad (2.4)$$

Therefore for $V = 5$V and $V_{GM} = 0{\cdot}5$W, $R = 12{\cdot}5\Omega$ for 100% duty cycle and 6·25Ω for a 50% cycle.

R is the minimum value of the trigger source impedance. Plotting the above two load lines on *Figure 2.7* they are seen to be tangential to their respective maximum gate power curves. Gate source impedance is normally chosen to give an operating load line between this line and the area bounded by the minimum firing gate voltage and current. An example is considered later.

The only other thyristor characteristics to be considered here are turn-on and turn-off times. Turn-on time is the time required to

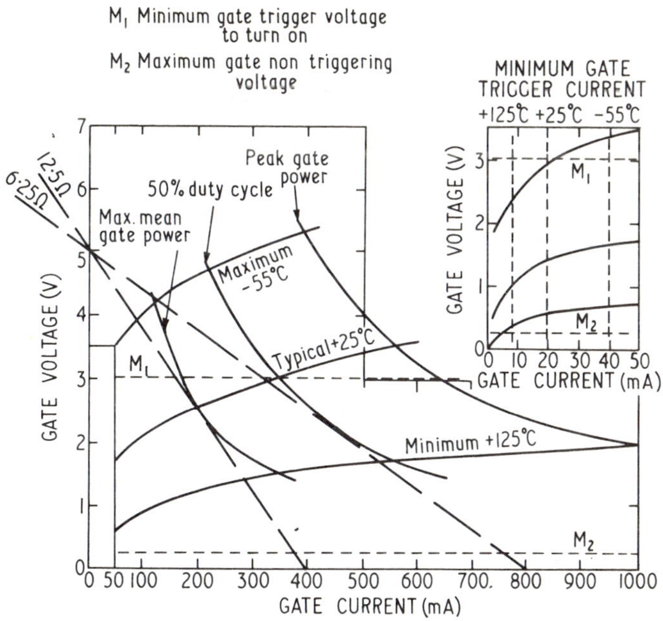

Figure 2.7. Thyristor gate characteristics

energise the thyristor (RL) before its gate drive can be removed and it becomes self-latching. Turn-off time is the duration for which the current has to be blocked in RL before it switches from the energised to de-energised state. Both characteristics are affected by several factors normally shown on auxiliary graphs in the data sheets.

DESIGN EXAMPLES

Perhaps the most elementary thyristor circuit is an a.c. static switch controlling a resistive load. This is shown in *Figure 2.8* with its associated gate driver. It is described in greater detail in Chapter 3. When the voltage at A is positive TR2 is on, and TR1 off, so that the thyristors remain non-conducting. Transistor TR2 goes off when A falls to the zero voltage rail. TR1 then functions as a blocking oscillator providing a continuous positive gate pulse to CSR1 and CSR2, which are therefore on.

Let us look at the design requirements for thyristors CSR1 and CSR2. If *R* is 40Ω then with a supply of 240V r.m.s. each thyristor carries a half sine wave of current, having a peak value of 8·5A.

Therefore their mean (180° conduction) rating is $8{\cdot}5/\pi$ or 2·8A. Clearly the CRS3 thyristor described before would be suitable. The peak voltage across each device is $240 \times \sqrt{2} = 340$V. CRS3/40 AF having a 400V crest working voltage rating gives an adequate safety margin.

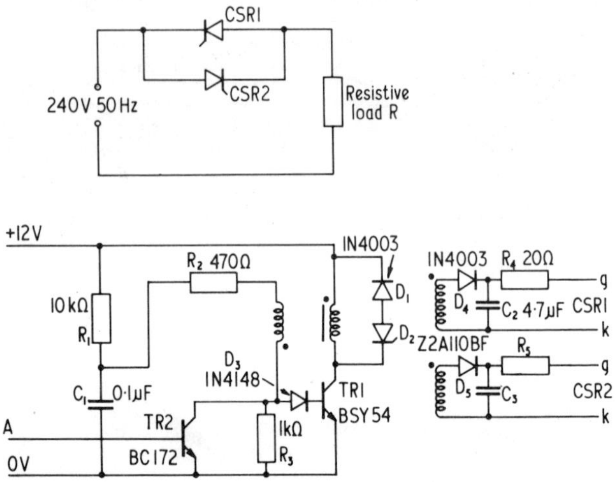

Figure 2.8. Static contactor

The gate drive transformer used has a 2:1:1:1 turns ratio. Therefore open-circuit drive voltage is 6V. The source resistor must be so chosen that its load line passes above the 3V, 20mA area shown in *Figure 2.7*, while never cutting the maximum mean gate power curve. A line joining 6V and 300mA fulfils these requirements so that the source resistor is $6/0{\cdot}3 = 20\Omega$. This is greater than the minimum value given by equation (2.4).

A thyristor beacon flasher, operating from a 24V d.c. supply is illustrated in *Figure 2.9*.

Thyristor CSR1 controls a 100W lamp. Since it is operating from a d.c. source it is necessary to force-commutate this thyristor when the lamp is to be turned off. This is done by CSR2. Suppose thyristor CSR1 is conducting. Then capacitor C_1 is charged to approximately 24V with plate *a* positive through R_7. If CSR2 is fired this capacitor charge appears as reverse voltage across CSR1. It is turned off and C_1 begins to charge via the load with plate *b* positive. Thyristor CSR1 will remain off so long as it is reverse-biased for a duration in excess of its turn-off time. The value of C_1 to achieve this can be calculated as follows.

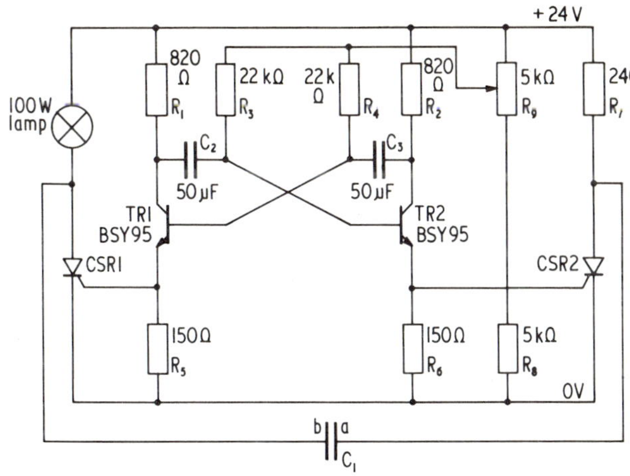

Figure 2.9. D.C. beacon flasher

Suppose C_1 is charged to V_0 with plate *a* positive prior to CSR2 being fired. If the supply is V volts and R is the resistance of the lamp, the capacitor voltage during this period is given by:

$$V_C = V\left[1-\exp\left(\frac{-t}{RC_1}\right)\right]-V_0\exp\left(\frac{-t}{RC_1}\right) \qquad (2.5)$$

When V_C equals zero the turn-off period ends so from equation (2.5):

$$C_1 = \frac{t_0}{R\log\left(1+\frac{V_0}{V}\right)} \qquad (2.6)$$

where t_0 is the thyristor turn-off time.

In our example $V_0 = V = 24$V. If thyristor CSR1 is a CRS3 device, its typical turn-off time is 17µs as given by data sheets. A maximum figure is not available but may be taken as 50µs. Since $R = 5{\cdot}76\Omega$ (100W at 24V) the minimum value of C_1, calculated from equation (2.6), is 12·6µF. Any convenient value above this figure can be used.

Both CSR1 and CSR2 have to withstand the supply voltage. Therefore, 50V devices would be suitable. To allow for low frequency operation their d.c. ratings should equal the peak value, *i.e.* 24/5·76 = 4·2A (a CRS3 could be used) and 24/240 = 0·1A. Note that CSR2 will have to carry a peak surge current of 48/5·76 or 8·4A, due to the discharge of C_1 when the thyristor is initially fired.

Gate drive requirements are best calculated by considering open-

and short-circuit conditions. For instance, with the thyristor gate open-circuited the drive voltage at the emitters of transistors TR1 and TR2 is $24 \times 150/970 = 3{\cdot}7$V. Assuming the gate to be short-circuit its current is $24/820 = 29$mA. Both these are above the minimum at 25°C specified in *Figure 2.7*.

MANUFACTURE AND ENCAPSULATION

The thyristor is a four-layer semiconductor device. The thickness of the dice is approximately 0·015in and its diameter varies from 0·1 to 1in, depending on its current carrying requirements. Although germanium may be used, all commercially available thyristors are of silicon. The complete slice is of monocrystalline material, *i.e.* the silicon atoms are arranged in a continuous crystalline structure, even though there are four *p-n-p-n* layers of different impurity doping.

Basically there are three techniques for introducing the impurities into a silicon slice during manufacture of the *p-n-p-n* structure. These are:

(i) **Diffusion method**

The silicon wafer is placed in a gas containing a certain percentage of doping material. Under high temperatures this impurity diffuses into the silicon to the desired depth.

(ii) **Alloying method**

The material with which the silicon slice is to be doped is contained in a metal layer. This is brought into contact with the silicon and the assembly is heated. The metal fuses with the silicon wafer and the zone beneath the metal disc is converted by the doping substance.

(iii) **Epitaxial method**

The doping gas is mixed with an organic gaseous silicon compound which is made to decompose close to the silicon wafer. At relatively high temperature the silicon is precipitated from the gas, in the form of a monocrystalline layer having a doping concentration corresponding to that in the decomposing gas.

These three methods are used, singly or in combination, to make

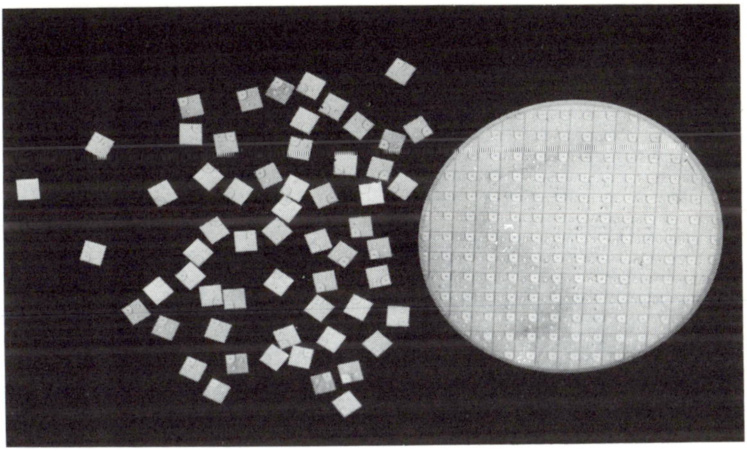

Plate 1. *As in other semiconductors, thyristors are fabricated from a silicon slice on which the separate layers are formed by diffusion or alloying. The dice is then scribed and broken into individual thyristor die. The larger the thyristor current rating the bigger the die size and the smaller the number of die per dice. The photograph shows a die design in which the gate section is situated in one corner (by courtesy of ITT Semiconductors, Foots Cray)*

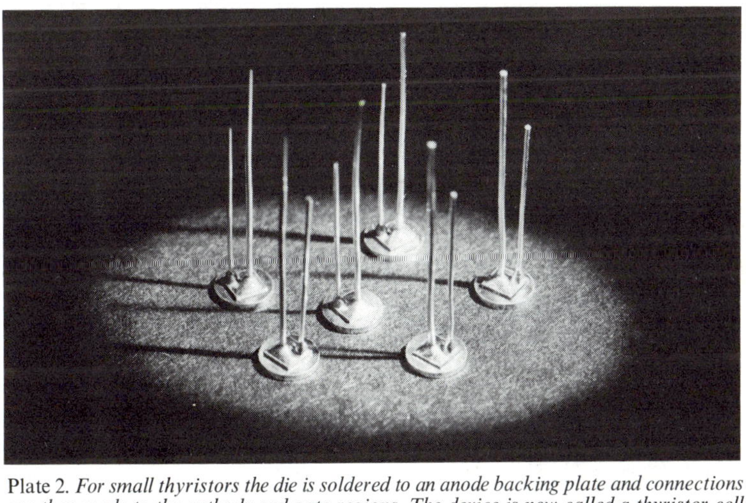

Plate 2. *For small thyristors the die is soldered to an anode backing plate and connections are then made to the cathode and gate regions. The device is now called a thyristor cell and is ready for encapsulation (by courtesy of ITT Semiconductors, Foots Cray)*

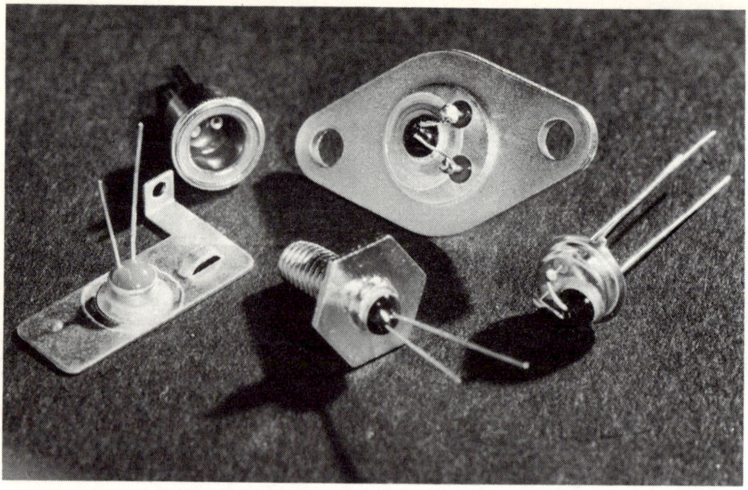

Plate 3. *The photograph illustrates a variety of thyristor cell encapsulations used on small devices. The cheapest configuration is the 'skill' package shown on the left. Others illustrated are stud, T03 and T05 (by courtesy of ITT Semiconductors, Foots Cray)*

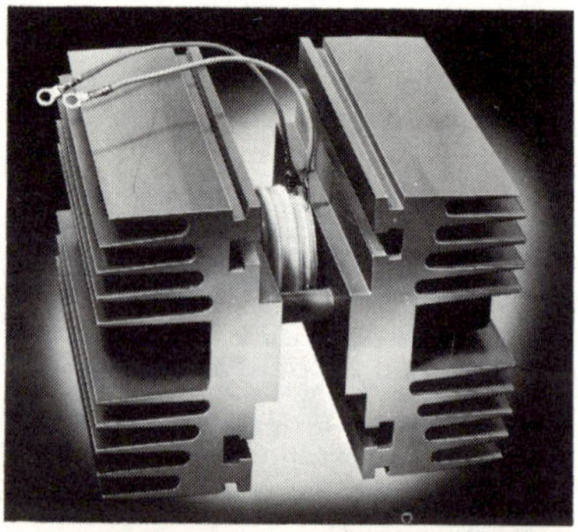

Plate 4 (above). *Large thyristors are assembled by compression bonding and encapsulated in 'hockey puck' cases, which allow both sides of the device to be cooled. The thyristor shown here can regulate over 1 MW of power (by courtesy of Westinghouse Brake and Signal Co. Ltd.)*

Plate 5 (left). *Thyristors and diodes can be connected in series or parallel to increase their ratings. The photograph shows a high voltage rectifier valve which is built from four basic modules. Each module can control 5 kV at 5 A (by courtesy of ITT Components Group Europe)* up the four-layer thyristor dice. The five stages in the production of an all-diffused construction are as follows.

Stage 1

Starting with a slice of *n*-type material, an oxide coating is formed and two *p*-layers are introduced by diffusion of gallium through the oxide. This diffusion, arranged to penetrate a controlled depth, occurs simultaneously from both sides of the slice as in *Figure 2.10*(*a*).

Stage 2

The oxide is selectively removed by a photographic process, exposing, where required, the gallium doped *p*-layer. In order to do this an emulsion sensitive to ultra violet light is applied to the oxide gallium diffused slice. Using a photographic negative, which is in effect a mask, ultra violet light is allowed to fall onto the emulsion where the oxide coating is to be retained. The solvent used to remove the emulsion will not affect those parts which have been exposed to the ultra violet light. Acid etching is used to dissolve the oxide not protected by the emulsion.

Stage 3

Phosphorus is deposited at a low temperature on the surface that has retained an oxide mask, as a preparatory stage in the forming of an *n*-type layer.

Stage 4

The reverse side of the slice is coated with boron to give a low resistance contact. Both boron and phosphorus are diffused into the silicon at high temperature. The oxide prevents phosphorus diffusing into the surface, and the *n*-layer occurs only where the oxide has been selectively removed (*Figures 2.10*(*b*) and (*c*)). This process is carried out in an oxygen atmosphere and results in a further build-up of oxide on the slice.

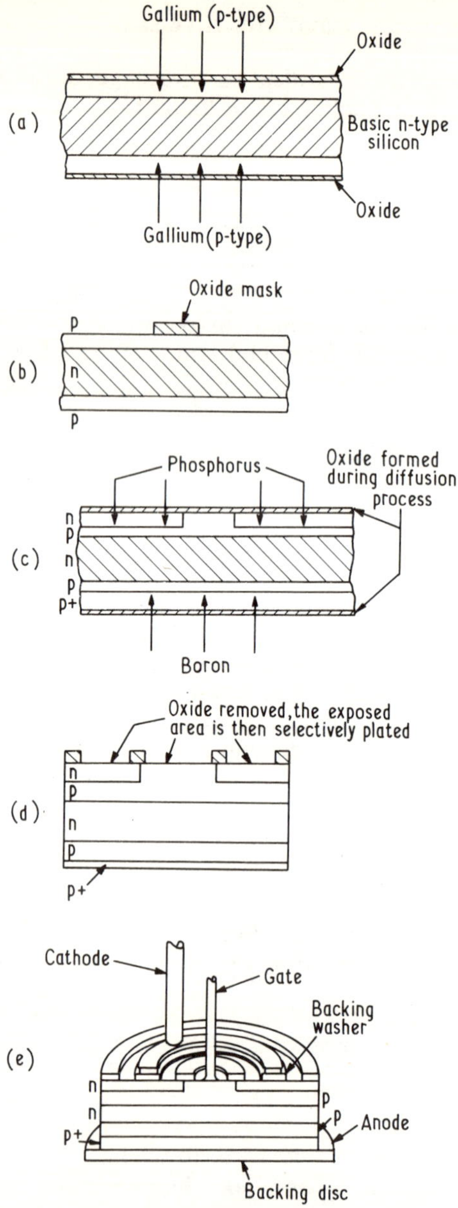

Figure 2.10. Stages in the fabrication of a thyristor
(a) Diffusion of gallium through oxide coatings
(b) Oxide selectively removed
(c) Diffusion of boron and phosphorus
(d) Removal of oxide
(e) Connections to cell

Stage 5

The oxide is again selectively removed to provide contact surfaces (*Figure 2.10(d)*). After plating, electrical contacts are soldered to the *p-n-p-n* cell. Connections to the gate, cathode and anode are as shown in *Figure 2.10(e)*. The cell is etched and protected ready for encapsulation.

The housing in which the silicon cell is encapsulated has to perform several functions:

(i) seal the internal structure from ambient conditions
(ii) provide mechanical strength
(iii) give adequate insulation between gate, cathode and anode terminals
(iv) provide convenient means for allowing current flow from the dice
(v) enable the heat generated in the silicon to be conducted away to the ambient.

External construction of the thyristor varies from TO-5 casing to stud type and Press-Pak or 'hockey-puck' mounting. The silicon is normally hard soldered to two thermally matched back-up plates,

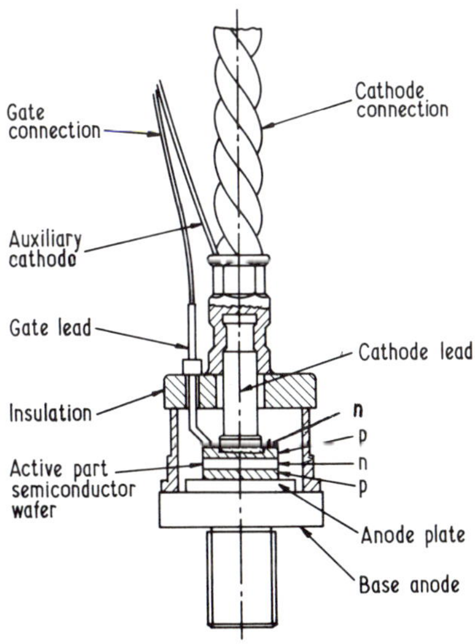

Figure 2.11. Cross-section through an encapsulated device

one of which (usually the anode) is soldered to a copper stud which provides the mass for conducting away the generated heat. Hard soldering is preferred to soft solders which are liable to fracture under thermal cycling. It has been shown that the number of cycles to failure N, for a slice of diameter D (inches) cycled through a temperature range ΔT is given, for hard soldered joints, by

$$N = \left(\frac{C}{D\Delta T}\right)^2 \tag{2.7}$$

where C is a constant of about 3 500.

Soft solder joints are easier to make and are used for less severe operating conditions.

Recently, compression bonded encapsulation has been used for larger devices. These do not use solder joints but rely on spring pressures to maintain the silicon in electrical and thermal contact with the case. These devices can be housed in stud or Press-Pak units, the latter having the added advantage of two cooling areas but being more difficult to mount. *Figure 2.11* shows the cross-section through a typical stud-type thyristor which uses hard solder bonds. The silicon slice has been exaggerated in size to show it more clearly.

OPERATING THEORY

Figure 2.12 illustrates the thyristor in its four layers. *p*-type material has holes as majority carriers and electrons as minority carriers. The reverse is true for *n*-type semiconductor. Therefore with the anode positive to the cathode, junctions J_1 and J_3 are forward-biased and J_2 is reverse-biased. The *p-n-p-n* structure can conveniently be represented by *p-n-p* and *n-p-n* devices as shown in *Figure 2.13*. Clearly, there are now two gate points and the combination can be turned on by putting current in at G_2 or taking it out at G_1. This latter technique has poor gain and is not normally used.

In *Figure 2.13* if α_1 and α_2 are current gains of the two transistors, as indicated, and I_A is the current across junction J_2 then for a

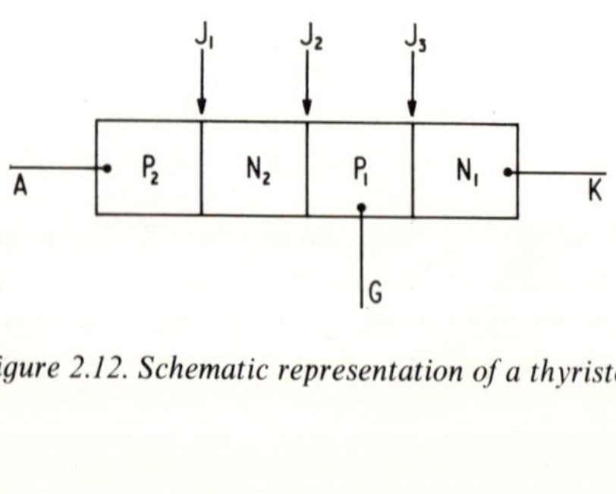

Figure 2.12. Schematic representation of a thyristor

leakage current of I_{co}:

$$I_A = \alpha_1 I_A + \alpha_2 I_A + I_{co}$$

or

$$\therefore I_A = \frac{I_{co}}{1-(\alpha_1+\alpha_2)} \tag{2.8}$$

Normally the two gains are small so that the device current is only slightly greater than I_{co}. The current can be increased (*i.e.* the thyristor turned on) by two methods:

(i) Passing gate current, which will cause emitter current I_E to increase and α to approach unity.

(ii) Increasing the supply voltage which increases the effective V_{CE} of the transistors and increases α again to unity, due to avalanche current build-up in the device.

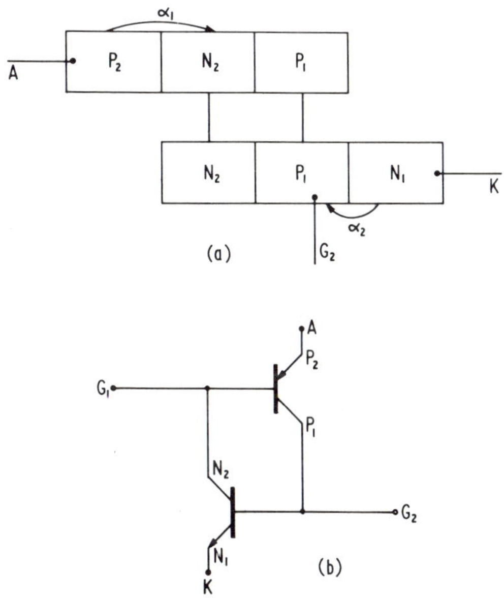

Figure 2.13. Two transistor analogy (a) schematic representation (b) discrete transistor connection

Once the thyristor is conducting, its junctions are heavily doped with holes and electrons. To turn the device off, the polarity of anode and cathode is reversed. This causes the carriers from J_1 and J_3 to be swept away to the supply, resulting in a reverse recovery current. However, the charge at junction J_2 can be removed only by gradual recombination, a process largely independent of external circuit conditions. When this has been completed (in about 20μs) the thyristor can again withstand forward voltage.

To increase the voltage rating of a thyristor it is usual to increase the thickness of the control layer. This, however, leads to devices with greater voltage drops and higher dissipation, as well as increasing the hole-electron recombination period. *Figure 2.14* shows a typical dice configuration in which the wafer has been bevelled to increase creepage distances and so boost the voltage ratings.

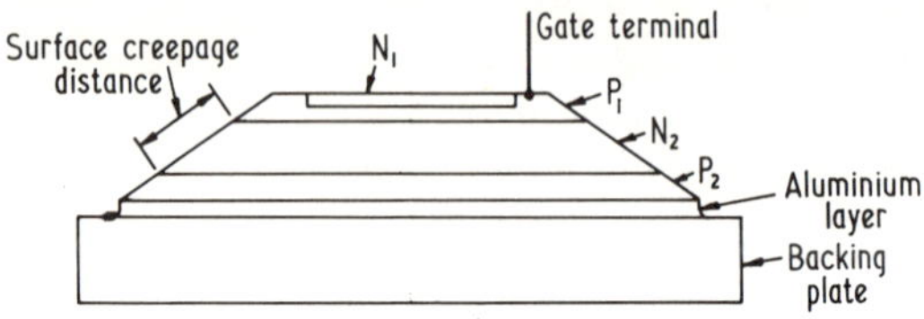

Figure 2.14. Transistor dice with bevelled edge

Even though a thyristor is able to withstand high voltages, it could turn on due to *dv/dt* effects. These cause large currents to flow through the capacitance of the depletion layers, the current crossing junction J_3 as gate current and turning the thyristor on. This is overcome by using 'shorted emitter' techniques, as in *Figure 2.15*, where the majority of the *dv/dt* current is shorted straight to the cathode, and does not flow through the *n*-layer. With this method, voltage rises of 1 000V/μs can be overcome.

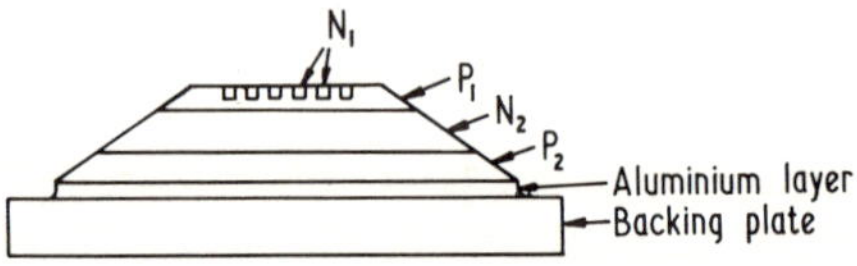

Figure 2.15. 'Shorted-emitter' thyristor construction

Another development worth noting is a technique to increase the *di/dt* capability of thyristors. When a device is initially triggered, most of the anode current tends to be squeezed through a small turned-on portion near the gate, causing localized heating and burn out. This effect is worsened by the fact that all thyristors have an 'emitter lip' between the cathode *n*-layer and the cathode plate. This lip presents impedance to current flow, as in *Figure 2.16(a)* so that if

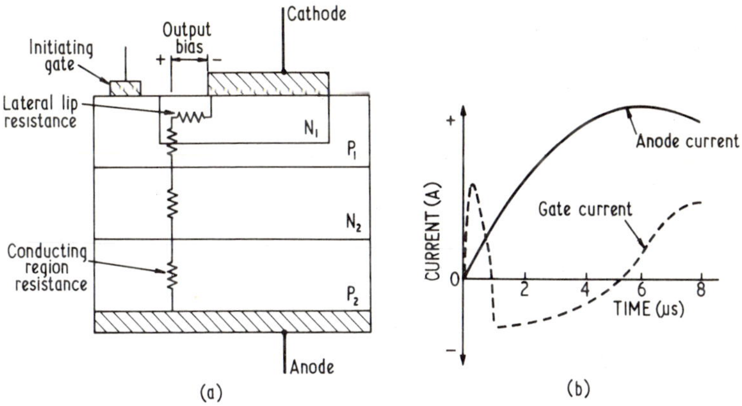

Figure 2.16. Effect of emitter lip in a thyristor (a) cell cross-section (b) current/time characteristic

the anode current is increasing rapidly a voltage is developed across it which opposes the gate signal. If the gate voltage is low, it may even be reversed, as in *Figure 2.16(b)*, which would reduce the turn-on time still further. Regenerative gate thyristors employ a main gate purely to initiate device conduction. The resistance of the emitter lip is increased to give a relatively large voltage drop which is then picked off, as in *Figure 2.17*, to feed several points on the *p*-layer and

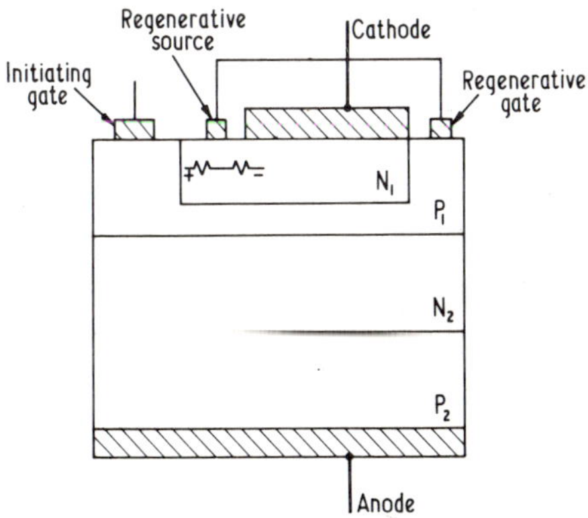

Figure 2.17. Regenerative gate construction

so provide a gate signal to ensure a more uniform and rapid device turn on. Current rises of over 1 000A/µs are now claimed with these thyristors.

COOLING

During operation of a thyristor, heat can be generated due to several phenomena:

(i) power loss during forward conduction
(ii) leakage current when the thyristor is blocking in the forward or reverse direction
(iii) switching losses which occur during turn-on and turn-off
(iv) gate losses, normally small when single-pulse firing techniques are adopted, but can be significant for d.c. or pulse train signals.

The heating of the thyristor causes uneven expansion between dice and its solder joints, which can lead to fracture if there is excessive cycling. Hard solder and compression bonding techniques are then essential forms of encapsulation. It is also necessary to dissipate the heat generated in the device, to maintain the temperature of the semiconductor junction below its peak rated value.

Referring again to the CRS3 thyristor, *Figure 2.2* gives the maximum working junction temperature to be 125°C, whereas the stud (or case) temperature varies with the conduction angle, as in *Figure 2.3*. Therefore in order to conduct 3A d.c. through the CRS3 thyristor, its stud temperature must be maintained below 107·8°C. A very useful concept in calculating cooling requirements of thyristors is thermal resistance. This is the resistance which the device presents to heat flow, and has units of degrees Celsius per watt. Therefore if P watts are flowing between a thyristor junction at T_j°C and the case at T_c°C, the thermal resistance between junction and case is

$$\theta_{jc} = \frac{T_j - T_c}{P} \text{ °C/W} \qquad (2.9)$$

Similarly there can be a thermal resistance between the case of the thyristor, at T_c°C, and the heatsink, at T_h°C, given by θ_{ch} and between the heatsink and the ambient θ_{ha} where:

$$\theta_{ch} = \frac{T_c - T_h}{P} \text{ °C/W} \qquad (2.10)$$

and

$$\theta_{ha} = \frac{T_h - T_a}{P} \text{ °C/W} \qquad (2.11)$$

The power loss across the device is found from curves given in data sheets, such as the one for the CRS3 thyristor which is reproduced in *Figure 2.18*. Therefore in order to conduct 3A d.c., the thyristor case temperature must be maintained at 107·8°C. The thermal resistance between junction and case is 6°C/W (*Figure 2.6*) so that this gives, from equation (2.9), a maximum junction temperature of 125°C ($P = 3{\cdot}2$W at 3A d.c., from *Figure 2.18*) which agrees with the device data sheets. Normally the thermal resistance between case and heatsink can be made unity by ensuring a good thermal contact.

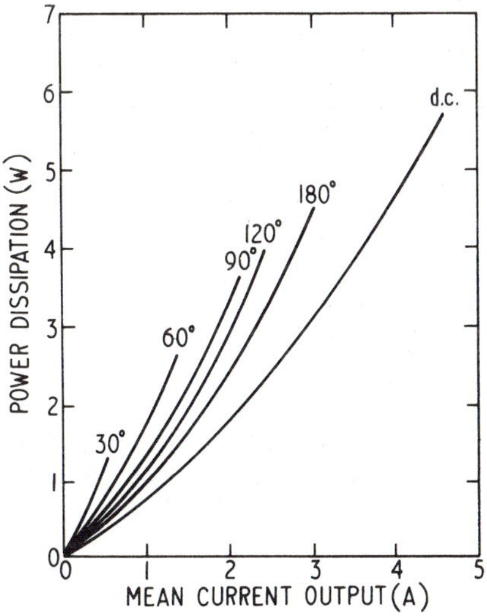

Figure 2.18. Power loss curves from a thyristor data sheet

Therefore for a working ambient temperature of 45°C the heatsink must have a thermal resistance of (equation 2.11)

$$\frac{107{\cdot}8-45}{3{\cdot}2} = 19{\cdot}6°\text{C/W}$$

Commercially available heatsinks have a stated thermal resistance, dependent on heatsink size, and can readily be selected. Alternatively, graphs are provided, as in *Figure 2.19*, for the heatsink under various conditions. Here forced air cooling is used. From these curves the value of θ_{ha} can readily be obtained. For instance with an air flow of 200 ft/min the temperature gradient in the heatsink for 50W dissipation is 95°C so that $\theta_{ha} = 1{\cdot}9$°C/W.

So far the discussions have dealt exclusively with instances in which there is steady-state power loss in the thyristor. Often, however, only intermittent operation is required. *Figure 2.20* shows the effect of a step increase in power on the junction temperature. Although a thyristor is physically small it presents a finite thermal

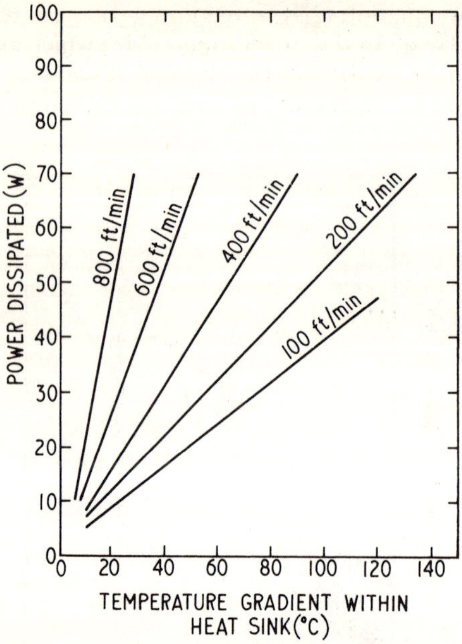

Figure 2.19. Heatsink design characteristics (2·5in type H forced air cooled)

mass so that the junction temperature increases gradually. Since thermal resistance is defined as the ratio of the rise in temperature to the power increase, this impedance will build up with time, as in *Figure 2.21*. This is referred to as the transient thermal resistance of a thyristor. Only after a definite time does the transient thermal resistance reach the d.c. impedance value.

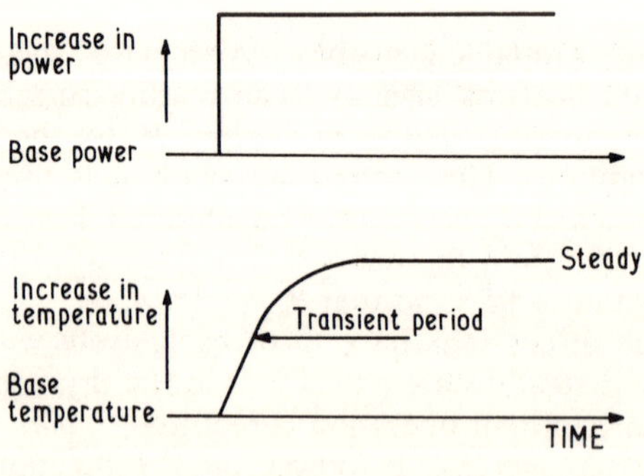

Figure 2.20. Thermal inertia of a transistor

The transient thermal resistance should be used in instances of pulsed operation, where the duration of the pulse is such that the thyristor junction does not reach its steady state value. Therefore in

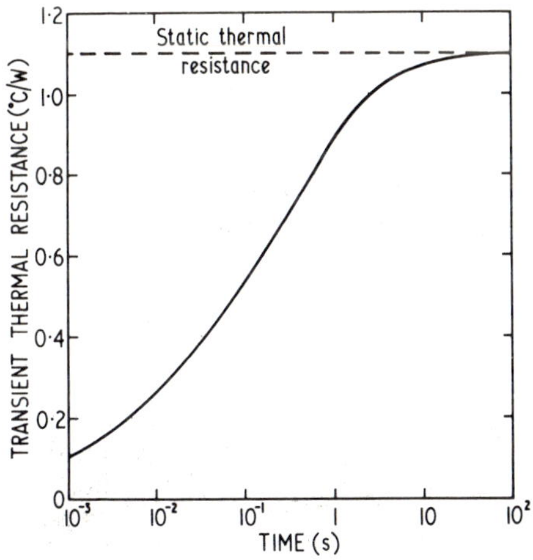

Figure 2.21. Typical transient thermal resistance curve

Figure 2.22(a), for a single pulse of P_m watts, the junction temperature rise above the case is given by:

$$T_{jc} = P_m \theta_T \tag{2.12}$$

where θ_T is the transient thermal resistance at time t (pulse duration). Similarly sinusoidal pulses as in *Figure 2.22(b)* can be approximated to a rectangular pulse with the same mean power, and the value of P_m used in equation (2.12).

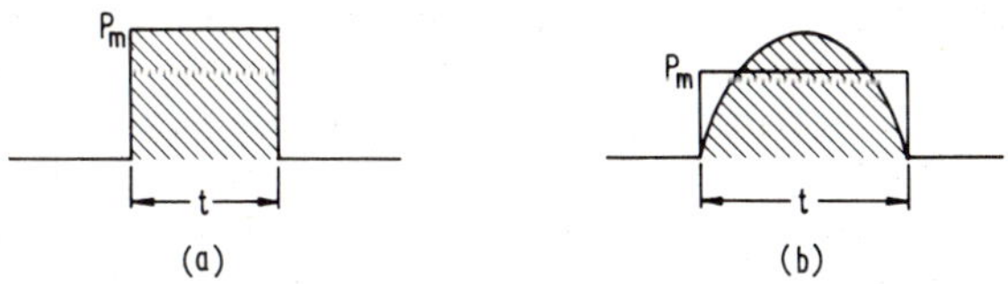

Figure 2.22. Power dissipation under various single-pulse shapes (a) square pulse (b) sinusoidal pulse

For the train of pulses shown in *Figure 2.23* the mean power dissipation P is used with the steady-state thermal resistance and the peak power P_m with the transient thermal resistance. This gives

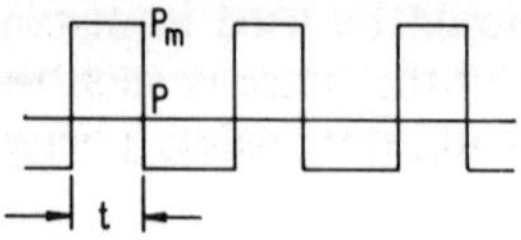

Figure 2.23. Power dissipation under a stream of pulses

the highest rise in junction temperature above the case as

$$T_{jc} = P\theta_{jc} + P_m\theta_T \qquad (2.13)$$

Calculations for a train of irregular pulses can get rather involved and should be approximated to a regular train, if possible.

OVERCURRENT PROTECTION

There are several factors which determine the choice of an overcurrent, or fault current, protection system for thyristor circuits. The most important of these is perhaps whether the supply source is 'stiff' or 'soft'. A 'soft' supply has series impedance, which would limit the rate of current rise during a circuit fault. Such a fault could take the form of a supply short circuit through the thyristor. A 'stiff' supply can cause a rapid rise in thyristor current in a fraction of a cycle, destroying the device.

When running from a 'soft' source two alternatives are open to the designer when considering fault current. If there is considerable improbability of a sustained fault over several cycles it is often possible to calculate the fault current and choose a thyristor with a surge current rating adequate to carry it. Often this will mean using a thyristor larger than that required to operate on the rated full load current, but no other fault current protection circuits are required. The second alternative is to open the faulty circuit by electromechanical contactors or thermal overcurrent trips. It is important to appreciate that these devices take at least several milliseconds to operate and the thyristor will have to carry the full current during this interval. Therefore they can be used only on supplies with sufficient regulation to limit the initial current build-up. A common technique is to use one thyristor, which is larger than the rest in the system, to short-circuit the supply lines when the device currents reach a preset maximum value, and so operate contact breakers in the line. Now only one thyristor need be rated to carry the initial current surge. Such a system is also commonly used for overvoltage protection, as discussed in the next section.

Stiff supplies require special consideration. The rapid fault current build-up means that circuit isolation must be quickly achieved. Fast-acting semiconductor fuses are now almost invariably used. *Figure 2.24* shows the protective action of such a fuse. At the commencement of a fault the current builds up rapidly, the rate of increase

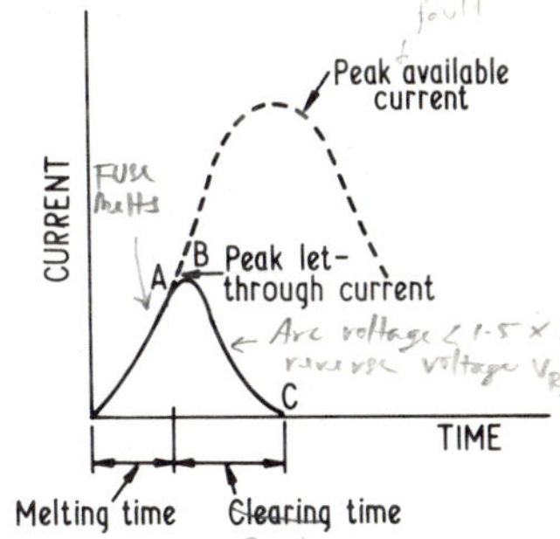

Figure 2.24. High-speed fuse characteristics

being limited by the relatively small circuit impedance. If no action were taken the current would rise to the peak available value before starting to decay (a half sine wave of supply voltage is assumed). However, long before this current is reached the fuse melts, at A. The current increases slightly to B before the energy dissipated in the fuse arc causes it to decrease to zero at C. It is important that the rate of decrease of current in the arc is not too severe or it can give rise to high voltage transients (Ldi/dt) in circuit inductances, which would destroy the thyristors. In a well designed fuse, melting and arcing times are approximately equal.

Semiconductors and fast acting fuses both exhibit the property of withstanding almost constant values of i^2t (ampere^2-second) below about one cycle. This greatly simplifies the design of protective fuses since all that is basically required is to ensure that the i^2t rating of the fuse is less than the thyristor it is to protect, so that it will fail first. If the fuse is in the primary of a transformer and the thyristor cell in the secondary, then the i^2t of fuse and thyristor must be related by:

$$i^2t_{(\text{fuse})} \leqslant \left(\frac{V_{\text{sec}}}{V_{\text{prim}}}\right)^2 \times i^2t_{(\text{thyristor})} \qquad (2.14)$$

The i^2t of a fuse is affected by circuit voltage and the prospective fault current. This is shown by the curves in *Figures 2.25* and *2.26*. As an example in use of these characteristics suppose a fuse is needed to protect a circuit operating at 300V and 300A, the asymmetrical peak current being 13·5kVA. Taking a GS450/400 (450V, 400A r.m.s.) fuse, the normal i^2t is 800 000A^{2}s with an asymmetrical peak current of 200kA. *Figure 2.25* gives a reduced i^2t of 280 000 at 300V and drawing a line in *Figure 2.26* passing through the 200kA, 280 000A^{2}s point,

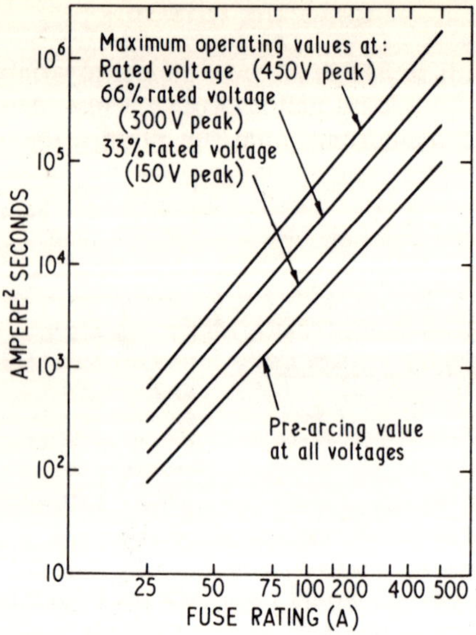

Figure 2.25. Values of i^2t at maximum prospective current of 200kA peak asymmetrical for type GS450 fuse links

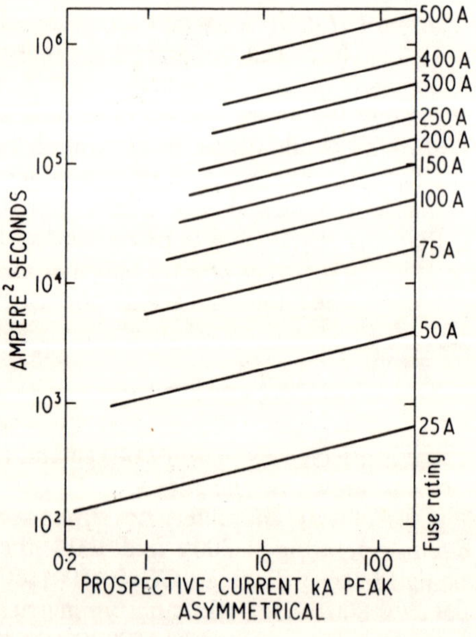

Figure 2.26. Variation of total i^2t related to prospective peak asymmetrical current at maximum rated voltage (450V p.i.v.) for type GS450 fuse links

the value of i^2t at 13·5kVA is 150000. This is now the rating of the fuse under the circuit conditions described. It should be less than that of the thyristors being protected. It is possible to reduce the i^2t of the fuse by choosing one with a higher voltage rating. However, the excess of fuse rating over supply voltage could give a very abrupt arc quenching and generate large voltage transients. A better method of reducing the fuse i^2t is to decrease the prospective peak asymmetrical current by increasing circuit reactance. However, it is important to ensure that the fault current reaches a sufficient value to blow the fuse within a relatively short time or the thyristor may be damaged. This is illustrated very clearly by *Figure 2.27* where it is

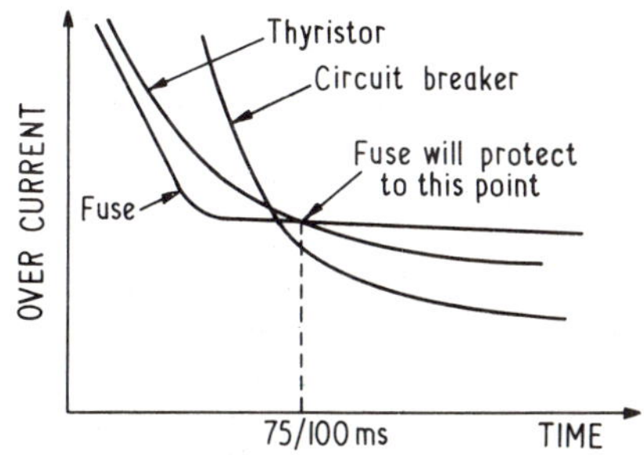

Figure 2.27. Characteristics of several protective devices

seen that for stiff supplies (rapid current rise) a fast-acting semiconductor fuse will protect a thyristor, whereas for soft supplies circuit breakers must be used.

OVERVOLTAGE PROTECTION

Voltage transients may be generated in circuits owing to several causes, the principal of which are:

(i) Inrush current to the transformer primary. Oscillations can now occur in the secondary, due to distributed capacitance and leakage reactance, which gives twice the peak supply voltage across devices.

(ii) Stray capacitance between primary and secondary of a step-down transformer. This causes voltage transients at the secondary when the primary is switched on.

(iii) Interruption of current into the primary of a transformer, the resulting rapid decay of flux inducing a large transient voltage into the secondary winding.

(iv) Sudden interruption of inductive load current, for instance due to acting of overcurrent protective devices, which gives large *Ldi/dt* voltage transients.

(v) Charge storage effects in semiconductors, and notably in the free-wheeling diode. The storage current does not cause any transients but it ceases so rapidly when the junction recovers, that the resulting voltage spikes can be severe.

Voltage transients, if they exceed the peak forward rating of the thyristor, could destroy the device. Even if this rating is not exceeded faulty firing could occur, which in many cases would lead to device overcurrents and operation of current limiting fuses. In the reverse direction a transient will cause relatively large currents to flow in a small area of the thyristor junction, causing localised heating and device failure.

It is necessary to ensure that as many sources of voltage transients are eliminated from a thyristor circuit as possible. For instance, it is preferable to switch the secondary of transformers rather than the primary, to ensure that breaking devices have a relatively long arcing time (the voltage transient being dissipated in the arc), and to use fast recovery semiconductor devices (to limit charge storage currents). When elimination of voltage transients is not feasible special steps must be taken to minimise their destructive effects. The most commonly used method is by connecting *R-C* circuits across the input lines, as in *Figure 2.28*(*a*). The circuit in *Figure 2.28*(*b*) is sometimes preferable since an electrolytic capacitor, which could be made physically small, can be used. A bleeder resistor discharges the capacitor when the circuit is de-energised. Instead of using input filtering, *R-C* circuits can also be be connected across individual devices.

Apart from using *R-C* suppressors it is possible to employ avalanche devices which are largely self-protecting. When an overvoltage occurs the semiconductor breaks over and dissipates the transient energy across itself. Several Zener devices are available which can be connected in the thyristor circuit to produce the same effect. *Figure 2.29* shows the connection of SafeTstaCs* to protect devices in a three-phase bridge circuit. The performance of these devices is largely given by manufacturers' data sheets, like that in *Figure 2.30*, which defines the dynamic characteristic of one plate of the selenium SafeTstaC. As a design example suppose a single-phase full-wave bridge rectifier is supplying 200A d.c. from 200V r.m.s., the input transformer having a magnetising current of 4·0A referred to the secondary. Then from *Figure 2.30*, since the maximum discharge current of the SafeTstaC is to be 4A, plate size 930 can be used. Each plate is rated at 25V r.m.s., so for 200V eight plates are

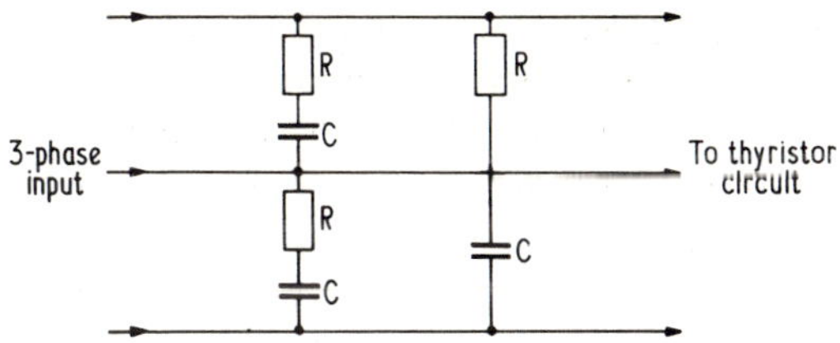

(a)

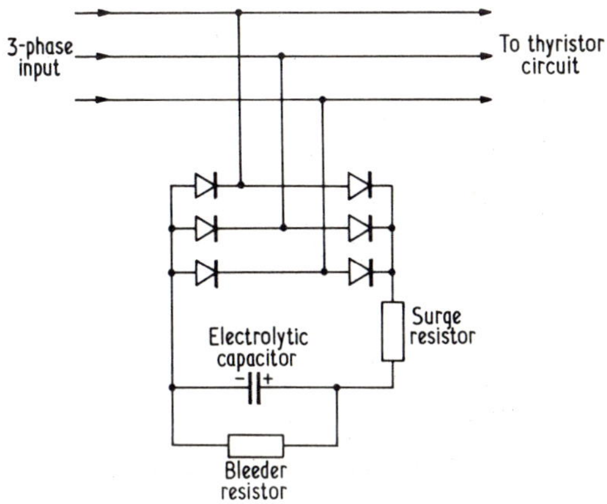

(b)

Figure 2.28. Overvoltage protection by R-C circuits
(a) individual suppressors across lines
(b) single R-C suppressor

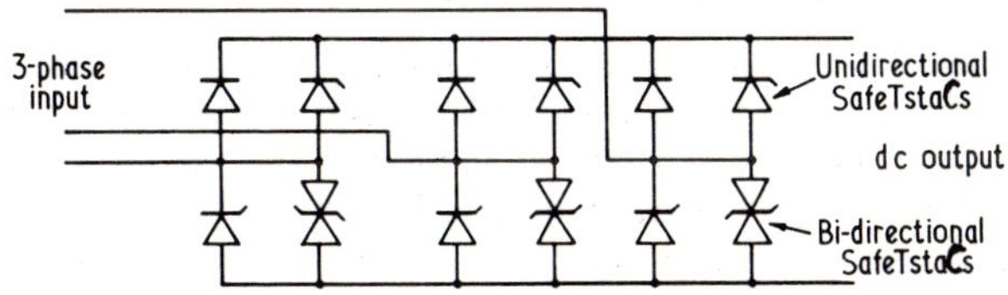

*Figure 2.29. Overvoltage protection using SafeTstaCs**

*Trade name: STC Ltd.

connected in series. The maximum discharge voltage per plate at the peak discharge current is 50V (the suppressed voltage from *Figure 2.30* is seen to be 47·5V but 50V is used here) so that the semiconductor devices must have a peak rating of at least $50 \times 8 = 400$V to be

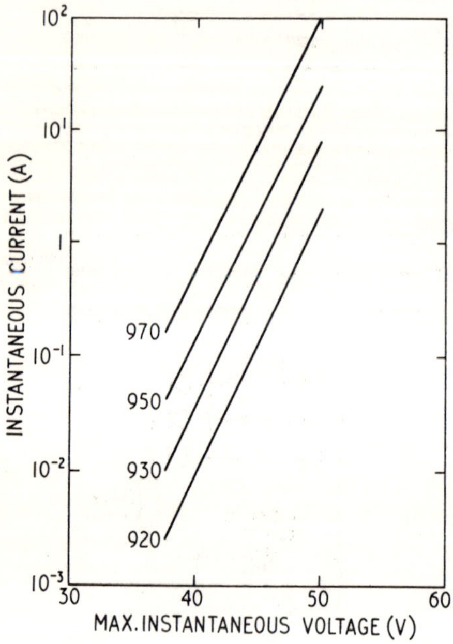

Figure 2.30. Dynamic characteristics of SafeTstaCs plates

adequately protected. Although surge suppressors, as described above, protect the circuit against peak voltages, they are not effective in guarding against *dv/dt* which can cause misfiring of a thyristor. *R-C* circuits should now also be used.

SERIES AND PARALLEL CONNECTIONS

Improvements in semiconductor techniques have led to thyristors with current ratings in excess of 500A and several thousand volts blocking capability. Given a projected increase over the next decade of almost 300% in both ratings it is likely that only very special applications would require that several thyristors be used together to increase the voltage or current obtainable from one device.

Figure 2.31(a) shows two thyrsistors connected in series to share a voltage V. If the peak rated voltage of each thyristor is V_{pk} it is hoped that the maximum permissible value of V can approach $2V_{pk}$. However since the two thyristors are in series they must share the same leakage current I_L. Therefore if they had blocking characteristics, as in *Figure 2.31(b)*, clearly thyristor CSR2 will be operating very close to its rated voltage while CSR1 only blocks a fraction of this voltage. Therefore the peak value of V is severely limited.

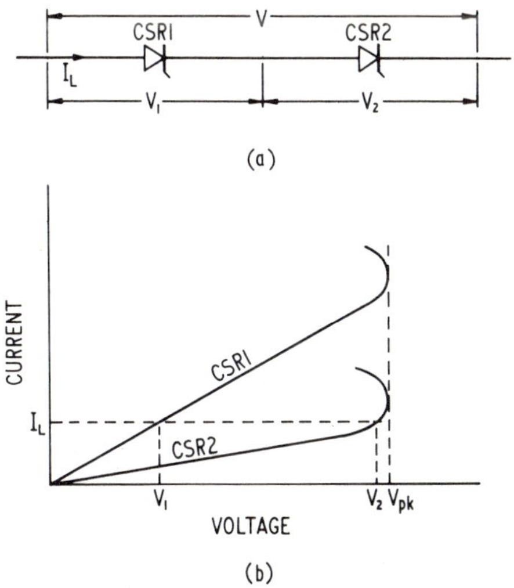

Figure 2.31. Unequal voltage distribution in a series string
(a) series-connected thyristor
(b) spread in device characteristics

Forced sharing of voltage amongst a series string of thyristors can be accomplished by using sharing resistors as in *Figure 2.32(a)*. To be effective the current through these resistors must be large enough to swamp the inequality in the thyristor leakage currents. If R_{pk} denotes the maximum value of this resistance, in the limiting condition the worst case of unequal sharing occurs when one thyristor, say CSR1, has negligible leakage whereas the remaining have maximum leakage current, I_L. If there are n cells in series then

$$V - V_1 = V_2(n-1) \tag{2.15}$$

and

$$V_1 = \left(I_L + \frac{V_2}{R_{pk}}\right) R_{pk} \tag{2.16}$$

Solving equations (2.15) and (2.16) gives:

$$R_{pk} = \frac{nV_1 - V}{(n-1)I_L} \tag{2.17}$$

where V_1 is the peak rated voltage of the thyristors, and V is the maximum voltage that can be safely applied to the series string when R_{pk} is used.

Series resistors are adequate in forced sharing of voltage between a string of thyristors in their static state. However when the thyristors are being turned off or turned on, sharing resistors are ineffective in

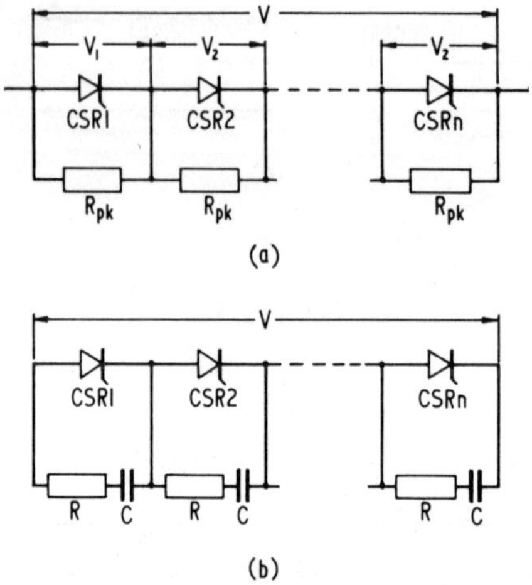

Figure 2.32. Additions for greater equality in sharing d.c. and transient voltages (a) d.c. voltage sharing (b) transient voltage sharing

ensuring equal voltages across devices. The thyristors which recover their blocking state fastest, or which turn on last, will see the full voltage of a series string, often with disastrous effects. This transient inequality can be protected against by using a capacitor across each device, as in *Figure 2.32(b)*. This slows down the rate of change of voltage across devices turning off or triggering on. A resistor R is now placed in series with C to limit the large discharge currents which would otherwise occur when the thyristors are turned on. The value of C is normally calculated using the following empirical formula:

$$C = \frac{5I_F}{V_1} \tag{2.18}$$

where I_F is the peak forward current flowing through the thyristor before commutation occurs, and V_1 is the maximum voltage rating of the thyristors.

Several thyristors connected in parallel all have the same voltage across them when conducting, but due to unequal characteristics they carry different currents. This is illustrated in *Figure 2.33*. Once again it is possible to swamp the inequality of the device voltage

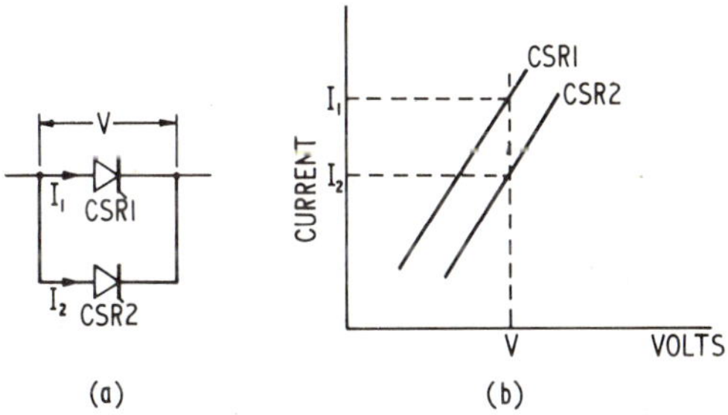

Figure 2.33. Unequal current distribution in paralleled thyristors (a) paralleled thyristors (b) spread in forward characteristics

drops, when conducting, by series resistances. (Series inductances may be used but are effective only during transient conditions when the current is changing.) This is shown in *Figure 2.34*. If V_1 and V_2 are maximum and minimum voltage drops across the thyristors, and I_1 and I_2 the permitted variation in the shared current, then:

$$V = I_1R + V_1$$
$$= I_2R + V_2$$

Therefore

$$R = \frac{V_2 - V_1}{I_1 - I_2}$$

Series resistances force d.c. more equally through paralleled thyristors but fail to compensate for unequal latching current, holding

Figure 2.34. Series resistance for forced current sharing

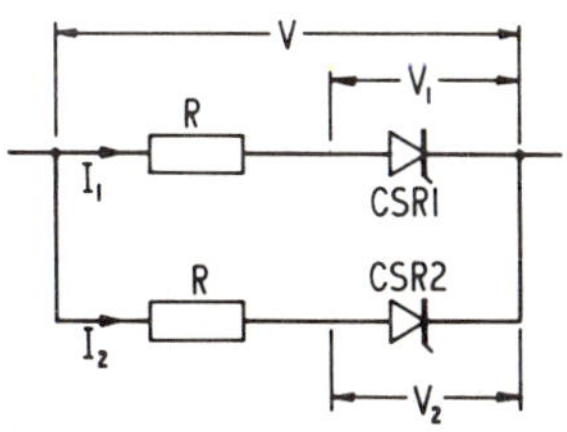

current or turn-on times. For instance, in a parallel-connected set, if one device turns on before the rest, it will be forced to momentarily carry the full load current. This need not be serious so long as the *di/dt* or surge rating of the thyristor is not exceeded. If required, a small reactor can be connected in series with each device to compensate for unequal turn-on times.

If the latching currents of a parallel set of thyristors differ, it could mean that for pulse firing some thyristors will reach their latching currents before the end of the pulse, whereas others will not. This will mean that only these devices will conduct and support the total load current. Clearly the firing signal must be long enough to prevent this. Similarly if the thyristor currents decrease momentarily, devices with holding currents below this value will turn off. When the load current again increases to its full load value the off thyristors will not turn on unless refired, and some devices may be overloaded. It is essential to ensure that the load current does not fall to too low a value.

Although series connection is sometimes used, paralleling thyristors is inefficient (loss across series impedance) and not common. In rare instances when it is necessary, specially matched pairs, with nearly equal volt drops, can normally be obtained from thyristor manufacturers.

The firing circuit required to operate thyristors in series must have several isolated outputs, as in *Figure 2.35*(*a*), each capable of

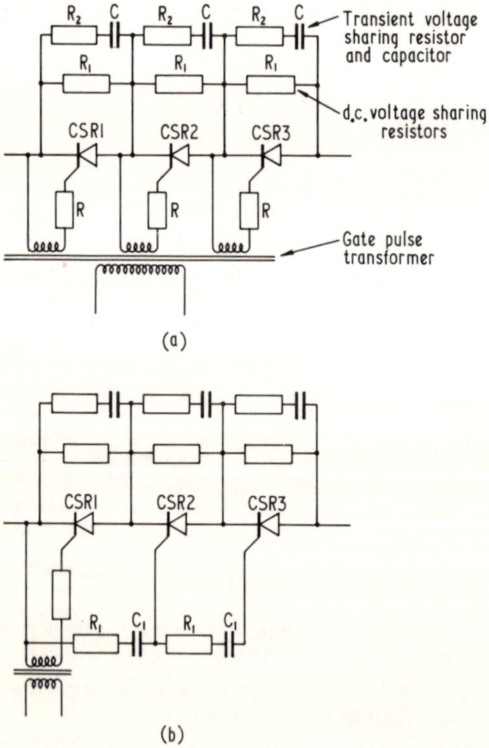

Figure 2.35. Firing series thyristors (a) simultaneous firing (b) slave firing

providing the required thyristor gate drive. Series resistors R are essential to prevent a thyristor with low gate impedance shunting current away from the remaining devices. Although this method of simultaneous firing is most successful, it is also possible to use slave firing as in *Figure 2.35(b)* where only CSR1 is directly fired, the remaining thyristors in the string turning on when the preceding device conducts.

With parallel thyristors, simultaneous and slave firing may also be used, as in *Figure 2.36.* Although simultaneous firing does not require several isolated gate drives, the cathodes of all thyristors now being commoned, it is still necessary to include drive resistors R_1 to prevent shunting by a low-impedance gate. Slave firing turns

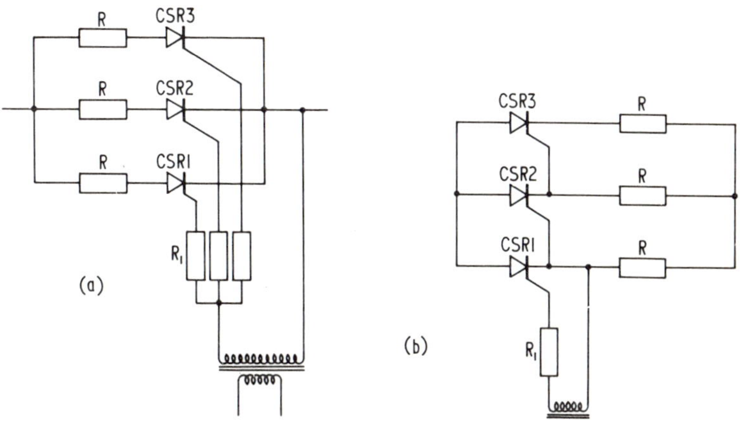

Figure 2.36. Connections for triggering parallel-connected thyristors (a) simultaneous firing (b) slave firing

on a thyristor when the preceding device begins conduction, as for a series connected string. Since successive conduction is an essential feature of slave firing systems, it means that the last device being turned on will momentarily support the full string voltage, or that the first device of a parallel group will conduct the total load current for a brief instance. This being so, slave firing cannot be used when more than about three devices are being operated in series or parallel.

CONCLUSIONS

Although the electromechanical analogy of a thyristor has been freely used to explain its parameters, it must not be forgotten that it is essentially a semiconductor device. Data sheets give the characteristics of the thyristor, which define its performance under certain operating conditions, and its rating, which must not be exceeded if damage is to be avoided. Transient voltages and currents which exceed the rated values can destroy the thyristor which must therefore be protected. Generally, as with most semiconductors, heat is the critical parameter and adequate heatsinking is essential even under normal operating conditions. Where the required rating of the thyristor exceeds that obtained with a single device, several can be connected in series or parallel.

3

Static switching

The most basic applications of a thyristor are those in which its use is essentially that of a switch connecting a load to its supply. These circuits are generally referred to as static contactors, owing to the absence of moving parts. In this chapter, typical d.c. and a.c. contactor systems will be examined and reference made to electronic control circuits which often enable them to perform a variety of functions.

A static contactor has several desirable features. There are no moving parts to wear or burn out. It is mechanically robust and noise free in operation. It can operate very quickly, in about 0·1ms (compared to about 50ms for a mechanical contactor) and is readily adaptable to sophisticated electronic control. This can include zero level switching (such that the a.c. contactor closes at voltage zero and so prevents interference in the lines), gradual build up in voltage, rapid overload isolation, etc.

However, the static contactor is not used as frequently as its mechanical counterpart, and for very good reasons. It is generally more expensive and physically larger. It is more prone to failure due to current and voltage overloads, and it does not provide complete isolation between the supply and the load. There are several applications where it has found acceptance. For example in explosive atmospheres where arcing of contacts would be dangerous, or where it could generate intolerable interference, as in computer installations. Contactors mounted in inaccessible conditions are of the static type because they need no maintenance. If the contactor is also required to operate frequently, or to respond to voltage signals generated by some other process, then again static contactors could be used.

Contactors can be used to operate on a.c. or d.c. Generally, mechanical contactors can have the same construction for both these sources, only their ratings being affected. However, since a thyristor can be turned off only when the current through it falls below a certain value, static contactors using them differ in circuit detail when operating on a.c. or d.c.

A.C. CONTACTORS

A.C. contactors can operate from single- or three-phase lines. Each can be divided into three parts, the power circuit comprising thyristors, the firing circuit used to turn on the thyristors, and control and protection circuits which switch the firing circuits on or off to perform a certain circuit function.

Power circuits

Figure 3.1 shows the simplest and most frequently used arrangement for a full-wave single-phase contactor. With CSR1 and CSR2 nonconducting, the load is effectively isolated from the supply lines A and B, if the leakage of the devices is ignored. When CSR1 and CSR2 are fired they can be considered as short circuits and the load is connected to the supply.

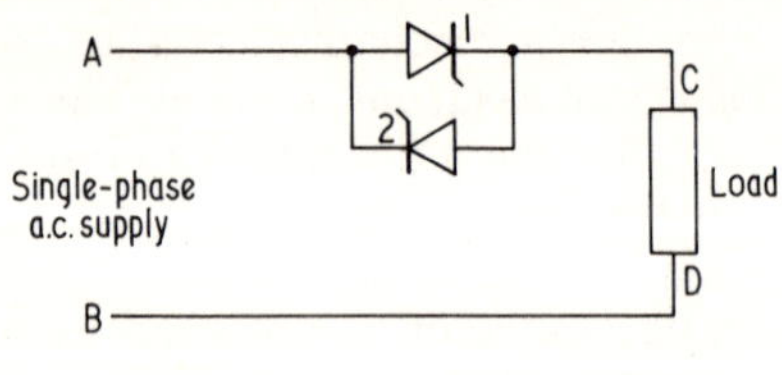

Figure 3.1. Single-phase contactor power configuration

Figure 3.2 shows a three-phase static contactor. Only one thyristor and a reverse diode are required in each line since one of the lines must conduct in the forward direction before any load current can flow. As in single-phase contactors, with no thyristor gate drive, the

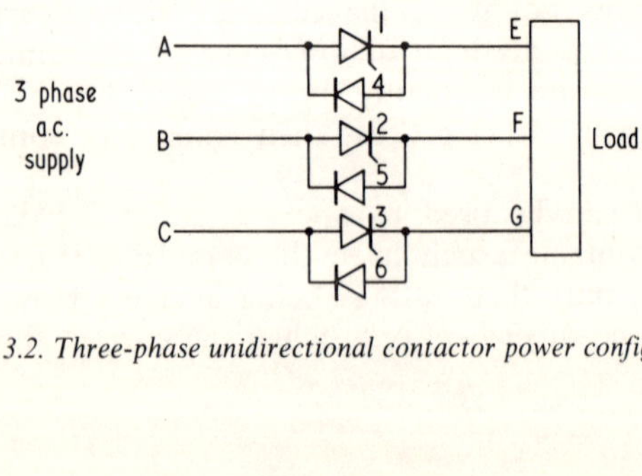

Figure 3.2. Three-phase unidirectional contactor power configuration

load is isolated or off. When the thyristors are turned on the load is connected to the supply.

If it is required to reverse the supply to the load, for instance where the load is a motor and its direction of rotation is to be changed, the circuit of *Figure 3.3* must be used. With CSR1 to CSR6 on and CSR7 to CSR10 off, line A feeds E, B feeds F and C feeds G. To reverse the supply to the load CSR1, CSR2 and CSR7 to CSR10 are on and CSR3 to CSR6 are off. Now A feeds E, B feeds G and C feeds F. It is necessary to use back-to-back thyristors in all the lines since if, for instance, 10 was a diode then firing CSR3 would cause a short circuit across lines B and C.

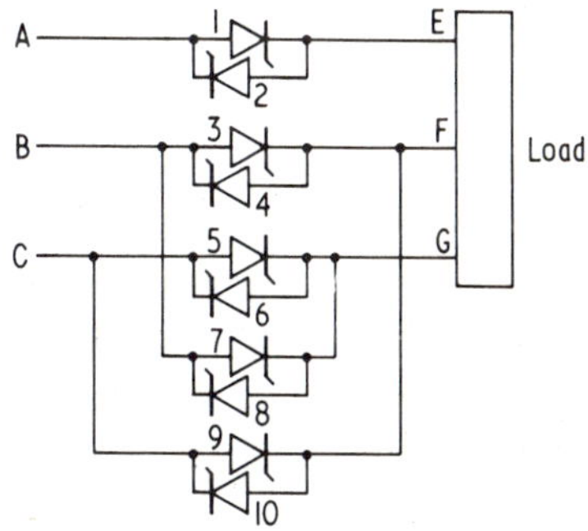

Figure 3.3. Three-phase reversing contactor power configuration

The devices in the circuits of *Figures 3.1* to *3.3* must all be rated to withstand the peak line voltage. For 240V single-phase and 440V three-phase supplies this means peak voltages of 340V and 630V respectively. To allow for line transients, it is usual to employ 500V and 800V devices. Additionally, surge suppressors must be used between lines and across devices if the transients are troublesome. In this context it is perhaps worth noting that the circuit of *Figure 3.2* is protected against reverse voltage transients by the diodes. Over-voltage in the forward direction could cause a thyristor to break over into conduction for a half cycle, but now it need not be damaged, and the load may be such as not to be appreciably affected by the half cycle of power.

The current rating of the devices can be calculated as follows, with reference to the single-phase circuit. Assuming the line current to the load to be given by $I_{pk} \sin \theta$, when the thyristors are conducting, then the r.m.s. line current is:

$$I_{rms} = \left[\frac{1}{\pi} \int_0^{\pi} I_{pk}{}^2 \sin^2\theta \, d\theta \right]^{1/2} = \frac{I_{pk}}{\sqrt{2}} \qquad (3.1)$$

The thyristors carry current for only half a cycle each so their mean rating is given by I_T as:

$$I_T = \frac{1}{2\pi}\int_0^{\pi} I_{pk} \sin\theta \, d\theta = \frac{I_{pk}}{\pi}$$

and from equation (3.1):

$$I_T = \frac{\sqrt{2}}{\pi} I_{rms} = 0{\cdot}45 I_{rms} \quad (3.2)$$

Therefore each thyristor must have a mean current rating of 0·45 times the r.m.s. line load current.

Equation (3.2) applies equally well to the thyristors and diodes of the three-phase circuits in *Figures 3.2* and *3.3* since these also carry half a cycle of the line current each.

Thyristor firing circuits

If the load were resistive it would be sufficient to fire each thyristor with a single pulse, at the start of each cycle, the pulse width being larger than the turn-on time of the thyristor. The waveforms for the single-phase contactors are shown in *Figure 3.4* and are equally applicable for the three-phase circuits. For inductive loads *Figure 3.4(e)* shows that there can be up to 90°E lag between voltage and current. Therefore a pulse to CSR1 at the start of the positive half cycle, as shown in (*b*) would be ineffective in turning the device on since load current is still flowing through CSR2 and CSR1 is

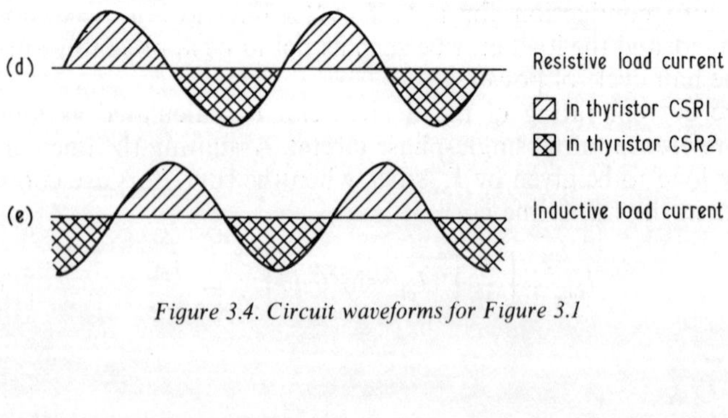

Figure 3.4. Circuit waveforms for Figure 3.1

reverse-biased. Only when this current has decayed to zero will CSR1 be able to conduct.

To enable the contactor to work on inductive loads it is therefore necessary to maintain continuous gate drive for 90°E. However, it is far simpler to fire CSR1 and CSR2 simultaneously and continuously for the duration of the contactor on period. This eliminates any requirements for synchronising the gate pulse to the supply. The only disadvantage to this is that gate pulses are now applied to a thyristor when it is reverse-biased, so increasing its leakage current and hence dissipation. However, since the reverse voltage is only equal to the forward voltage drop of the other conducting thyristor this is not important.

Figure 3.5 shows a firing circuit for *Figure 3.2*. TR1, T_1 forms a free-running blocking oscillator with capacitor C_1 in its base. This determines the frequency of the oscillator (about 10kHz) and prevents the output transformer T_1 from saturating, which would result in excessive dissipation of transistor TR1. The output from the three secondaries is half wave rectified and then smoothed before being fed through limiting resistors to the thyristors. Zener D3 and diode D2 are chosen such that the reverse voltage across the primary of transformer T_1 during the off period of TR1 is approximately equal to the on voltage when TR1 is conducting. This provides equal energising and de-energising potentials for T_1 and gives an almost 1:1 output voltage waveform, so reducing the output smoothing requirements.

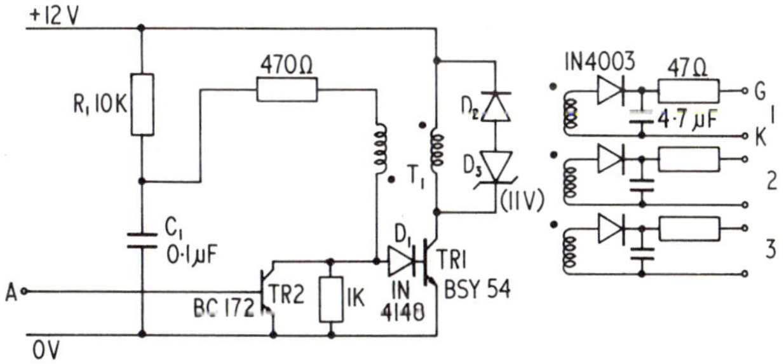

Figure 3.5. Blocking oscillator firing circuit for Figure 3.2

R_1 is a starting resistor which provides the initial base current to TR1 initiating its conduction, this action then being regenerative due to the feedback winding on T_1.

Transistor TR2 acts as a control for the blocking oscillator. With TR2 conducting TR1 is held off and thyristors CSR1, CSR2 and CSR3 are non-conducting. Diode D1 ensures a bias to TR1 holding it off. When TR2 is off the blocking oscillator is free running and functions normally.

This circuit responds to negative logic, that is to say when input at A goes negative (to 0V) there is an output from the contactor; when the input goes positive the contactor is off. Conversion to positive logic requires TR2 to be replaced by a *p-n-p* transistor, A being fed to its base, the emitter being connected to D1 and collector to the zero volt rail.

An identical circuit to *Figure 3.5* can be used for *Figure 3.1* but T_1 would now need only two output secondaries. *Figure 3.3* would require two separate oscillators, each with six output secondaries, one oscillator firing thyristors CSR1 to CSR6 and the other CSR1, CSR2 and CSR7 to CSR10.

Protection circuitry

The circuit of *Figure 3.5* is the basic thyristor driver which responds to voltage signals at its input A. These signals can come directly from the operator in the form of 'on' or 'off' commands. Superimposed on these, through suitable gating circuitry, can be signals from various protection circuits, as described in the subsections below.

Motor over-temperature protection

In *Figure 3.6* the p.t.c. thermistor is embedded in the motor whose temperature is to be monitored. The temperature at which the thermistor changes state from a low to high resistance determines the maximum operating point. Under normal running conditions R_2 is low in value compared to R_1 and TR1 is off. Diode D1 is open

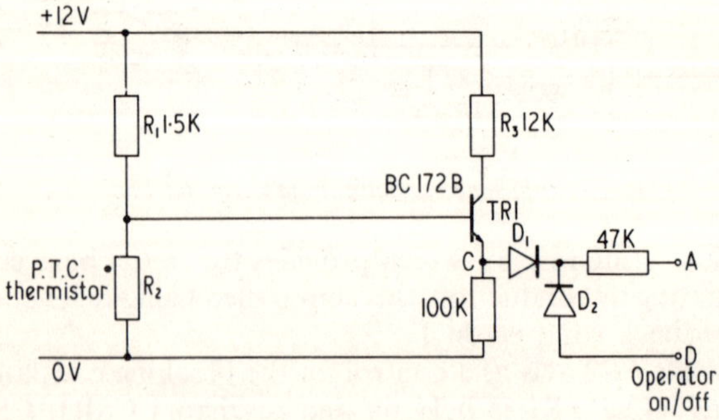

Figure 3.6. Thermistor protection

circuit and the potential at A is determined by the operator input at D. Assuming that D is negative then point A, which corresponds to the same letter of *Figure 3.5*, is low and the oscillator is on. If the motor now overheats R_2 changes to a high-resistance state, TR1 is driven on and current is fed via R_3 to A turning the blocking oscillator in *Figure 3.5* off and disconnecting the drive to the motor. When the motor has cooled sufficiently the thermistor again changes state to a low resistance turning TR1 off and enabling the operator to control the contactor by the input at D.

Only one protection circuit is required for both single- and three-phase contactors. For the three-phase reversible circuit two separate outputs are taken from C (*Figure 3.6*) and fed to the two oscillators through diodes.

Overload protection

Figure 3.7 shows a simple circuit which gives overload protection with a remarkably good inverse time characteristic. Lines 1 and 2 are fed from a current transformer connected in series with the load. It is assumed that this transformer has a linear characteristic so that the voltage at 1 and 2 is proportional to the load current. This is rectified and fed to a potentiometer, the setting of which determines the operating current of the overload circuit. This enables a standard current transformer to be used for a range of contactors.

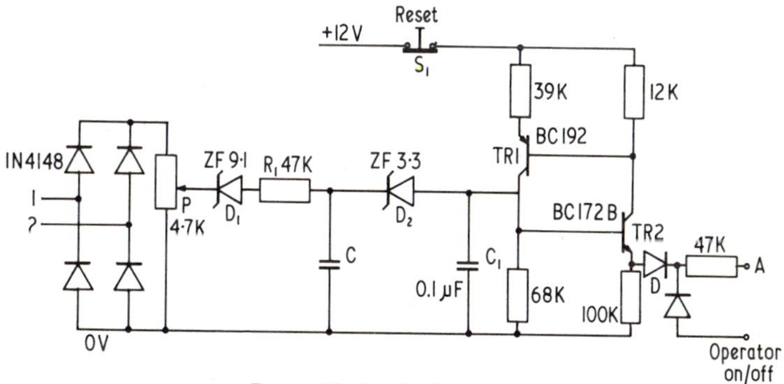

Figure 3.7. Overload protection

When the voltage at P exceeds that of D1 capacitor C commences to charge via R_1, the charging rate being increased by an increase in input voltage. When C reaches the voltage of D2 and the base-emitter of TR2, it turns this transistor on. Transistors TR1 and TR2 form a self-latching complementary Schmitt trigger circuit such that turning TR2 on turns TR1 on which keeps TR2 conducting even when the voltage on C has fallen below that of D2. It can be reset by momentarily breaking the supply at S_1. If this latching circuit were not used, then motor oscillations could result where overload would turn the contactor off, which would reduce the load and turn the thyristors on again, which would again result in an overload and turn the thyristors off, and so on.

When TR2 turns on, current is fed to A, as in the overtemperature protection circuit, overriding the operator's control and turning the contactor off. C_1 has a much smaller value than C and serves only in removing switch-on transients from the base of TR2 which would tend to turn it on.

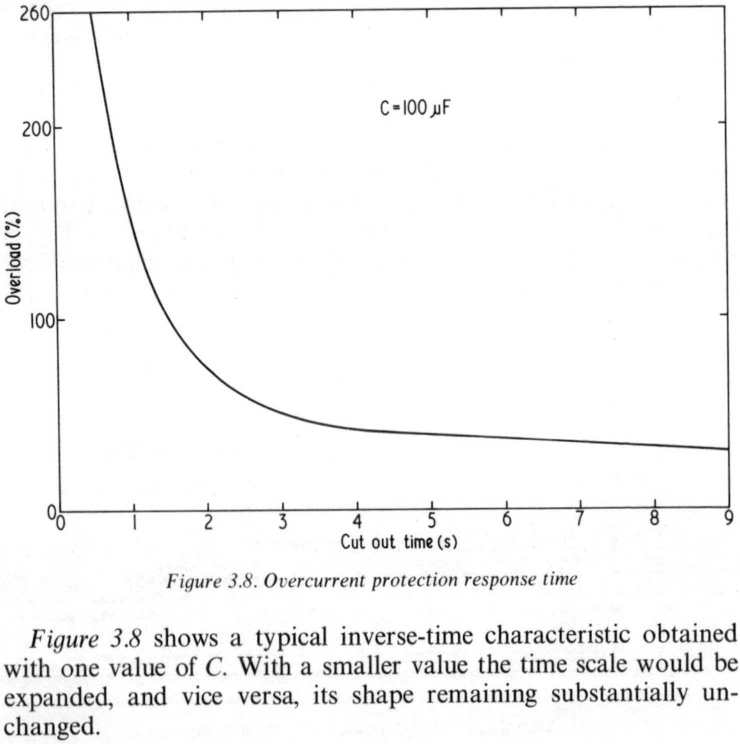

Figure 3.8. Overcurrent protection response time

Figure 3.8 shows a typical inverse-time characteristic obtained with one value of C. With a smaller value the time scale would be expanded, and vice versa, its shape remaining substantially unchanged.

Zero-voltage switching

By zero-voltage switching is meant opening and closing of contacts when the supply voltage is close to zero. This prevents sudden changes in voltage across the load, which can give steep waveforms in the current and so generate interference. With mechanical contactors it is impossible to determine the operation time accurately since it is relatively long, and can change substantially between successive cycles. However, this problem is not present with static contactors using thyristors.

A thyristor automatically turns off at current zero so the problem of breaking load currents and generating interference does not exist. Care should, however, be taken to switch on at a voltage-zero. This can be done relatively simply by the circuit shown in *Figure 3.9*. It should be noticed that the use of this results in a circuit with positive logic. When the operator input goes positive diode D1 is reverse-biased and, assuming TR2 to be off, TR1 is driven on. This results in the oscillator input at A going negative (to zero volts) and allowing it to free run. When the operator input goes negative TR1 is off (irrespective of the state of TR2) and A is positive inhibiting the oscillator. In effect TR1 is an inverting stage.

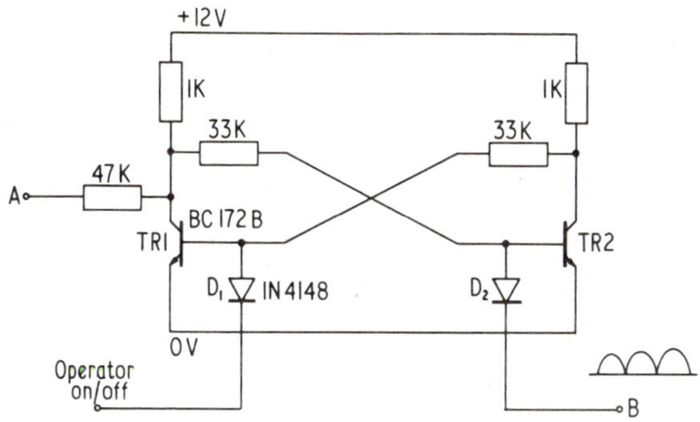

Figure 3.9. Zero-voltage switching

A second transistor TR2, coupled to TR1, has its base supplied at B from a voltage which corresponds to the full wave rectified a.c. input to the contactor. Normally TR2 is on so that TR1 is off irrespective of the state of the operator input. However, when the voltage at B falls to zero, TR2 goes off momentarily. If the operator input is also at zero, nothing will happen, TR1 remaining off. However, at this instant if the operator input is positive TR1 will come on allowing the thyristor to be fired. TR2 will now be off for the rest of the half cycle (the operator input still being positive) and remain off when the input a.c. voltage again goes through zero. At any moment in the cycle the operator can turn TR1 off, so removing the thyristor gate drive. However, the thyristors will conduct until the load current decays to zero. When this is done TR2 will come on so that TR1 cannot be turned on till the next voltage zero.

Reversible contactor protection

The reversible contactor of *Figure 3.3* is driven by two drivers, one firing thyristors CSR1 to CSR6 and the other CSR1, CSR2 and CSR7 to CSR10. Each driver is controlled by an input as at A in *Figure 3.5*. These can be called A_1 and A_2. If both these inputs are taken negative, both drivers will be on and the ten thyristors in *Figure 3.3* fired. This would result in a short circuit between lines B and C. To prevent this the circuit of *Figure 3.10* is used, the operator

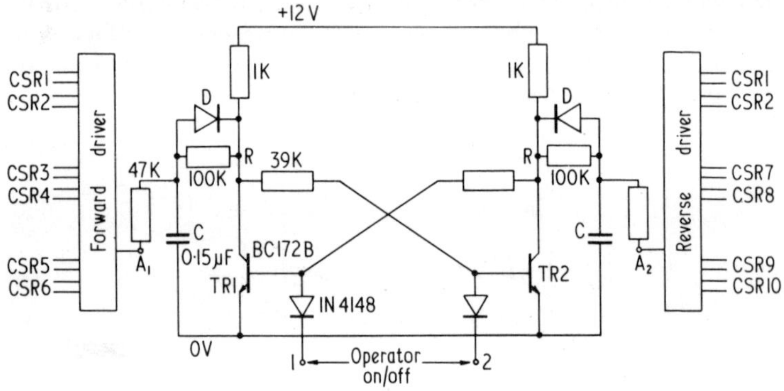

Figure 3.10. Reversible contactor protection

inputs at 1 and 2 not feeding A_1 and A_2 but into the bases of TR1 and TR2. Note that as in *Figure 3.9* this inverts the logic so that the contactor now turns on for positive operator input. If any of the operator inputs (say 1) goes positive then its corresponding transistor (TR1) is on provided that the other (TR2) is off. Therefore the forward driver is running and the reverse drive is off. If the other operator input (2) now also goes positive TR2 is still off since TR1 is on. The reverse driver will come on only when the first input (1) goes to zero turning TR1 off and stopping the forward driver. *R* and *C* are chosen such that a delay greater than half a cycle of a.c. input (10ms for a 50Hz system) exists on switch on. This enables the conducting set of thyristors to be extinguished before the other set is fired. Diode D resets *C* rapidly when the transistor turns on. Therefore this circuit allows that driver to run whose operator input goes positive first. If both inputs are negative together then TR1 and TR2 are both off and both drivers are non-operational.

Gradual start and current limit

The voltage applied by the contactor to the load can be gradually increased by advancing the firing point *a* over successive half cycles on first start-up, as in *Figure 3.11*. Also if the load current is sensed,

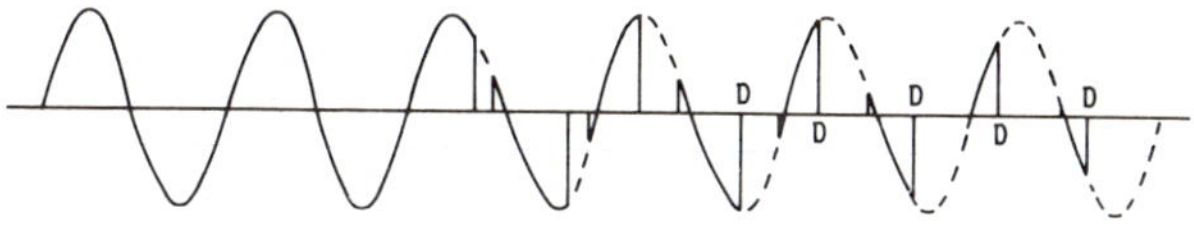

Figure 3.11. Phase control

say by a current transformer, the firing point can be retarded if this exceeds a preset value. However, the contactor is no longer a contactor in the truest sense of the word but more a variable voltage device as described in Chapter 1.

D.C. CONTACTORS

Once conduction has been initiated in a thyristor it remains on till the current decays to zero. This is accomplished naturally in an a.c. circuit owing to the reversal of the supply voltage. For d.c. contactors the voltage reversal must be artificially provided across the thyristor for a time in excess of its rated turn-off time.

Power circuits

Perhaps the simplest of d.c. contactor circuits is that shown in *Figure 3.12*. Thyristor CSR1 is fired to connect the load across the supply. This also permits *C* to charge through *R* with plate 1 positive. To turn the contactor off CSR2 is fired. *C* is connected across CSR1 and is reverse-biased. In effect *C* provides an auxiliary path

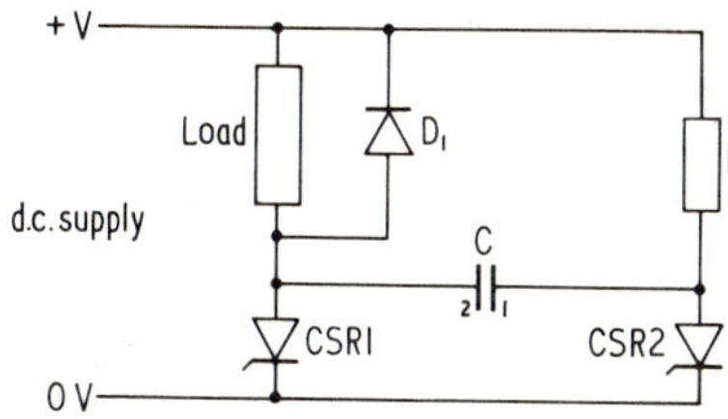

Figure 3.12. Simple d.c. contactor

for the load current and consequently commences charging with plate 2 positive. Provided the time during which CSR1 is reverse-biased is greater than its turn off time it will remain off. When CSR1 is next fired CSR2 turns off since C is connected across it. This contactor has two disadvantages.

(i) It is inefficient since CSR2 is conducting during the off period and energy is lost in R. This resistor can be made large valued but then (ii) below is accentuated.

(ii) After CSR1 is turned on sufficient time must be allowed for C to charge to the supply voltage, through R, before CSR2 can be fired to turn it off. Therefore the contactor has a top limit to its operating frequency. The greater the value of R the lower this frequency.

The contactor shown in *Figure 3.13* is capable of much higher operating frequencies even under conditions of high load impedance. If the load is substantially constant, diode D2 may be omitted. Initially C charges through D3 to the d.c. supply with plate

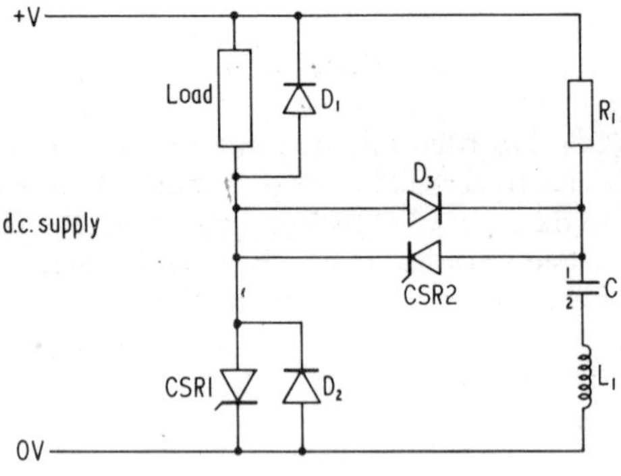

Figure 3.13. Modified d.c. contactor

1 positive. CSR1 is now turned on to supply load current. During this period C would discharge owing to the leakage currents of CSR2 and D3 had this not been compensated for by the current through R_1. Therefore this resistor can have a large value since it has only to pass the leakage current of two devices. To turn CSR1 off, CSR2 is fired. Capacitor C resonates with L through CSR1 and CSR2 and the voltage on C reverses, plate 2 being positive. Once again C resonates with L, through D2 and D3, turning CSR1 and CSR2 off. *Figure 3.14* shows the voltage on capacitor C assuming negligible resonant losses.

The mean current of thyristor CSR1 is given by $I_{L(max)}$ with a peak of $V\sqrt{(C/L_1)}$ where $I_{L(max)}$ is the maximum load current and V the supply voltage. The peak voltage is $V+I_{L(max)}\sqrt{(L_1/C)}$.

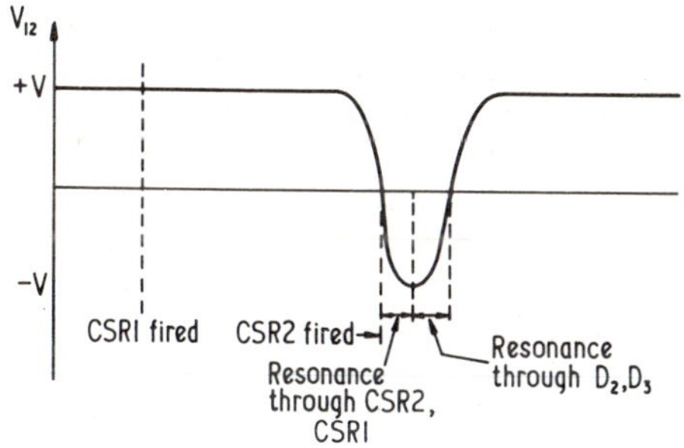

Figure 3.14. Waveform across capacitor C

The maximum turn-off time seen by CSR1 is:

$$\sqrt{(L_1C)}\left\{\frac{\pi}{2}-\sin^{-1}\left[\frac{I_{L(max)}}{2V}\sqrt{\frac{L_1}{C}}\right]+\cos^{-1}\left[\frac{I_{L(max)}}{V}\sqrt{\frac{L_1}{C}}\right]\right\}$$

The mean rating of CSR2 is low, the peak value being $V\sqrt{(C/L_1)}$. Its voltage rating must exceed V.

Diode D1 passes a peak current of $I_{L(max)}$ and sees a voltage of $2V_B+I_{L(max)}\sqrt{(L_1/C)}$.

The mean current of diode D2 is low the peak value being $V\sqrt{(C/L_1)}$ with a peak voltage $V+I_{L(max)}\sqrt{(L_1/C)}$.

The mean current of diode D3 is also small the peak current being $V\sqrt{(C/L_1)}$ at a voltage of V.

Alternative circuits which may be used for d.c. switching are illustrated in Chapter 5.

Thyristor firing and control circuits

Although the blocking oscillator circuit of *Figure 3.5* can again form the basic firing circuit it is usually sufficient to apply a single pulse of width large enough to allow the load current to rise beyond the holding current of thyristor CSR1. The circuit is as shown in *Figure 3.15*. A positive input at A would turn TR1 on for a time determined by the *R-C* time constant, the pulse at the secondary of transformer T_1 feeding the thyristor to be fired. When the voltage at A goes negative TR1 is held off and *C* reset. If TR1 is replaced by a *p-n-p* transistor an output pulse is obtained when the input goes negative, the polarity of the logic being reversed from that of *Figure 3.15*.

Figure 3.16 shows an on/off circuit suitable for use with the contactor of *Figure 3.13*. Amplifier TR1 feeds two driver stages TR2 and

TR3. When the operator input goes negative TR1 is off and TR2 turns on for a time determined by R_1, R_2, C_1. Thyristor CSR1 turns on and supplies load current. When the input at A goes positive TR1 comes on, resetting C_1, and turning TR3 on for a time determined by C_3 and R_3. Thyristor CSR2 is fired turning CSR1 off. When the input again goes negative CSR1 is refired and C_3 reset.

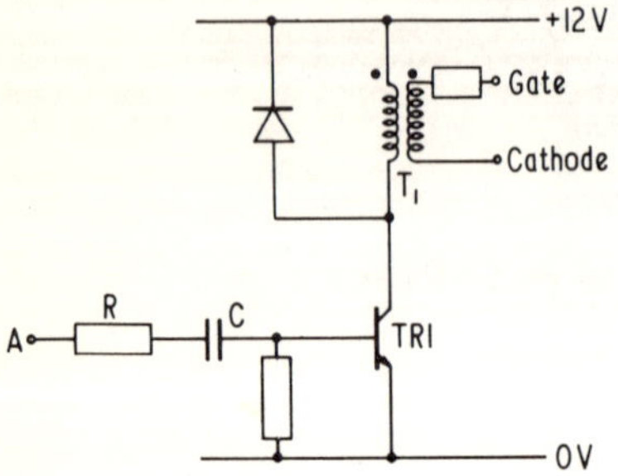

Figure 3.15. Basic thyristor firing circuits

Another input at B allows protection circuitry to be used as was done with a.c. contactors. The thermistor protection circuit is directly applicable, the input from C (*Figure 3.6*) feeding B of *Figure 3.16*. When the motor temperature exceeds the maximum level, B goes positive turning TR1 on and firing CSR2.

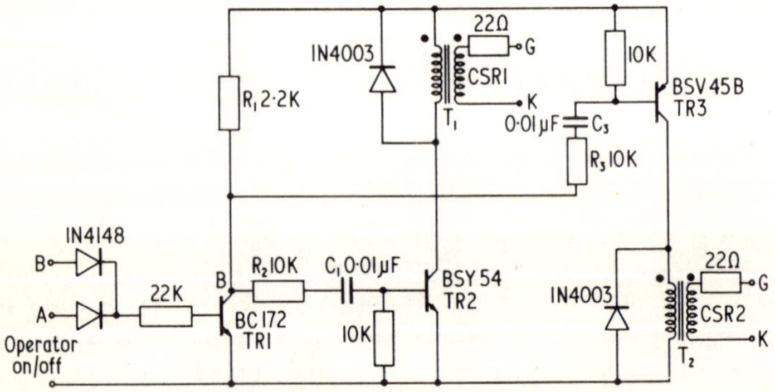

Figure 3.16. On/off control for Figure 3.13

The overcurrent circuit of *Figure 3.7* can also be used. Point D (*Figure 3.7*) now feeds B of *Figure 3.16*. Potentiometer P (*Figure 3.7*) is fed from a d.c. current transformer or a d.c. amplifier which senses the voltage across a shunt resistor placed in the d.c. contactor load circuit. In either case the sensing element must have a linear output.

In the event of an overload B again goes positive, firing CSR2 and turning the contactor off. This state is maintained until the contactor is reset.

Another useful circuit which gives gradual build up of motor voltage is shown in *Figure 3.17.* The circuit operates on a fixed

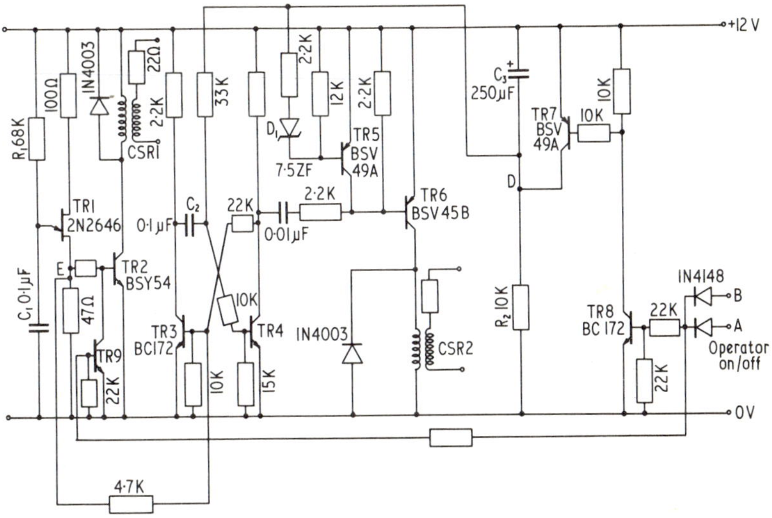

Figure 3.17. Slow run up d.c. contactor control

frequency variable mark-space model, the 'on' time of thyristor CSR1 in any cycle gradually increasing until it is eventually conducting continuously. The operating frequency is chosen to be about 50Hz. The load waveforms are shown in *Figure 318.*

With the operator input positive (off) TR8, TR9 and TR7 are on. Therefore C_3 is discharged and TR2 off. Unijunction transistor TR1 operates at a frequency determined by R_1, C_1, triggering the monostable TR3, TR4 and therefore TR6. Since point D is at +12V the

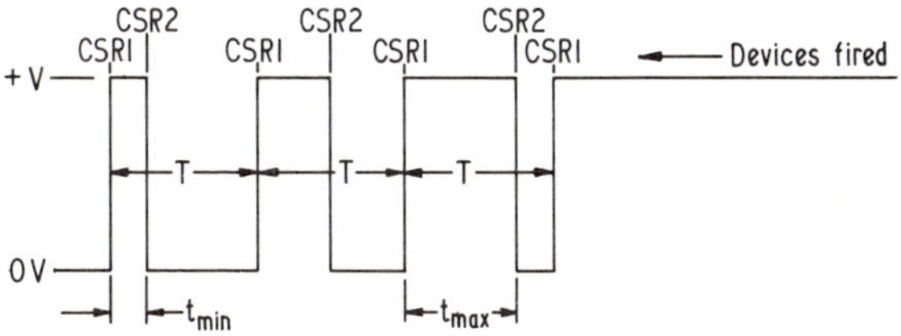

Figure 3.18. Load voltage during slow run up

monostable period is small, *i.e.* TR6 turns on soon after the pulse at E.

When the operator input goes negative TR8, TR9 and TR7 go off. The pulse at E turns TR2 on firing thyristor CSR1. At the same time TR3 is turned on. On the reset of the monostable, when TR4 comes on, TR6 turns on firing thyristor CSR2. The load voltage is on for time t_{min}. As C_3 charges through R_2 point D moves towards 0V which increases the monostable delay time, so increasing the time between the firing of CSR1 and CSR2 in any period T, and therefore the load voltage (*Figure 3.18*). When D reaches a preset voltage D1 breaks over turning TR5 on and so holding TR6 off. Therefore CSR1 is fired and the load is continuously applied. The maximum time t_{max} at which this occurs is about 80% of the period T. The input at B can again be used for protection circuitry.

It is worth noting that point D could be fed from a current sensing amplifier so that the contactor could run up to full load under a preset maximum current instead of a time delay as shown. Relatively cheap electronic control circuitry can therefore replace bulky series resistor starters in large installations.

CONCLUSIONS

Mechanical contactors have been with us for many years and are not likely to be replaced in the forseeable future. However, owing to its versatility, there is often a strong case for considering static contactors where sophisticated control is required. The temptation to use something for novelty value must, however, be avoided. Far better to have a simple machine that sells than an up-to-the-minute innovation that has priced itself out of the market.

4

A.C. line control

The power flowing in an a.c. line can conveniently be controlled by series resistors. Such a method is inefficient so that when better performance is required transformers and variacs are used.

The thyristor a.c. line regulator was introduced in Chapter 1. This is relatively very efficient since it works on the principle of blocking unwanted power rather than dissipating it across the control device. It also gives a system which is physically smaller and lighter than conventional methods. This is specially true when the power being handled is large, as is evident when comparing a 1kV 100A thyristor a.c. regulator with an equivalent auto-transformer.

Phase-control techniques give the simplest a.c. regulating system and are the best known. However, there are three other methods by which thyristor control of a.c. lines is possible, namely synchronous tap-changing, zero voltage switching and a.c. chopper control. These are discussed below.

PHASE CONTROL

Figure 4.1 shows three typical thyristor single-phase control systems with the circuit waveforms. Referring to *Figure 4.1*(ii) and (iv), at time t_0 input line A goes positive to B. Since CSR1 is off, no load current can flow and the load voltage is zero. The supply voltage now appears across the section AC. When CSR1 is fired at t_1 current flows through it and diode D2 to the load. Similarly in the negative half cycle, the firing of CSR2 is delayed by α from the zero voltage point. It is evident now why this system is called 'phase control'. It controls the phase (α in *Figure 4.1*(iv)) between the start of the supply voltage and the start of the load current, in order to vary the power flowing to the load.

The circuit of *Figure 4.1*(ii) uses two extra diodes over *Figure 4.1*(i). This increases its cost and reduces the efficiency since for any conduction path there is now a series thyristor and diode. It has the advantage that the cathodes of the thyristors are commoned so the gate drive circuit may be simplified.

Figure 4.1(iii) shows an alternative system which uses only one thyristor and a diode bridge. The waveforms of *Figure 4.1*(iv) still apply, although it must be kept in mind that the voltage across the thyristor is now never negative (due to the action of the diode bridge). Therefore with line A positive, CSR1 conducts from t_1 to t_2. At t_2 the load voltage is zero and the thyristor must turn off. If this does not happen, then as soon as the supply reverses, the voltage across the thyristor will become positive, turning it on, and delay period t_2 to t_3 will be lost. This loss of control is most likely to occur on inductive loads.

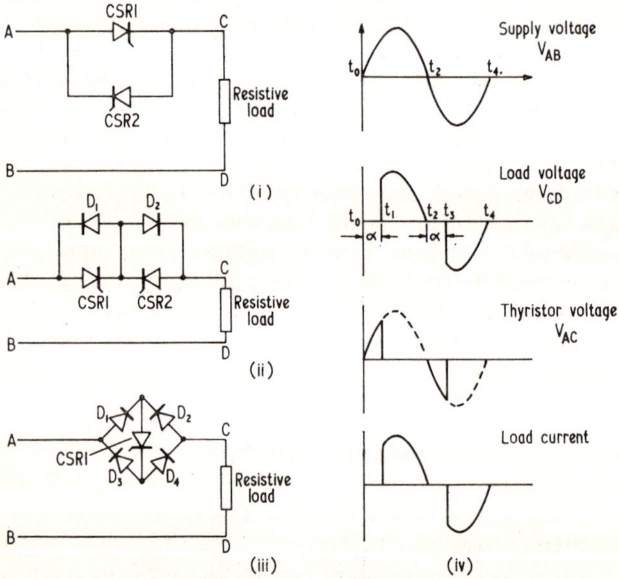

Figure 4.1. Thyristor phase control for resistive load

Another disadvantage of the arrangement of *Figure 4.1*(iii) is that there are voltage losses across three devices (two diodes and one thyristor) in any direction, so that the efficiency is lower than in *Figure 4.1*(i) or (ii). For high-voltage systems this may not be important. For instance, a loss of 2V (1V across a thyristor and 0·5V across each diode) in 250V is an efficiency reduction of only 0·8%. This circuit can sometimes prove cheaper than *Figure 4.1*(i) (diode bridge weighed against a thyristor) and has the advantage of a simpler firing circuit since only one thyristor is used.

The circuit of *Figure 4.1*(i) is generally better suited to most applications than the other two, and is the one commonly used.

To derive the rating of the devices let:

E = r.m.s. input voltage of a.c. lines
R = load resistance
α = firing angle delay

Then the r.m.s. current rating of each thyristor and diode in *Figure 4.1*(i) and (ii) is given by:

$$I_T = \left[\frac{1}{2\pi}\int_\alpha^\pi \left(\frac{\sqrt{2}E}{R}\right)^2 \sin^2\theta \, d\theta\right]^{1/2}$$

$$= \frac{E}{\sqrt{2}R}\left[1 - \frac{\alpha}{\pi} + \frac{\sin 2\alpha}{2\pi}\right]^{1/2} \qquad (4.1)$$

For *Figure 4.1*(iii) the diodes conduct only in each half cycle so that their rating is given by equation (4.1). However, the thyristor conducts during both half cycles so equation (4.1) must be multiplied by a factor of $\sqrt{2}$.

It was mentioned in Chapter 2 that thyristor ratings can conveniently be obtained on an r.m.s. basis except at very low conduction angles. At these control settings the ratio of peak to r.m.s. current is large and care must be taken not to exceed the repetitive peak current. All devices must be rated to withstand a peak repetitive voltage of $E\sqrt{2}$.

The r.m.s. load voltage and currents are similarly given by:

$$V_L = \left[\frac{1}{\pi}\int_\alpha^\pi E\sqrt{2}\sin^2\theta \, d\theta\right]^{1/2}$$

$$= E\left[1 - \frac{\alpha}{\pi} + \frac{\sin 2\alpha}{2\pi}\right]^{1/2} \qquad (4.2)$$

and

$$I_L = \frac{E}{R}\left[1 - \frac{\alpha}{\pi} + \frac{\sin 2\alpha}{2\pi}\right]^{1/2} \qquad (4.3)$$

As seen from equations (4.2) and (4.3), values for $\alpha = 0$ are those of the sine wave input.

The load voltage and current waveforms shown in *Figure 4.1*(iv) are rich in harmonics. *Figure 4.2* gives a Fourier analysis of the spectrum up to the seventh. Where load harmonics must be minimised a.c. chopper regulation is preferred. This is dealt with later.

It was mentioned in Chapter 1 that for inductive loads the control thyristors do not cease conduction when the input voltage reverses.

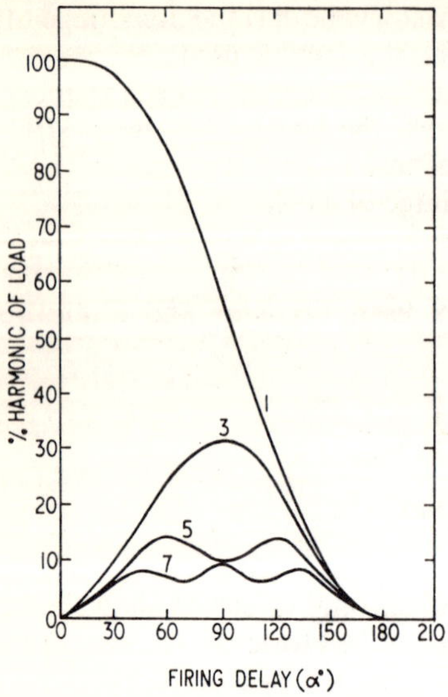

Figure 4.2. Harmonic content of resistive-load waveform

They are kept on by energy stored in the load, which is fed back to the supply. *Figure 4.3* shows the modified waveforms for a series resistance-inductance load of power factor angle ϕ. CSR1 (*Figure 4.1*(i)) is fired at time t_1. This causes the load current to flow (note that the rising edge of the current is slower than for purely resistive load). CSR1 is kept conducting until t_{21}, energy flowing from the

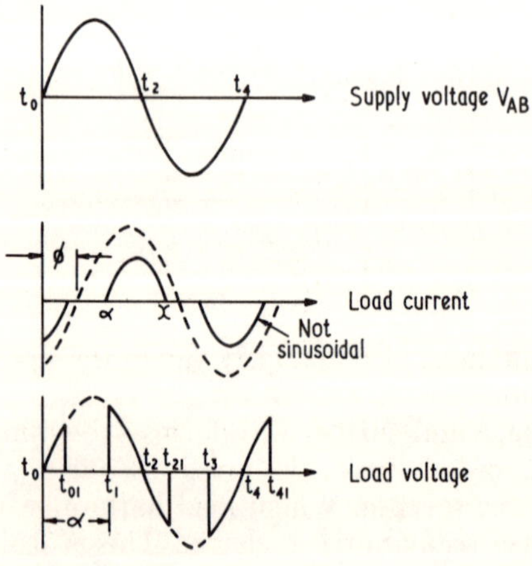

Figure 4.3. Phase-control circuit waveforms with a series resistance-inductance load

load to the supply from t_2 to t_{21}. The load voltage remains at zero until CSR2 is fired at t_3. Clearly if $\alpha \leqslant \phi$, load current and voltage are sinusoidal.

When the load consists of a parallel resistance-inductance combination the circuit waveforms are modified as in *Figure 4.4*. When CSR1 is fired at t_2 the load current rapidly increases in the resistive component and more slowly in the inductance. At t_3 the load voltage reverses but CSR1 is kept conducting until t_4. The thyristor then turns off. The load voltage, however, does not fall abruptly to zero, as in *Figure 4.3*, but follows a gradual decay as the inductive energy is dissipated in the resistive load. Depending on the load power factor and the firing angle, the load current may not have decayed to zero before CSR2 is fired at t_5 as shown in *Figure 4.4*.

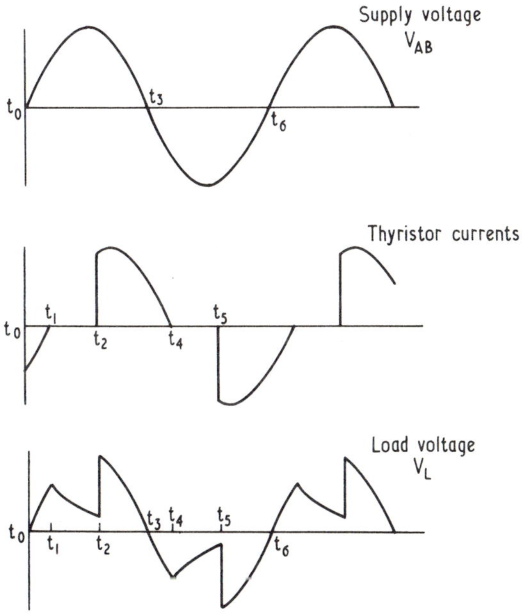

Figure 4.4. Phase-control circuit with a parallel resistance-inductance load

Generally the performance factors of a circuit do not differ appreciably between resistive and inductive loads. For instance the harmonic curves of *Figure 4.2* are still approximately correct on inductive loads, but the harmonic content decreases with reducing power factors. The peak of the third harmonic will also now shift to slightly higher firing angles. The r.m.s. thyristor current and load voltage for a series resistance-inductance load are given by

$$I_T = \frac{E}{Z}\left[\frac{1}{2}\left\{\frac{x-\alpha}{\pi} - \frac{\sin(x-\alpha)}{\pi\cos\phi}\cos(\alpha+x+\phi)\right\}\right]^{1/2} \quad (4.4)$$

$$V_L = E\left[\frac{x-\alpha-\sin(x-\alpha)\cos(x+\alpha)}{\pi}\right]^{1/2} \quad (4.5)$$

Z is the load impedance and the cut-off angle x in *Figure 4.3* is given by

$$\sin(x-\phi)-\sin(\alpha-\phi)\exp\{-\cot\phi(x-\alpha)\} = 0 \quad (4.6)$$

Figure 4.5 shows graphs of the variation of the r.m.s. load voltage for resistive and inductive loads.

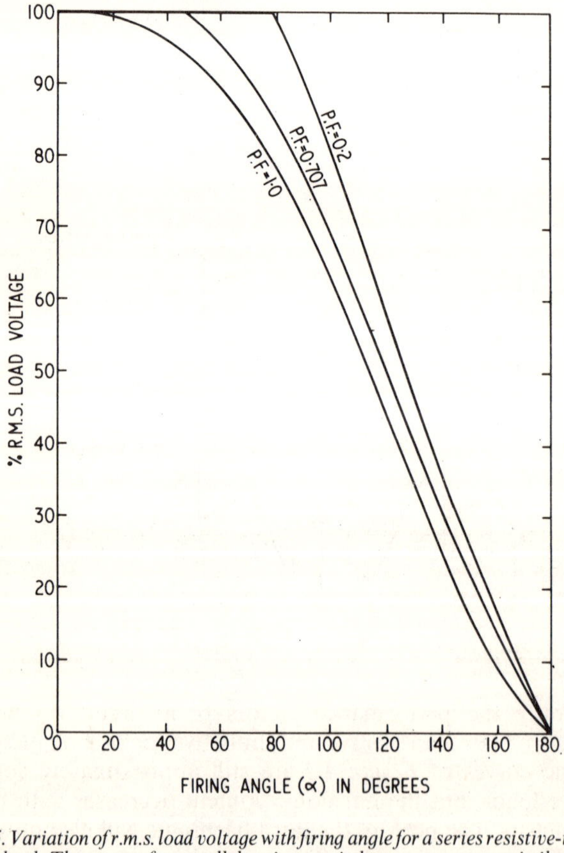

Figure 4.5. Variation of r.m.s. load voltage with firing angle for a series resistive-inductance load. The curves for parallel resistance-inductance are very similar

There are four effects which it is essential to guard against when working into non-unity power factor loads. These concern:

(1) *di/dt*. When the load is capacitive a rapid rise of charging current can occur on initially firing the thyristors. This *di/dt* effect can destroy the thyristor by causing local hot points even though the r.m.s. current rating has not been exceeded. When the load is a transformer stray capacitance across it would produce the same effect. In all such cases a linear or saturable reactor in series with the thyristor can be used to slow down the rate of current change.

(2) *dv/dt*. Referring to *Figure 4.1*(i) and *Figure 4.3* it can be seen that CSR1 conducts from period t_1 to t_{21}. The voltage across CSR2 is now negligible. At t_{21} thyristor CSR1 goes off and the voltage across CSR2 rises rapidly to the value of input line voltage at this point. This *dv/dt* effect can cause the thyristor to switch on and conduct even in the absence of a gate signal. It is best protected against by a capacitance-resistance network across the devices.

(3) Voltage transients. Since the thyristor is a fast acting switch, large-valued voltage transients can occur when switching inductive loads. To illustrate this suppose a thyristor is controlling an inductive load of 1H. Load current falls to a value below the device holding current (say 10mA) and the thyristor turns off (say in 1μs). This would result in a load transient of 10kV! It can be damped by *R-C* circuits across the thyristor which effectively increase its turn-off time.

(4) Firing-circuit considerations. These will be dealt with in the next section.

The single-phase circuits shown in *Figure 4.1* can be connected into three-phase lines in a variety of ways. Although any of the three basic systems shown in *Figure 4.1* may be used, that illustrated in *Figure 4.1*(i) is the most popularly used. *Figure 4.6* shows a few examples. The connection of *Figure 4.6*(i) resembles the single-phase system almost exactly and the same equations and circuit waveforms are applicable. It produces the lowest-voltage harmonics over most of the control range but is restricted for use with open-star connected loads only. The circuit of *Figure 4.6*(ii) is also similar in operation to the single-phase systems. Thyristors are fired in pairs, spaced 120° apart, in order to give the required phase sequence output, the load current flowing from a more positive to a negative phase. A typical sequence of conduction could be B1 Y2 R1, Y2 R1 B2, R1 B2 Y1, B2 Y1 R2, Y1 R2 B1 and R2 B1 Y2 over a 360° period.

The control system shown in *Figure 4.6*(iii) uses one thyristor and a diode in each line. It has no counterpart in a single-phase circuit since the diode would then supply load current during a complete half cycle. However, for three-phase systems the diode cannot conduct unless a thyristor in a more positive line has been fired. This circuit is simpler to use since only the thyristors require gate pulses, the diodes conducting automatically.

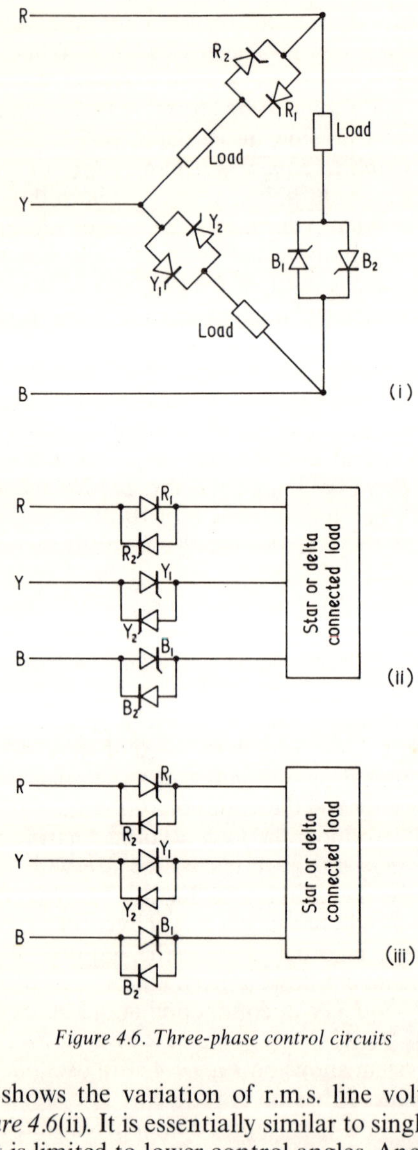

Figure 4.6. Three-phase control circuits

Figure 4.7 shows the variation of r.m.s. line voltage with firing angle for *Figure 4.6*(ii). It is essentially similar to single-phase operating curves but is limited to lower control angles. Another peculiarity

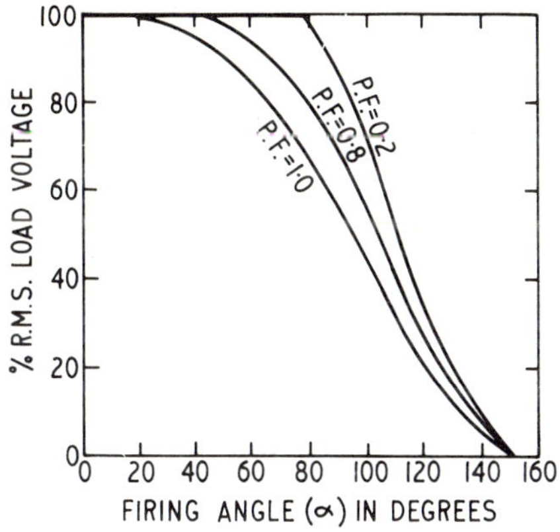

Figure 4.7. Variation of r.m.s. load voltage with firing angle for the circuit of Figure 4.6(ii)

of three-phase systems is that the neutral point voltage will vary with the conduction angle owing to unbalance in the instantaneous phase voltages. When a neutral line is present, a current will flow. *Figure 4.8* illustrates the waveform of this current for the arrangement of

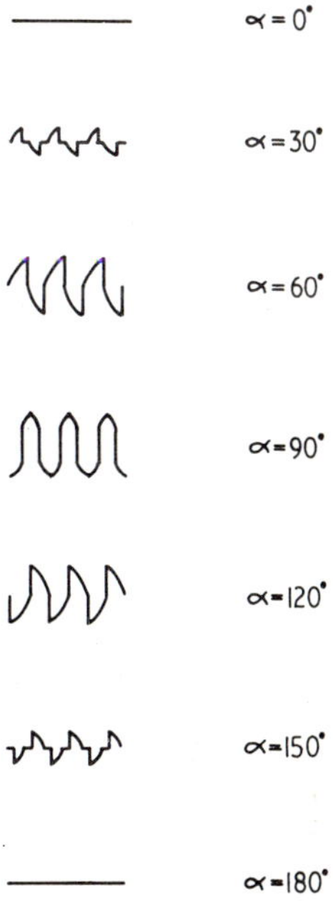

Figure 4.8. Neutral current waveforms for the circuit of Figure 4.6(ii) with a star-connected resistive load

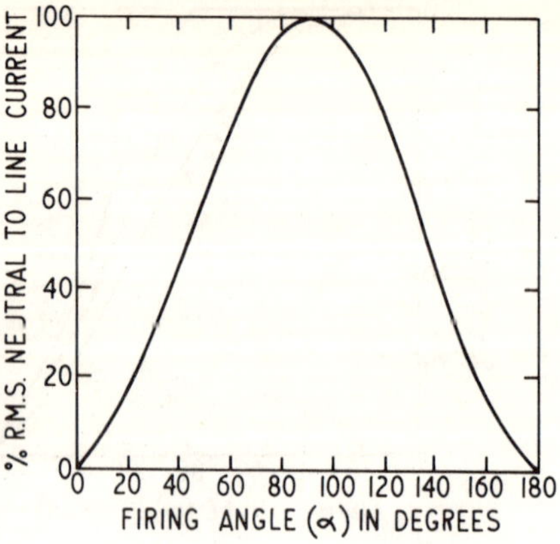

Figure 4.9. Neutral current for the circuit of Figure 4.6(ii) with a star-connected four-wire resistive load

Figure 4.6(ii) when a four-wire star-connected load is used. Clearly the unbalance is worst at a delay angle of 90°. This is illustrated in *Figure 4.9* where the neutral current is then seen to be equal to the line current.

FIRING CIRCUITS

The essential features of a firing circuit for phase control are shown in *Figure 4.10*. The detector senses the zero voltage points of the input lines or some other reference. The delay circuit provides a variable setting from this reference, after which the gate drivers are energised to fire the thyristors.

Figure 4.10. Phase-control firing circuit block diagram

The detector could be used to sense the voltage across the thyristors and begin the delay from the instant the previous thyristor goes off. However, for inductive loads this can lead to asymmetry in the firing points. For instance in *Figure 4.1*(i) suppose the load is inductive and that due to momentary unbalance CSR1 is kept conducting slightly longer than CSR2. This will mean that the timing delay for CSR2 will start from a later point, still further increasing the asymmetry in the two thyristor gate drives. This effect will build up over several cycles until eventually CSR1 is almost fully on and CSR2 fully off. Therefore sensing the input a.c. lines directly is preferred. With inductive loads it is also essential to apply gate pulses to the thyristors for a time in excess of the load phase angle. Therefore in *Figure 4.3* CSR2 is conducting at t_0. If CSR1 is fired at this point ($\alpha = 0$) it will not turn on since it is reverse-biased. At t_{01} thyristor CSR2 goes off. However, CSR1 will not come on unless it is refired or the gate pulse which was applied at t_0 is maintained up to this point.

Practical circuits based on *Figure 4.10* are numerous. *Figure 4.11* shows a simple controller for resistive loads. When line A goes positive capacitor C_1 charges through the load, D2, R_1 and R_2. This charging rate is controlled by the setting of R_2. When it reaches the break-over voltage of trigger diode TD1 a pulse is applied to the gate of CSR1 turning it on. Similarly in the next half cycle C_2 discharges

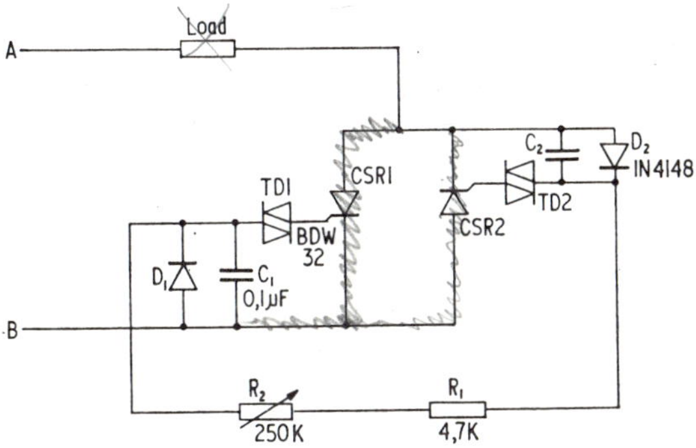

Figure 4.11. Thyristor pulse firing circuit

through TD2 firing CSR2. Sensing in this circuit is obtained by D2 and D1, only one diode being forward-biased in each half cycle. The delay components are R_1, R_2, C_1, C_2. Gate drivers are TD1 and TD2 with their associated capacitor discharge pulse. This circuit is suitable only for resistive loads. *Figure 4.12* gives an alternative circuit which applies a continuous gate pulse to the thyristors over their whole conduction period. From commencement of each half cycle either TR1 or TR2 is forward-biased by its corresponding Zener voltage. This therefore represents the detector. Delay is obtained by the well known phase-shift circuit of C_1 and R_1. Depending on the setting of R_1, transformer T_2 is turned on supplying power to the corresponding gate drive transformer and hence the thyristor. Once the gate signal has commenced, it remains on until the end of the

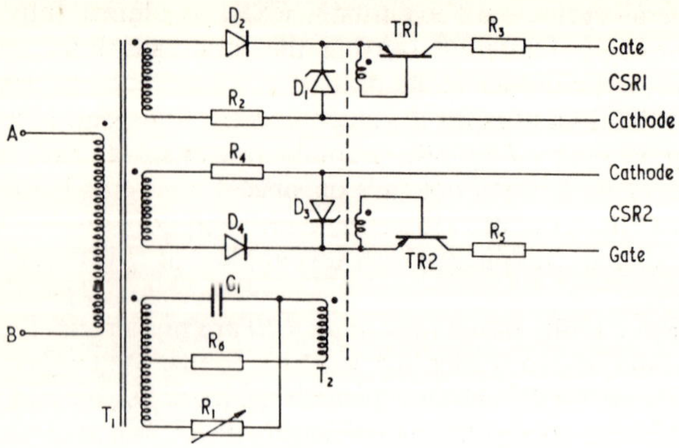

Figure 4.12. Firing circuit suitable for inductive loads

half cycle. If required the transistors could be replaced by low current thyristors.

Further firing circuits will be looked at in Chapter 6 where phase control is considered in circuits providing controlled rectification.

A.C. CHOPPER REGULATION

Regulation of the a.c. power input to a load can be accomplished by phase control methods as described in the previous section. This has several disadvantages, one of which is the high harmonic content in the output. This is specially evident at large delay angles. An alternative to phase control is a.c. chopper regulation which is illustrated in *Figure 4.13*.

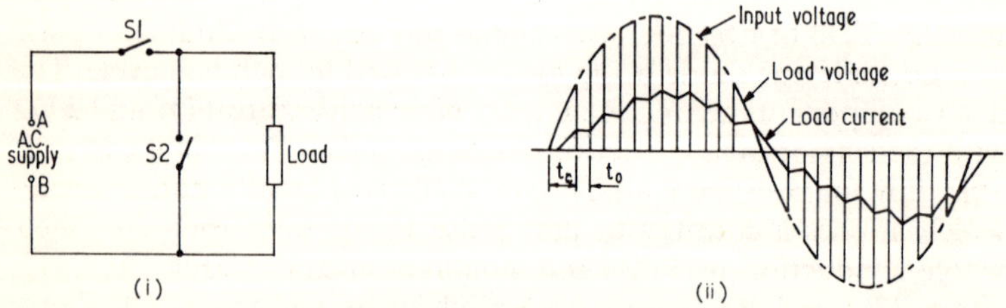

Figure 4.13. A.C. chopper regulator

When switch S_1 is closed, for period t_c, power is supplied to the load from the a.c. supply, and the magnitude of the load voltage is equal to the a.c. source provided the drop across S_1 is neglected. For an inductive load S_2 must be closed when S_1 opens. The load current freewheels in S_2 and its voltage is low. Assuming that S_1 is open for period t_0 the load waveforms are as in *Figure 4.13*(ii). The mean load voltage is controlled by changing the on period during an operating cycle, *i.e.* $t_c/(t_c+t_0)$. The output voltage function is given by

$$V = \sqrt{2}kE \sin \omega_1 t + \sum_n \left[\frac{\sqrt{2}E}{n} \sin nk\{\sin(\omega_1 + n\omega_2)t - \sin(\omega_1 - n\omega_2)t\} \right] \quad (4.7)$$

where k is ratio of the chopper pulse width to the chopper period (*i.e.* $t_c/(t_c+t_0)$), $\sqrt{2}E \sin \omega_1 t$ is the input a.c. voltage, and $2\pi/\omega_2$ is the chopper period. The relative harmonics are plotted in *Figure 4.14*.

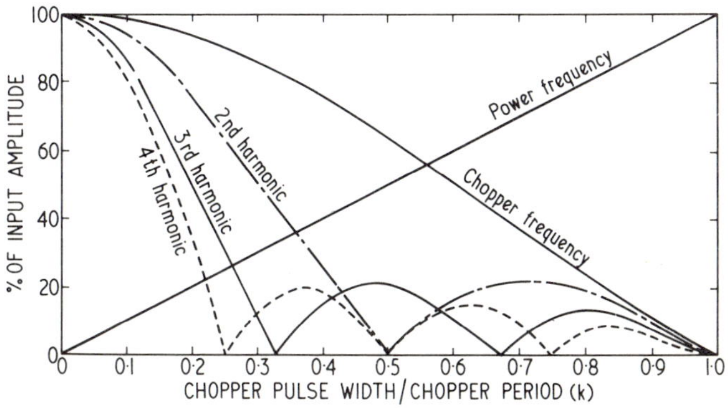

Figure 4.14. Harmonic content of a chopped sine wave

From this it is seen that the power frequency output varies linearly with k. The lowest harmonic is approximately equal to the chopper frequency and can be filtered out relatively easily for high chopping frequencies.

A.C. chopper regulation has two basic advantages over phase-control techniques:

(i) The speed of response is faster. This is so since a chopper can be turned on or off at any instant in the a.c. cycle, whereas in phase controlled circuits once the thyristors have been turned on they will go off only when the supply voltage has reversed.

(ii) The harmonic content is lower. This is evident when comparing the load current waveforms given in *Figure 4.13* with those in *Figure 4.3*. For low voltages the phase control circuit produces a series of short current pulses, spaced 180° apart, whereas for chopper regulation the waveform can still be made to approximate to a sine wave, provided the chopping frequency is relatively high. *Figure 4.15* shows the plot of effective power factor (assuming a resistive load) against duty cycle, which is $t_c/(t_c+t_0)$ for chopper regulators and $(\pi-\alpha)/\pi$ for phase-control circuits. This graph shows that the chopper regulator has a higher power factor below 50% duty cycle, whereas above this value the phase-controlled waveform is superior. However the difference at these higher voltage settings is not large so that overall a chopper regulator gives a better performance.

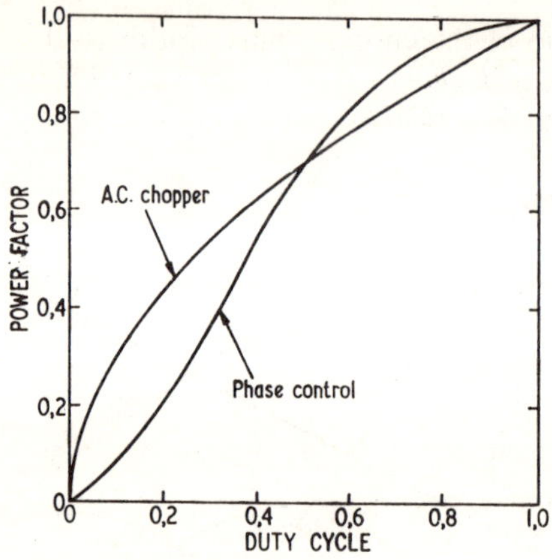

Figure 4.15. Power-factor curves

Disadvantages of chopper regulators are principally:

(i) higher cost due to more elaborate control and power circuitry

(ii) greater radio-frequency interference generation since the chopping is performed several times per cycle. This is discussed in the next section.

Since the thyristors in a chopper regulator need to be turned off before the end of a half cycle they must be forced commutated. Several forced-commutation circuits may be used and these are discused in Chapters 5 and 9. *Figure 4.16* shows a popular arrangement. Thyristors CSR1 and CSR2 perform the function of series switch S_1, CSR3 and CSR4 replace parallel switch S_2. Capacitors C_1, C_2 and centre-tapped inductor L_0 operate such that when a series thyristor is fired its corresponding parallel thyristor is turned off, and vice-versa. For instance suppose line A is positive to B and thyristor CSR2 is conducting. Capacitor C_2 charges to the peak supply voltage. To turn the series thyristor off CSR4 is fired. C_2 discharges through the lower half of L_0 which couples a pulse to CSR2 turning it off.

The rating of series thyristors on resistive load can be obtained by an integration of the voltage waveform. The r.m.s. value is given by

$$I_{T(r.m.s.)} = \frac{E}{R}\left[\frac{t_c}{2(t_c+t_0)}\right]^{1/2} \tag{4.8}$$

The rating of the shunt thyristors depends largely on the load power factor. They are usually made comparable to the series devices to allow for inductive loads operating on a 50% duty cycle.

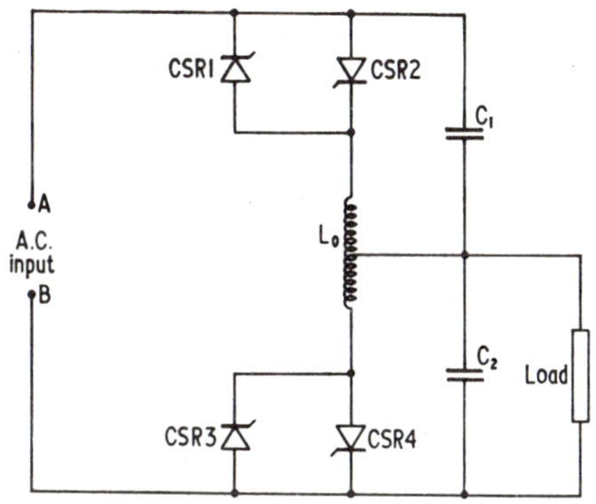

Figure 4.16. A.C. chopper regulator thyristor arrangement

A.C. chopper regulators are primarily used in applications which require a sine-wave output, since their harmonic content is then more easily filtered out than in a comparable phase-control circuit. Owing to the higher-frequency harmonics involved, the filter section can be relatively simple, as in *Figure 4.17*. It consists essentially of a series band-stop section, which removes the fundamental chopper fre-

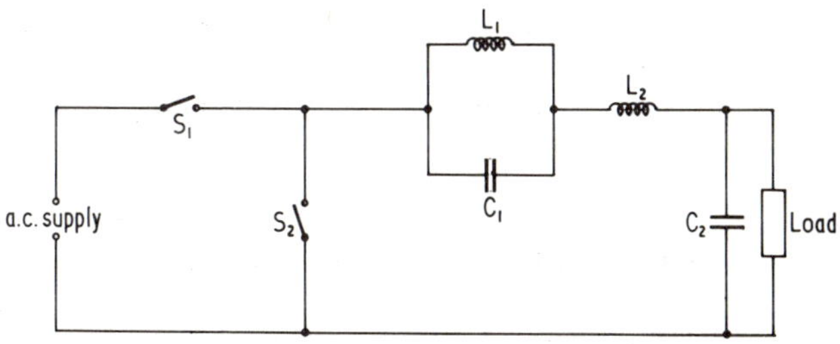

Figure 4.17. A.C. chopper regulator with a sinusoidal output

quencies, followed by a low-pass section, which filters out higher-order harmonics.

The control circuit for a chopper regulator, which is designed to produce a stabilised sine-wave output, is given in *Figure 4.18*. The a.c. supply is chopped before being transformed to the required voltage level. It is then filtered and fed to the load. The sensed voltage feedback from the load to the comparator may be a.c. or d.c. If a d.c. signal is required, the load voltage must first be rectified and smoothed. A d.c. reference voltage also feeds the comparator which produces an error signal to the control signal generator. This modifies the firing of the thyristors, so as to change the pulse width of the a.c. output and so correct the error between the output voltage and its reference.

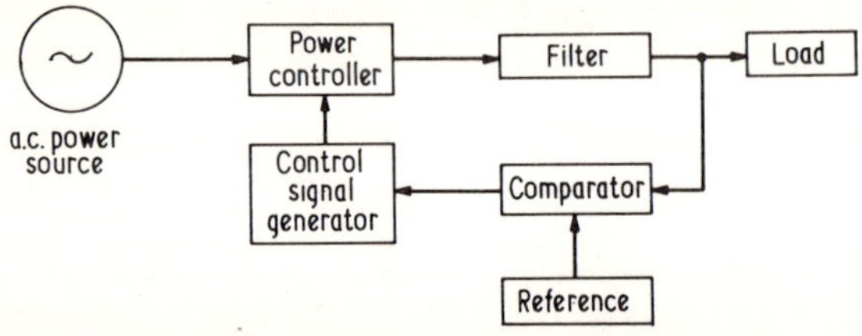

Figure 4.18. Control circuit for a.c. chopper regulator

This control method produces an output voltage whose mean or r.m.s. value is monitored with reference to a desired input. It has no control over the shape of the sine wave. The system also has a slow response speed owing to the delay introduced by signal rectification and smoothing. This disadvantage can be overcome by feeding a sine-wave a.c. reference (derived from the a.c. power source) to the comparator along with an a.c. signal fed back from the load. A comparison then occurs between instantaneous values of output and reference. If the output is greater the parallel switch is operated; if it is less the series switch is closed to deliver more power to the load. The chopping frequency will clearly now be variable. Apart from removing the delay in response speed this system also causes the instantaneous output to follow the a.c. reference voltage, to positive or negative limits which are determined by the gain of the overall system.

RADIO-FREQUENCY INTERFERENCE

Radio-frequency interference (or electromagnetic interference) is a well known phenomenon which has plagued many generations of engineers. Although it is commonly caused by circuits which produce contact sparking, such as relays and rotating machines with mechanical commutators, it may in fact be generated by any system in which the current or voltage changes rapidly. In thyristor circuits it is caused by the rapid turn on of a device in phase control or chopper regulators. It also occurs when the device current decays abruptly on reverting to the blocking state.

Figure 4.19 shows a typical control system in which the thyristor has been replaced by switch Sw. L_1 and L_2 are series line impedances which could be part of the supply transformer leakage reactance and

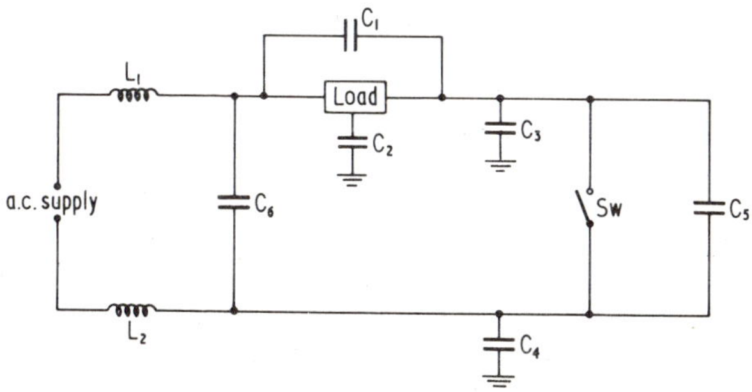

Figure 4.19. Stray capacitance and inductance in a typical thyristor control system

the self-inductance of cables between this and the load. C_1 and C_2 are stray capacitances across the load and from the load to earth. C_3 and C_4 are also stray capacitances to earth, and C_5 and C_6 appear between the lines. With Sw open, the capacitors are charged to the peak instantaneous supply voltage; when Sw closes, the capacitors discharge. This creates an oscillatory system which produces a wide spectrum of unwanted frequencies. Clearly the magnitude of the interference produced is determined by the peak energy stored in the capacitors at the instant at which Sw closes. It is therefore worst at $\alpha = 90°$ in a phase-controlled system.

Radio-frequency interference can be radiated or conducted. This does not mean that one can divide it into two distinct parts, since very often one effect gives rise to the other. Radiated interference is usually in the audio-frequency range and has serious effects on equipments which depend for their operation on radiated *E* and *H* waves, such as radio and television. Although these are normally tuned to accept a narrow band of frequencies and reject all else, they are incapable of differentiating between wanted and unwanted signals within this band. Conducted interference, on the other hand, travels down the power lines and can cause faulty operation of any mains-connected apparatus. It is especially serious when it influences the performance of instruments since these can then give false readings, and the operator may not even be aware of the presence of any interference. Some forms of interference are not objectionable in certain applications. For example switching off an electric light causes a short duration interference on a radio, which would often pass unnoticed. However the same interference could create havoc with a computer store, and would need special protection.

Although interference cannot be eliminated in thyristor chopping circuits, it can be prevented from affecting adjacent equipment. Suppression filters are now used, which consist basically of L and C components. It must, however, be kept in mind that effective suppression is a very specialised art and no hard and fast rules can be laid down which would be successful every time. One type of circuit which completely suppressed interference from one system may be quite useless in another. It is also essential to pay attention to detail. The exact location of the filter is often critical. Generally it should be placed directly at the source of interference and output and input leads should never be bundled together

Figure 4.20 shows some of the basic filter arrangements which are used for suppressing conducted interference. Where large attenuation is required, it is preferable to use several stages of smaller

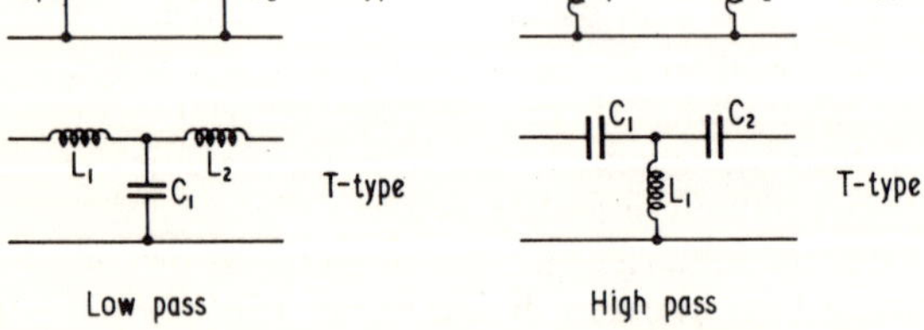

Figure 4.20. Basic filter configurations

filters rather than one large system. Low-pass filters are commonly used in a variety of modifications, as shown in *Figure 4.21*. The arrangement of *Figure 4.21*(i) is perhaps the most basic. However, as seen from *Figure 4.19* there are two modes of conducted interference. One is known as symmetrical interference and is due to leakage between the two lines. Asymmetrical interference on the other hand is caused by leakage between lines and earth. The filter shown in *Figure 4.21*(i) is not effective in suppressing asymmetrical interference and where this is a problem a split inductor is used as in *Figure 4.21*(ii). The polarity of the windings is such that the d.c. components of current through them are cancelled so that the core size is reduced. *Figure 4.21*(iii) and *4.21*(iv) introduce progressively more attentuation of interference signals at the expense of greater circuit complexity. Generally arrangements of the type shown in *Figure 4.21*(iv), where the capacitor is directly across the thyristor

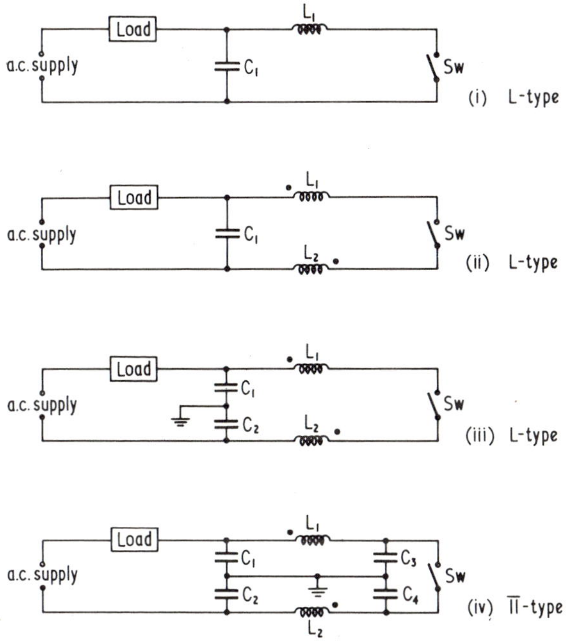

Figure 4.21. Modifications of the basic filter arrangements

switch, should be avoided. This capacitor would charge to a large voltage when the thyristor is off, and when it fired it would give a high current pulse through the device, which could destroy it. In selecting components for filters it is of course important that the inductors should have low stray capacitance, so that they should not be multi-layer, and the capacitance should have low series inductance. Ceramic capacitors are usually suitable provided their connecting leads are kept to a minimum length.

Although filters can be used to reduce the extent of conducted interference, they have no effect on the interference radiated directly from the thyristor and its associated circuitry. To eliminate this it is essential to shield these components. A solid sheet steel enclosure is usually the best but may sometimes prove too bulky. Metallic wire mesh shields having a 50% open area and greater than sixty strands of wire per wavelength are slightly lighter and almost as effective. Braided metallic wire shielding is rather poor although it does give a flexible enclosure. There should be no openings in the shields. Where these are unavoidable, due to ventilation or input-output lead requirements, it must be accepted that the amount of radiated interference escaping will be roughly proportional to the size of the opening. Shielding may be extended to the power cables for maximum protection since conducted interference down these leads would also give rise to radiation. In all cases, however, it is essential that the shield be connected to earth. Special attention must be paid to the bond used to ensure that it has a low impedance to radio-frequency interference currents. Permanent connections produced by welding or brazing are normally the best, although clean metal surfaces held together by nut and bolt or lock-washers give an equally effective joint.

ZERO-VOLTAGE SWITCHING

One of the disadvantages of phase-control and chopper regulation techniques is that the thyristor can be caused to conduct when there is a relatively large voltage across it. This results in a sharp increase in load current with the possible generation of radio-frequency interference. The effect of this can be minimised by the use of filters, as described in the last section. However, if the power levels being handled are large the filters can be bulky and expensive. Alternative techniques, known as zero-voltage switching, can then be used.

The principle of zero-voltage switching consists in firing a thyristor at the beginning of a half cycle or not at all. The load voltage build up then follows a sine wave and is not abrupt. Therefore the main cause of the interference is removed and no suppression filters are required. There are two methods for controlling the amount of power delivered to a load. These are illustrated in *Figure 4.22*. For burst firing the thyristor is either fully on or fully off for the duration of the sensing period. For instance if the temperature of an oven is being controlled, the thyristors will be on when the oven cools below a reference setting. The power supplied to the heaters will cause the oven temperature to rise. When it exceeds the desired value the thyristors turn off.

In proportional control (also known as cycle syncopation) the system works on a fixed period, measured in numbers of cycle or half cycles. *Figure 4.22(b)* illustrates a system in which this period is five cycles. To regulate the power the on to off duty cycle within this periodic time is varied. Once again this change can be in half cycles or in cycles. The important consideration is that the thyristor turns on as close to the start of the halfcycle as possible.

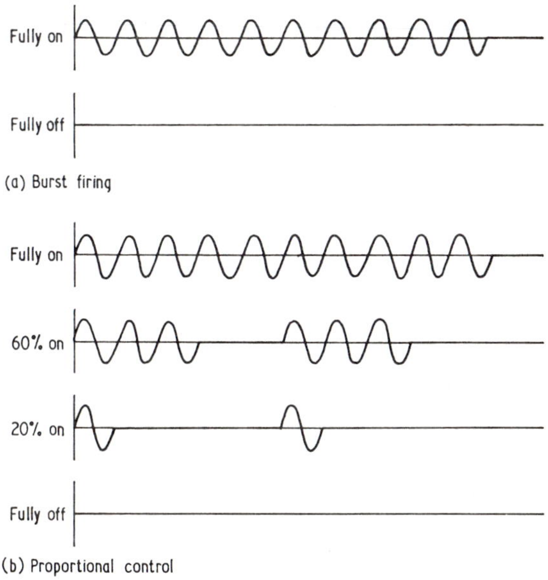

Figure 4.22. Operating modes for zero-voltage switching control

The difference between burst firing and proportional control is one of detail only, both being very similar in principle. Burst firing control systems have a large gain, as illustrated in *Figure 4.23*(i), so that the thyristors are either off or on. Although satisfactory for many applications this high gain can in some cases lead to problems of instability, commonly known as hunting. To overcome this the system gain is reduced, as in *Figure 4.23*(ii), so as to introduce a proportional band. Within this band the controller is able to anticipate the future state of the controlled parameter, and to adjust the duty cycle of the thyristors accordingly. The width of the proportional band is important. If it is made too large the system will be sluggish and reach its final state slowly; if it is too narrow there will be large overshoots about the required setting, as in burst firing.

Irrespective of the method of control used, the mean power and r.m.s. voltage are dependent on the thyristor duty cycle. If E is the r.m.s. input voltage, and P the power in the load with the supply uncontrolled, then the controlled voltage E_c and power P_c are given by

$$P_c = \frac{Pt_0}{T} \tag{4.9}$$

$$E_c = E\left[\frac{t_0}{T}\right]^{1/2} \tag{4.10}$$

where t_0 is the thyristor conduction period, expressed in units of time or number of half-cycles or cycles, and T is the operating period in the same units as t_0.

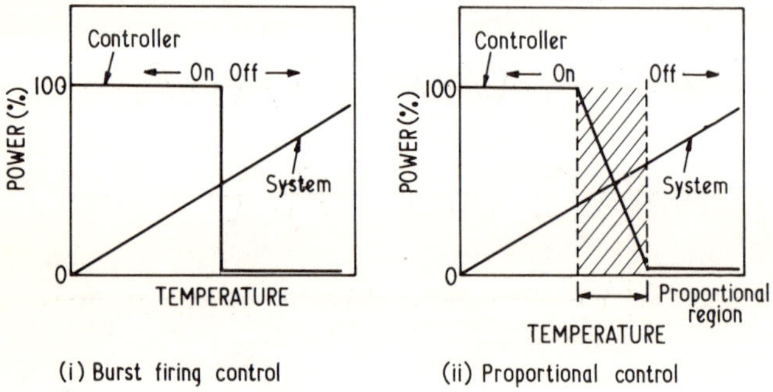

Figure 4.23. Control characteristics for zero-voltage switching systems

There are many circuits that may be used to control thyristors operating in a zero-voltage switching mode. They all have three basic sections:

(i) A power-supply stage to drive all internal amplifiers and feed the gate energy to the power thyristors.

(ii) A zero-voltage detecting stage, which senses the instant of zero supply voltage. This stage releases the power amplifiers for a short duration around this cross-over point so that they may trigger the thyristors if required.

(iii) An amplifier stage which magnifies the control signal to fire the thyristors.

Figure 4.24 shows a simple control circuit which has been used to provide whole cycles of power to a load. Thyristors CSR2 and CSR3 are the output devices, CSR1 performs the function of amplifying the signal on its gate, while the power supply and cross-over detection is derived directly from the a.c. input. Point C is fed from the control signal which determines the duty cycle of the power input. When C is positive thyristor CSR1 is on and both power thyristors CSR2 and CSR3 are held off. Suppose *C* now goes negative. Gate drive to CSR1 is removed but it will stay conducting for the rest of the positive half cycle (when line A is positive to B). When the supply voltage reverses, CSR1 goes off. During the next half cycle when line A again goes positive, a leading current is supplied to CSR2 via R_1 and C_1. Therefore this thyristor will turn on at the

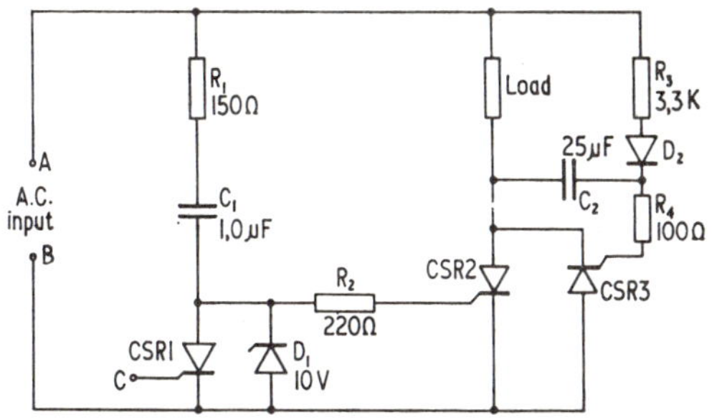

Figure 4.24. Zero-voltage switching control circuit

commencement of the half cycle. Thyristor CSR1 may be turned on soon after this instant, but will not be able to turn CSR2 off. With CSR2 conducting a half cycle of power is provided to the load. Simultaneously capacitor C_2 is charged via R_3 and D2. When the supply voltage reverses, C_2 discharges through R_4 and the gate cathode junction of CSR3, turning it on and providing the negative half-cycle load power. Therefore this method allows control of load voltage in full cycles of the supply, all switching occurring at zero voltage.

Owing to the increase in popularity of zero-voltage switching techniques, several manufacturers have produced monolithic integrated circuits, in standard TO5 or dual-in-line packages, which incorporate many of the functions required. *Figure 4.25* shows the internal construction of the PA424 (marketed by General Electric). The use of the device in a simple on-off (burst-firing) system is illustrated in *Figure 4.26*. A d.c. power supply is derived across an external capacitor C_E by internal diodes and external resistor R_E. This resistor, along with transistor TR3 and associated diodes, also performs zero-voltage sensing. The input to the unit is formed by long-tailed pair TR1, TR2, and the output amplifier is the Darlington pair TR5, TR6 which is controlled by inhibit transistor TR4.

When TR1 is on, transistors TR3 and TR4 are also driven on, so that no output is produced at pin 4. When TR1 goes off, TR3 and TR4 are still maintained on so long as the supply voltage is non-zero. It is only the combination of zero a.c. supply and TR1 off which

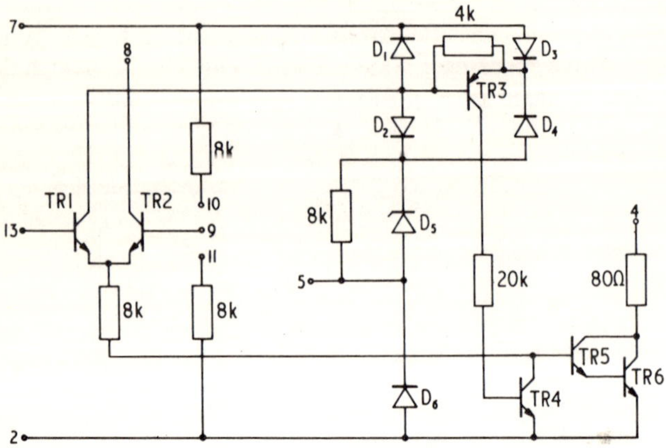

Figure 4.25. Circuit diagram of the PA424 integrated-circuit zero-voltage controller

allows TR4 to go off so that TR5 and TR6 turn on. These transistors therefore come on for a short duration (typically 100μs) around the zero-crossing point. If a longer pulse duration is required, to ensure that the thyristor reaches its holding current, the output from TR6 can be used to trigger a monostable circuit which then provides the required gate drive. Pin 4 can feed a thyristor firing pulse transformer either directly or through further amplifier stages.

The circuit shown in *Figure 4.26* is only capable of giving burst-firing signals, which are controlled by the relative magnitudes of R_A

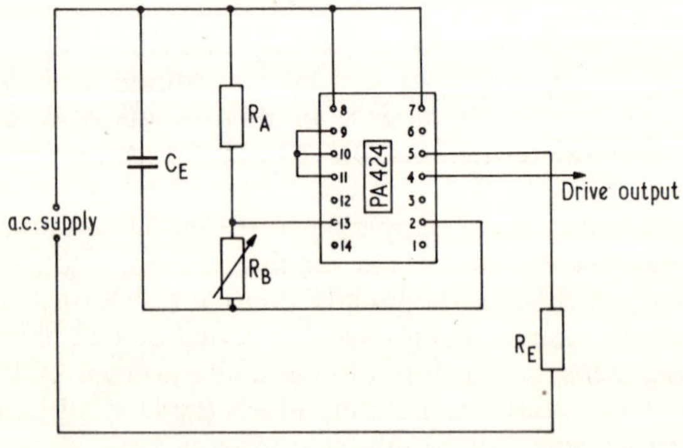

Figure 4.26. Connection of the PA424 to give burst firing control

and R_B. Here R_A could be some form of sensor, say a thermistor, while R_B represents the demand temperature. A modification for proportional control is shown in *Figure 4.27*. The magnitude of the proportional band is controlled by resistor R_D which injects a time varying current onto the d.c. bias produced by R_F and R_G. The period of this variation, which also determines the total periodic

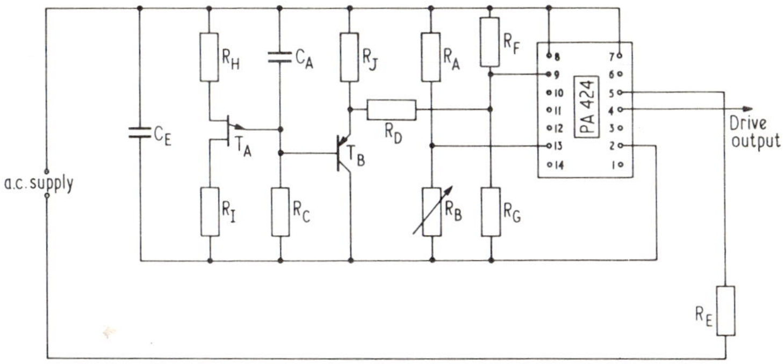

Figure 4.27. Connection of the PA424 to give proportional firing control

time in *Figure 4.22*, is determined by R_C and C_A. When C_A charges to the trigger voltage of complementary unijunction transistor T_A, it breaks over and discharges the capacitor. Therefore a sawtooth waveform appears at the emitter of T_A which is then amplified by emitter follower T_B before being fed to the remainder of the circuit. Whole cycle control is not suitable for low inertia loads, such as electric lamps, since these would show a noticeable change in control between pulses of power. The lower the load inertia the smaller must be the operating period for proportionate control systems. However, since changes can only occur in half-cycle blocks this would result in a coarse control system.

SYNCHRONOUS TAP CHANGER

In systems which have an output transformer and require only small adjustment of the voltage, synchronous tap changing is often preferable to ordinary phase control. *Figure 4.28* shows a typical system and its output waveforms.

At t_0 input line C goes positive and CSR3 is fired. After a delay α, thyristor CSR1 is turned on. This reverse biases CSR3 and turns it off. Output voltage jumps to the new value of input until at t_2 CSR1 goes off (resistive load assumed). CSR4 is fired at t_2 followed by

CSR2 at t_3 to complete the negative half cycle. By moving through one complete tap changing sequence synchronously with the supply voltage natural commutation of the thyristors is possible. The values

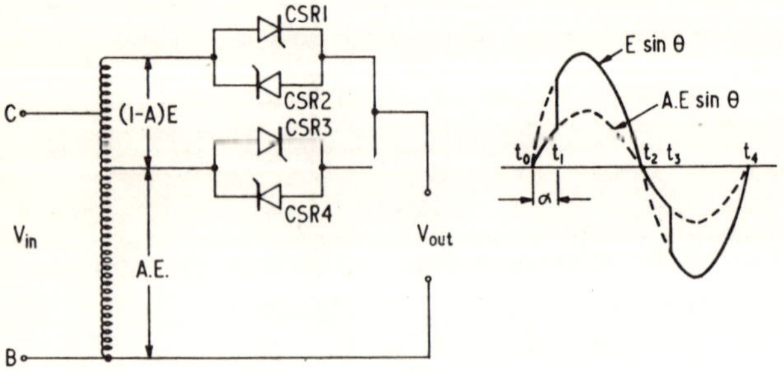

Figure 4.28. Synchronous tap changer

of A and α determine the output voltage as a function of input E. Although A is fixed for a given system, adjustment of the load is possible by varying α. The r.m.s. output voltage is obtained as:

$$V_0 = \left[\frac{1}{\pi}\int_0^{\alpha}(AE\sin^2\theta)^2\,d\theta + \int_{\alpha}^{\pi}(E\sin\theta)^2\,d\theta\right]^{1/2}$$

$$= \frac{E}{2}\left[\frac{1}{\pi}\{(1-A^2)(\sin 2\alpha - 2\alpha) + 2\pi\}\right]^{1/2} \tag{4.11}$$

The r.m.s. current rating of the thyristors depends on the conduction point α and is given by:

$$\frac{EA}{2R}\left[\frac{1}{\pi}\{2\alpha - \sin 2\alpha\}\right]^{1/2} \tag{4.12}$$

for thyristors CSR3, CSR4 and:

$$\frac{E}{2R}\left[\frac{1}{\pi}\{\sin 2\alpha - 2\alpha\} + 2\pi\}\right]^{1/2} \tag{4.13}$$

for thyristors CSR1 and CSR2, where R is the load resistance.

The volting rating of devices need only be $(1-A)E$ since with one or other arm always conducting only the tap voltage appears across them.

This control method produces a lower distortion than traditional phase-control techniques. Fourier series of the output is:

$$V_0 = \frac{a_0}{2} + \sum (a_n \cos n\omega t + b_n \sin n\omega t)$$

where

$$a_n = \frac{2E(1-A)}{\pi(1-n^2)}[\cos\alpha \cos n\alpha + n \sin\alpha \sin n\alpha - 1] \qquad (4.14)$$

$$b_n = \frac{2E(1-A)}{\pi(1-n^2)}[\cos\alpha \sin n\alpha - n \sin\alpha \cos n\alpha] \qquad (4.15)$$

for $n = 3, 5, 7, \ldots$

From this the magnitude of any harmonic for a fixed value of α and A can be found. Harmonics can be reduced by increasing the number of tapping points at the expense of increased circuit complexity. Fundamental coefficients are given by:

$$a_1 = \frac{2E(1+A)\sin^2\alpha}{2\pi} \qquad (4.16)$$

$$b_1 = \frac{2E(1-A)[(\alpha \sin^2\alpha - \cos^2\alpha) - \sin\alpha \cos\alpha]}{2\pi} \qquad (4.17)$$

The amplitude of the fundamental is:

$$[a_1^2 + b_1^2]^{1/2} \qquad (4.18)$$

The operation of the circuit shown in *Figure 4.28* becomes slightly more complicated on inductive load. For instance at t_2 thyristor CSR1 is conducting. Even though the line voltage now reverses this thyristor will be kept on by the stored energy of the load. To turn it off, and so limit the output voltage to $AE \sin\theta$, thyristor CSR3 is turned on. This means, however, that provided CSR3 is still on, thyristor CSR2 cannot be fired at t_3, as would normally be done, or it would short circuit through CSR3. Therefore although the control circuit for the thyristors is very similar to the usual phase-control systems described before, it is now necessary to sense the load current as well as the zero-voltage cross-over points and to inhibit the firing pulses if danger of short-circuit exists.

CONCLUSIONS

Of the four a.c. control methods discussed, phase-angle variation is the simplest and the most commonly used. Where the load power being controlled is large and radio frequency interference requirements stringent (as for domestic heating applications), zero-voltage switching is preferable provided the load has a relatively large inertia. A.C. chopper control finds application primarily in special systems where speed of response and minimisation of harmonics are essential. Where only small variations in output are required synchronous tap changing is often used.

5

D.C. line control

When discussing a.c. line control in Chapter 4 it was stated that the control of power between an a.c. source and a d.c. load can be accomplished by variable line impedances. This method is inefficient and inferior to that using thyristor control techniques. Similarly, d.c. line control can employ series thyristors to vary the power flow between a d.c. source and d.c. load. This is known as chopping. The principle has been discussed in Chapter 1 and is illustrated in *Figure 5.1*.

With CSR1 conducting the load voltage is equal to V_B (loss across

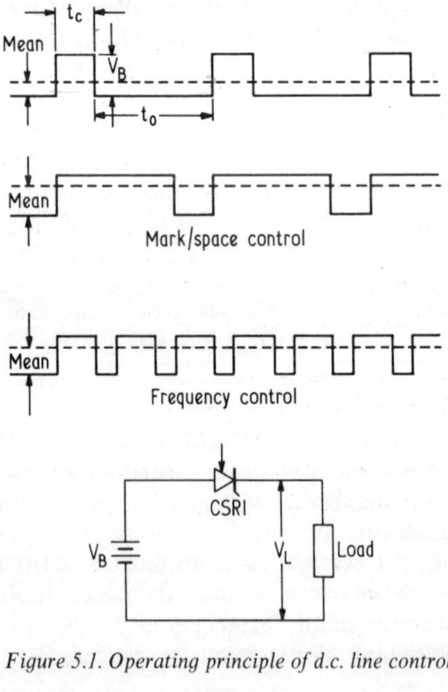

Figure 5.1. Operating principle of d.c. line control

the thyristor being assumed small). When CSR1 is non-conducting load voltage is zero. Varying the ratio of thyristor on to off times, either in a variable mark-space or controlled frequency mode, the mean load voltage is changed, being given by:

$$V_{L(mean)} = V_B \frac{t_c}{t_c + t_0} \tag{5.1}$$

An important difference between a.c. and d.c. line control should be noted. In a.c. systems the current source is alternating so that the thyristors turn off every time the supply voltage reverses. When the source is d.c. an external mechanism is needed to reverse bias the conducting thyristors, for a duration in excess of their turn-off time, to turn them off. This is known as 'forced commutation'. A large number of circuits have been developed for commutating thyristors but all can be grouped into four sections. This chapter will examine this classification system and discuss a few typical circuits. The appendix at the end of the chapter gives an example of one of the most frequent uses of d.c. line control.

THE CLASSIFICATION SYSTEM

Thyristor chopper circuits all operate in the modes illustrated in *Figure 5.1*. They differ only in the commutation method adopted to turn off the main thyristor, even though their turn-off pulse is usually supplied by a charged capacitor. Logically any classification method should place these circuits into different commutation groups. This chapter considers the following four divisions:

(1) *Parallel-capacitor commutation*, the charged capacitor being placed directly across the thyristor to be turned off as in *Figure 5.2(a)*. Switch Sw forms part of the commutation circuit.

(2) *Parallel capacitor-inductor commutation*, an inductor being connected in series with the commutation capacitor as in *Figure 5.2(b)*.

(3) *Series-capacitor commutation*, the commutation capacitor being connected in series with the thyristor to be turned off as in *Figure 5.2(c)*.

(4) *Coupled-pulse commutation*, the turn-off pulse being coupled to the thyristor through a transformer or auto-transformer as in *Figure 5.2(d)*.

Figure 5.2 indicates only the basic principles of the commutation methods. The auxiliary circuits used to prime the commutation capacitor, ready to turn off the main thyristor CSR1, are not shown.

In common with any engineering design problem, choice of a

thyristor chopper requires balancing of technical performance against cost. For a given price it is true that there is an optimum circuit. To help in drawing a technical comparison between various

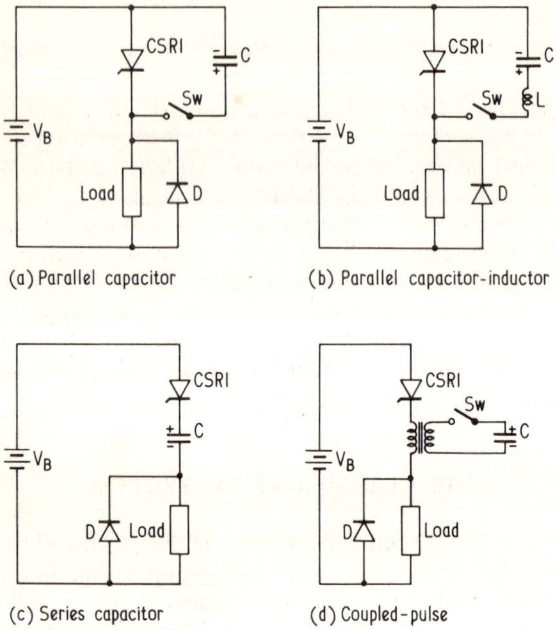

Figure 5.2. Chopper commutation methods

commutation systems it is necessary first to establish a check list of relevant technical factors. The following questions must be considered:

(1) Can the chopper be operated in a variable frequency or a fixed frequency variable mark-space mode, as desired?

(2) What are the limits on minimum and maximum output voltages?

(3) Is the commutation capacitor voltage inherently increased in proportion to the load current to be commutated? Such an increase would permit optimisation of capacitor size.

(4) If commutation were to be unsuccessful, say, due to an overload, would it be retried and be successful when the load reduces?

(5) Is the current rating of the main thyristor increased by the commutation circuit?

(6) Does a low impedance fault current path exist across the supply?

These factors will be considered again when comparing the different systems.

PARALLEL-CAPACITOR COMMUTATION

In the basic circuit shown in *Figure 5.2(a)* assume C to be charged to a voltage V_f with the polarity indicated. The load current just prior to Sw closing is $I_{L(max)}$ and is assumed constant during the short discharge period of C. If the reverse recovery current of CSR1 is neglected then:

$$I_{L(max)} = \frac{CV_f}{t_F} \tag{5.2}$$

where t_F is the time during which CSR1 is reverse-biased. For commutation to be successful, this time must be greater than the device turn off time at the rate of re-applied forward voltage given by:

$$\frac{dV}{dt} = \frac{I_{L(max)}}{C} \tag{5.3}$$

Equations (5.2) and (5.3) are applicable to all parallel-capacitor commutated circuits and enable the correct capacitor value to be chosen for a given load and device characteristic. However, the magnitude of V_f will be determined by the auxiliary commutation circuit adopted.

Parallel-capacitor commutated circuits can be further subdivided into two groups:

group 1 where the capacitor is primed ready for commutation by a charge path which does not include the load

group 2 where the capacitor charging current flows through the load.

The simplest example of group 1 commutation is shown in *Figure 5.3*. CSR1 is fired to initiate load current and at the same time charges C, via R_1, to V_B. CSR2 is fired to turn CSR1 off, the voltage

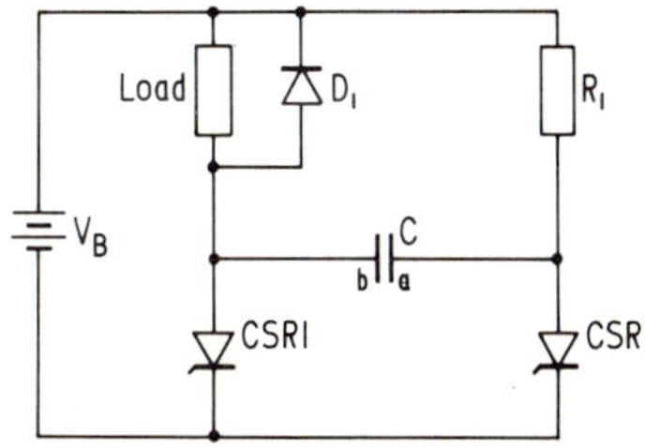

Figure 5.3. Elementary parallel-capacitor chopper of group 1

on C reversing such that when CSR1 is next fired CSR2 will turn off. The value of V_f is here equal to V_B.

This circuit is limited in efficiency and frequency of operation and is not suitable for high-power chopper applications.

A modification which overcomes these disadvantages is shown in *Figure 5.4*.

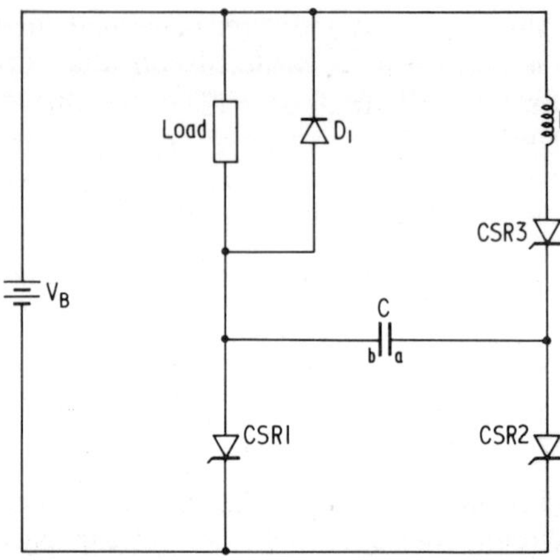

Figure 5.4. Addition of a resonant charge circuit to Figure 5.3

CSR1 is fired to commence the load cycle and simultaneously, or some time later, thyristor CSR3 is fired. C is charged through L_1 to V_f with plate *a* positive. To turn CSR1 off CSR2 is fired, the turn-off process being as in *Figure 5.2(a)*.

Assuming C to be discharged prior to CSR3 being turned on, the value of V_f is $2V_B$ (resonant losses in L_1 being neglected), so that the commutation capacitor size is half that required for *Figure 5.3*.

The following should be noted:

(1) The circuit of *Figure 5.4* can be operated in either a variable frequency or variable mark-space mode.

(2) 'Minimum on time t_c', which determines the minimum load voltage, is equal to $\pi\sqrt{(L_1C)}$. However, C is usually fixed by commutation requirements and the peak current through CSR3 and CSR1, given by $2V_B\sqrt{(C/L_1)}$ may be prohibitive for a small L_1. 'Minimum off time t_0' is load dependent. This determines the maximum load voltage. On light loads it could be large, seriously reducing the maximum output. This disadvantage is common to all circuits where the turn-off pulse flows through the load.

(3) Commutation voltage is not increased by load current (ignoring lead inductances) since D1 prevents plate *b* of capacitor *C* rising above the supply voltage under all load conditions.

(4) If CSR1 fails to turn off on firing CSR2, then when CSR3 is next fired *C* will recharge ready for another commutation attempt.

(5) The rating of CSR1 is increased by the charging pulse of *C* via CSR3.

(6) A fault condition causing CSR2 and CSR3 to be gated on simultaneously would result in a low-impedance path across the supply which could destroy both devices.

Chopper circuits of group 2 are the most frequently used. Although not necessarily of high technical merit they represent a good balance between performance and cost and are suitable for many applications.

For the circuit shown in *Figure 5.5* thyristor CSR2 is fired initially, charging *C* to voltage V_B with plate *b* positive. CSR1 is now turned on. *C* discharges via CSR1, D2, L_1 and resonates, recharging to V_f with plate *a* positive. For negligible resonant loss V_f is equal to V_B.

To turn CSR1 off CSR2 is fired. The voltage across CSR1 is reversed, *C* discharges through the load (current assumed constant) and recharges with plate *b* positive. CSR2 then turns off.

The following performance features can be noted about the circuit in *Figure 5.5*.

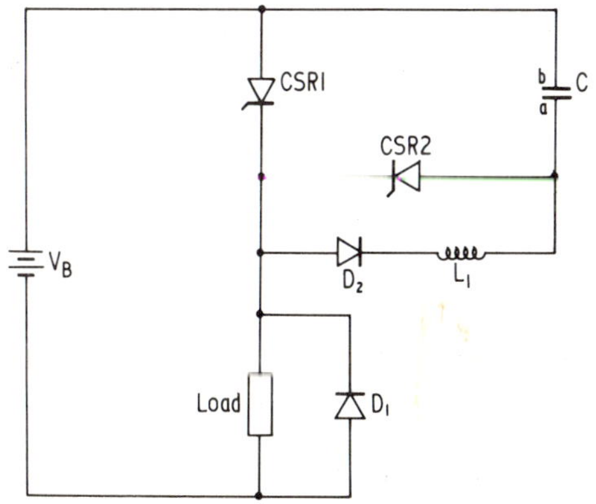

Figure 5.5. Parallel-capacitor commutation group 2

(1) Both variable frequency and variable mark-space operation modes are possible.

(2) The same comments made regarding minimum on and off times for *Figure 5.4* again apply here.

(3) Commutation voltage is not boosted by load current.

(4) If CSR2 is fired, plate *a* of the capacitor being positive and CSR1 conducting, and main thyristor CSR1 fails to turn off, *C* will discharge to zero and CSR2 will turn off. However, since CSR1 is still conducting *C* cannot recharge to V_B so commutation will not be re-attempted. 'Once-for-all' commutation failure has occurred and the chopper can be re-started only by first breaking the supply to CSR1, rendering it non-conducting, and then re-applying the supply.

(5) The rating of CSR1 is increased by the resonant capacitor charge.

(6) No low impedance path exists across the supply, assuming that D1 has not failed to short-circuit.

Figure 5.6 shows an alternative circuit to that in *Figure 5.5*. The discharge resonant pulse of the capacitor does not flow through

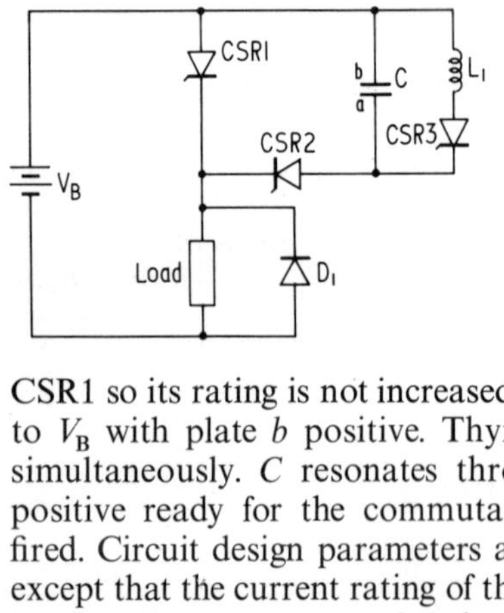

Figure 5.6. Avoidance of resonant current through CSR1

CSR1 so its rating is not increased. CSR2 is initially fired. *C* charges to V_B with plate *b* positive. Thyristors CSR1 and CSR3 are fired simultaneously. *C* resonates through L_1 recharging with plate *a* positive ready for the commutation interval when CSR2 is next fired. Circuit design parameters are identical to those in *Figure 5.5* except that the current rating of thyristor CSR1 is not now increased by the resonant current of *C* through L_1.

PARALLEL CAPACITOR-INDUCTOR COMMUTATION

Figure 5.7 illustrates the most basic commutation circuit in this group. On first switch on *C* charges through L_1 and the load with plate *b* positive. When CSR1 is fired the load cycle commences. Simultaneously *C* discharges through L_1 and recharges with plate *a* positive. The capacitor then again discharges through L_1 and the load, CSR1 turning off when capacitor current equals that of the load.

This circuit has a very restricted output voltage range. It can only be operated in a variable frequency mode with a thyristor conducting time of about $\pi\sqrt{(L_1 C)}$. Once turned off CSR1 can be re-fired only after C has recharged to V_B through the load. On light loads this time can be considerably greater than the on time.

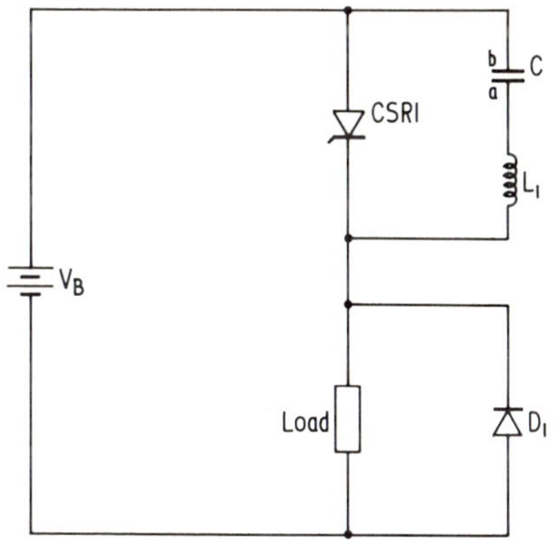

Figure 5.7. Basic parallel capacitor-inductor commutated circuit

The circuit of *Figure 5.8* gives a chopper with greatly improved performance. L_1 is now a saturable reactor. When CSR1 is fired the voltage of charged capacitor C is applied across L_1. It blocks the voltage for the time t_p required to drive it into positive saturation.

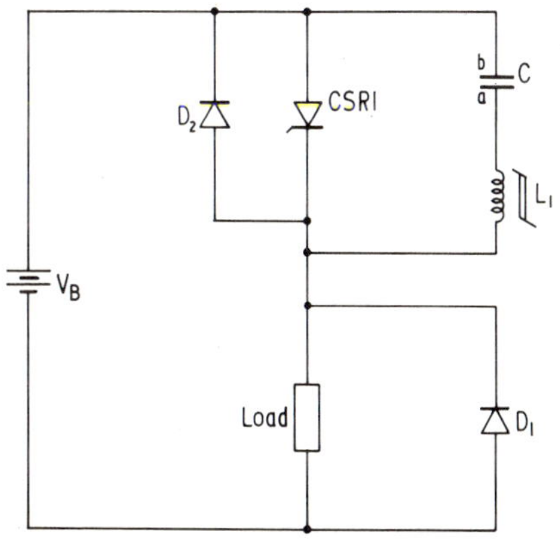

Figure 5.8. Modified saturable-reactor circuit

C then resonates with the after-saturation reactance of L_1 and recharges with plate a positive. Again L_1 blocks current for time t_n

required to drive the inductor from positive to negative saturation. Capacitor C then discharges through the load and diode D2 turning off CSR1. This circuit has increased the thyristor conduction time to $\pi\sqrt{(L_1C)}+t_p+t_n$ and reduced the minimum off time t_0 to $\pi\sqrt{(L_1C)}$ (when the load is an open circuit) so greatly increasing the maximum output voltage. It is still, however, only capable of variable frequency operation. *Figure 5.9* shows a modification which overcomes this disadvantage. Initially C charges through L_1, D3 and the load with plate a positive. The load cycle is commenced

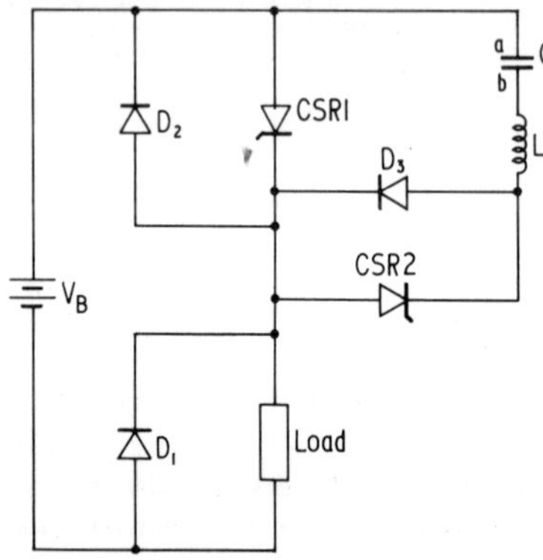

Figure 5.9. Modification to the basic parallel capacitor-inductor circuit for variable mark-space operation

when CSR1 is turned on. C cannot discharge since D3 is reverse-biased and CSR2 is non-conducting. To turn CSR1 off CSR2 is fired. C resonates through CSR2 and L_1 and discharges through L_1 and D3 turning CSR1 off. D2 limits the maximum discharge period, even with the load open circuit, as in *Figure 5.8*. The following points can be noted about the circuit of *Figure 5.9*:

(1) Both variable frequency and variable mark-space operation are possible.

(2) The minimum on and off times are $\pi\sqrt{(L_1C)}$.

(3) The commutation capacitor voltage is increased proportional to load current due to the transfer of energy from L_1 to C when CSR1 turns off. The capacitor voltage is therefore given by

$$V_B+I_{L(max)}\sqrt{(L_1/C)}$$

(4) A commutation failure prevents the circuit from re-attempting commutation.

(5) Rating of CSR1 is increased by the resonance of C through L_1.

(6) There is no low-impedance short circuit path across the supply.

The values of C and L_1 required to commutate a thyristor of turn-off time t_F in parallel capacitor-inductor circuits is given by:

$$C = \frac{2\sqrt{2t_F I_{L(max)}}}{\pi V_B} \tag{5.4}$$

$$L_1 = \frac{\sqrt{2t_F V_B}}{\pi I_{L(max)}} \tag{5.5}$$

From this it can be seen that parallel capacitor-inductor commutation is unsuitable for use in systems controlling large currents from low supply voltages. For example, to turn off 250A from 40V using thyristors with $t_F = 50\mu s$ would give $C = 280\mu F$ and $L_1 = 2{\cdot}5\mu H$. In all probability the inductance of the connecting leads would exceed this value.

SERIES-CAPACITOR COMMUTATION

Referring to the basic series-capacitor commutated circuit of *Figure 5.2(c)* it is seen that CSR1 will turn off once C has charged to V_B volts. Before CSR1 can be refired the capacitor must be reset. There are several ways in which this may be done. *Figure 5.10(a)* uses a parallel resistor R_1. However a better solution is illustrated in *Figure 5.10(b)*. CSR1 is fired to commence the load cycle. This causes C to charge through L_1 and turn CSR1 off. CSR2 is fired which resets C through L_2 after which CSR2 turns off and CSR1 can be refired.

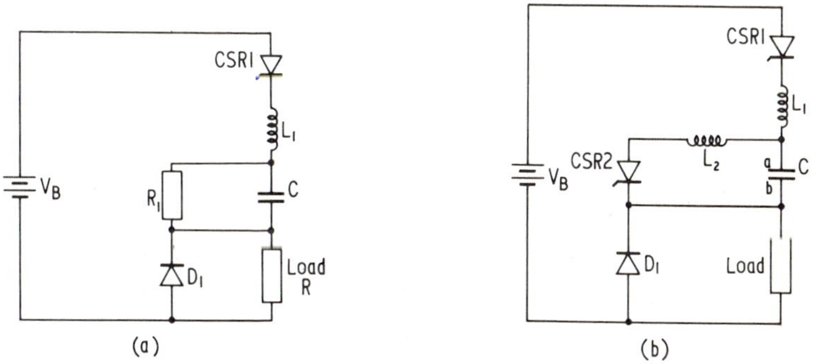

Figure 5.10. Series-capacitor commutated circuits

Series-capacitor commutation circuits have a very limited operating range, due to the capacitor reset time, and are rarely used as choppers. They find much greater application in sine wave inverter circuits and will be discussed further in Chapter 9.

COUPLED-PULSE COMMUTATION

A chopper circuit using an auto-transformer to couple the turn-off pulse to the thyristors is shown in *Figure 5.11*. CSR1 is the main thyristor and CSR2 an auxiliary device used to commutate CSR1. L_1 L_2 is a tapped auto-transformer. The circuit operation is as follows. CSR1 is fired to commence the load cycle. This also causes C_1 to charge through L_1 to supply voltage V_B. The clamping action of D2 prevents C_1 charging to a higher voltage. To turn CSR1 off CSR2 is fired. Capacitor C_1 discharges through L_2 coupling a voltage pulse via L_1 to CSR1 so turning it off. The voltage on C_1 falls to zero after which D1 conducts, carrying the current due to energy stored in L_2 and the load. CSR2 turns off when the current in L_2 has fallen to below the device holding value. C_2 is not normally required. By including it CSR1 can be fired before CSR2 has turned off since then C_2 discharges through L_1 coupling a pulse to CSR2 via L_2 and so turning it off. This allows a wider voltage control range.

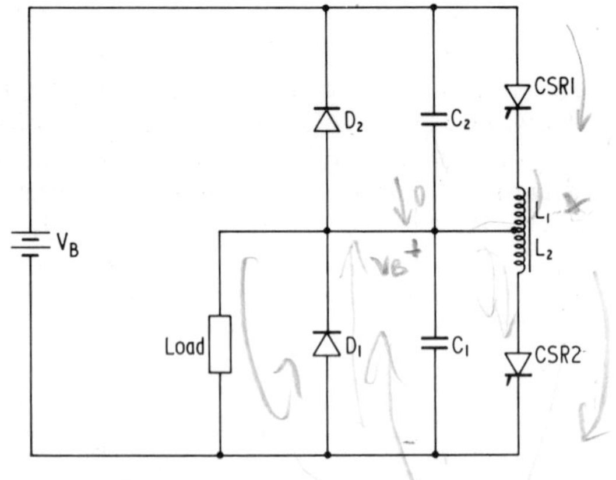

Figure 5.11. Coupled-pulse commutation

Assuming a symmetrical system with $C_1 = C_2 = C$ and $L_1 = L_2 = L$ the values of L and C can be found as (see Chapter 8):

$$L = \frac{2{\cdot}76 \times V_B \times t_F}{I_{L(max)}} \tag{5.6}$$

$$C = \frac{2{\cdot}15 \times t_F \times I_{L(max)}}{V_B} \tag{5.7}$$

This again shows the unsuitability of this commutation method for low-voltage high-current operation.

The following can be noted about the circuit shown in *Figure 5.11*:

(1) The circuit can be operated in a variable frequency or variable mark-space mode.

(2) The minimum on and off times are approximately given by $\pi\sqrt{(L_1C_1)}$ and $\pi\sqrt{(L_2C_2)}$ for low load current.

(3) Due to the clamping action of D1 and D2 commutation voltage is fixed at V_B irrespective of load current.

(4) An unsuccessful commutation would not allow the circuits to retry commutation.

(5) The charging pulse for C_1 and the discharge pulse for C_2 flow through CSR1 increasing its current rating.

(6) A low impedance failure path exists across the direct current supply.

CONCLUSIONS

There are a large number of chopper circuits that can be used. When selecting the best system for any application it is necessary to decide on the commutation class desired. Generally parallel-capacitor commutation circuits are the most flexible and are suitable for operating from a wide range of supply voltages. However, in applications where the supply voltage is high and the load current small, parallel capacitor-inductor commutation should be considered. These are generally simpler than the corresponding parallel-capacitor circuits. Coupled-pulse commutation is more expensive owing to the use of a transformer. It finds greater application in inverter circuits. Similarly series-capacitor commutation is frequently used in sine-wave inverters but is rarely used in chopper circuits.

Appendix to Chapter 5

An electric vehicle controller

Thyristor choppers provide an efficient and versatile method of controlling power flowing between a d.c. source and a d.c. load. The principles of choppers have been discussed earlier in this chapter. This appendix describes a complete chopper system which has been used for control of fork-lift trucks. Relatively few modifications would be required to convert this into a system for controlling an electric passenger car.

POWER CONTROLS

A chopper controls the mean load voltage by varying the proportion of the time in any cycle when it is connected to the supply V_B. If t_c and t_o are the load connected and load disconnected times, mean load voltage V_L is given by equation (5.1).

Since the thyristor is operating from a d.c. source it is necessary to force-commutate the device by reverse biasing for a period greater than its turn-off time. *Figure 5.12* shows a typical chopper

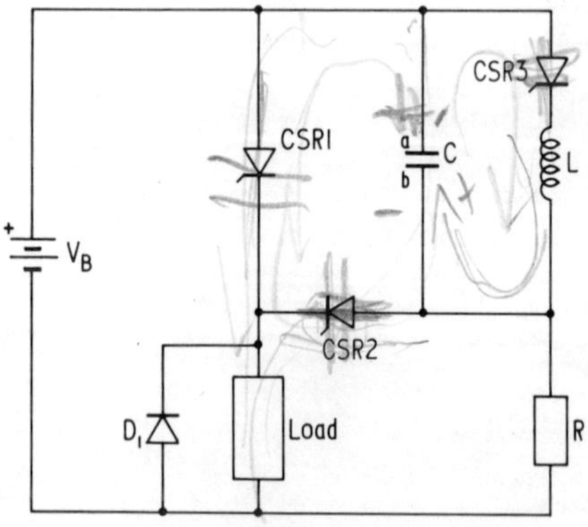

Figure 5.12. Chopper power circuit

circuit and *Figure 5.13* gives the voltage waveform across capacitor *C*.

On first switch on, CSR2 is fired. This causes *C* to charge to V_B volts with plate *a* positive to *b*. CSR2 now turns off. To commence the load cycle at t_1 CSR1 and CSR3 are fired. The motor voltage rises to $+V_B$ and *C* discharges through CSR3 and *L*. This circuit is oscillatory so that the voltage on *C* reverses and plate *b* charges positive to *a* by V_B (assuming no losses) at time t_2. The voltage on the capacitor is then maintained since CSR3 is now reverse-biased and CSR2 is off.

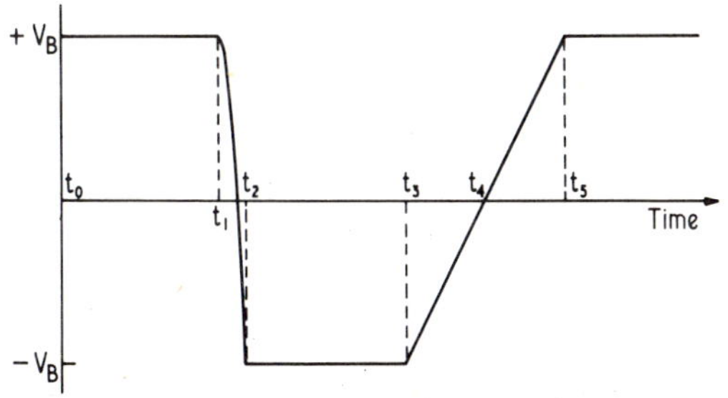

Figure 5.13. Commutating capacitor voltage

CSR2 is fired at time t_3 to commutate CSR1. This applies the reverse voltage of capacitor *C* across CSR1 turning it off. The capacitor discharges through the load and battery from t_3 to t_4 and recharges to $+V_B$ volts as before from t_4 to t_5. Time $(t_4 - t_3)$ must exceed the turn-off time of CSR1 for the maximum load current if commutation is to be successful.

The load voltage is shown in *Figure 5.14*. This is normally zero

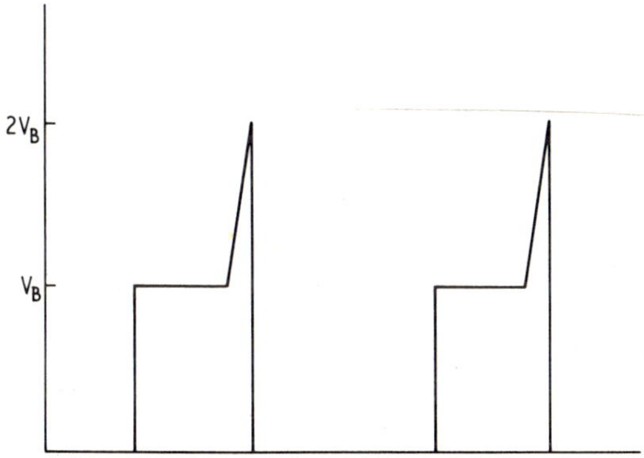

Figure 5.14. Load-voltage waveform for the circuit of Figure 5.12

rising to battery voltage when CSR1 is fired. When CSR2 is turned on to commutate thyristor CSR1, the voltage jumps to $2V_B$ decreasing to zero as C discharges and recharges in the reverse direction. The spikes of $2V_B$ are extremely short compared to the actual on time so that equation (5.1) still gives the mean voltage with reasonable accuracy.

For the circuit to operate satisfactorily minimum pulse width must not be reduced to zero. When CSR1 and CSR3 are fired sufficient time must be allowed for C to resonate through L and reverse its charge before CSR2 is again fired.

A limit is also placed on minimum off time. Assume CSR1 to be on and C to be charged with b positive to a. CSR2 is now fired turning CSR1 off. C discharges through the load and recharges to the supply voltage. CSR1 and CSR3 should not be refired until the charging of C has been completed and CSR2 turned off. This delay is obviously load-current dependent, being minimum on heavy loads.

It is usual to allow 1ms at either end of control so that if chopping at f Hz the output range is from $f/10$ per cent to $(100-f/10)$ per cent of battery voltage.

The chopper shown in *Figure 5.12* can be operated in a fully on mode. The gate pulses are removed from CSR3 so that CSR1 cannot

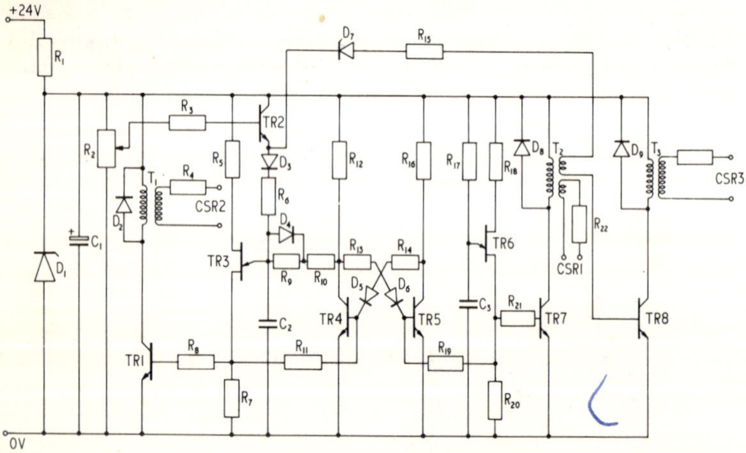

Figure 5.15. Thyristor firing circuit

be commutated and conducts during the whole period. It is essential to include resistor R which compensates for the leakage of charge from capacitor C and maintains it primed at V_B volts. This enables CSR1 to be commutated, when gate pulses are restored to CSR3, so that the load voltage can be reduced.

THYRISTOR FIRING CIRCUIT

Figure 5.15 illustrates the basic control circuit for the chopper of *Figure 5.12*. Supply is derived from a 24V battery tapping and stabilised by Zener D1. R_2 controls the chopper voltage. Suppose that it is set so that the potential at the emitter follower TR2 is a maximum. This means that when TR7 is turned on by free-wheeling oscillator TR6, the second winding on T_2 causes TR8 to be pulsed on, firing CSR3 simultaneously as CSR1.

When unijunction TR6 breaks over a pulse is fed to TR5, turning it on and turning off TR4. Capacitor C_2 commences to charge through high impedances R_9, R_{10} and also through R_6 which is comparatively low valued. The higher the voltage at D3 the sooner TR3 fires after TR6, which turns on TR1 and CSR2. As the control voltage reduces, the charging time for C_2 increases and so does the pulse width of the output. The operating frequency is fixed by R_{17} and C_3.

When the control voltage has reduced to a chosen value (*i.e.* output has reached 90% say of the battery voltage) it is insufficient to break over Zener D7 so that TR8 is not fired by T_2. Gate pulses are removed from CSR3 and CSR1 is maintained in a continuous conduction mode. When the control voltage again increases, pulses to CSR3 are restored and thyristor CSR1 is commutated each cycle. The voltage at which this jump between the controlled and fully-on mode occurs is set by the value of Zener D7.

STOPPING AND REVERSING

Three factors should be considered when designing a system capable of stopping and reversing an electric vehicle:

(1) Braking must be smooth and controlled. A system of dynamic braking is normally employed.

(2) Reversal should include a time delay or be current controlled to prevent it from being too rapid.

(3) Preselection should be included for reversal. This means that if the vehicle is moving in a certain direction and a reverse direction is selected on the control panel, this will not be effected until the accelerator pedal is recycled. This facility is often required in materials handling.

Figure 5.16 shows an extension of the basic power circuit of *Figure 5.12*. Battery V_B supplies power to the load which is a series motor. CSR1 is the main thyristor controlling this while *C*, *L*, CSR2 and CSR3 are its associated commutation circuitry. The inductive energy of the motor is carried by free-wheeling diode D1. Across the armature is connected diode D2 and resistor R_{DB} which constitute the dynamic braking circuit; R_{CL} is a low-value current-sensing resistor.

The field of the motor is connected between two reversing contactors. When both forward F and reverse R contactors are de-energised, contacts 1 and 3 are closed and 2 and 4 opened as illustrated in *Figure 5.16*. No current can flow through the motor even if CSR1 is triggered.

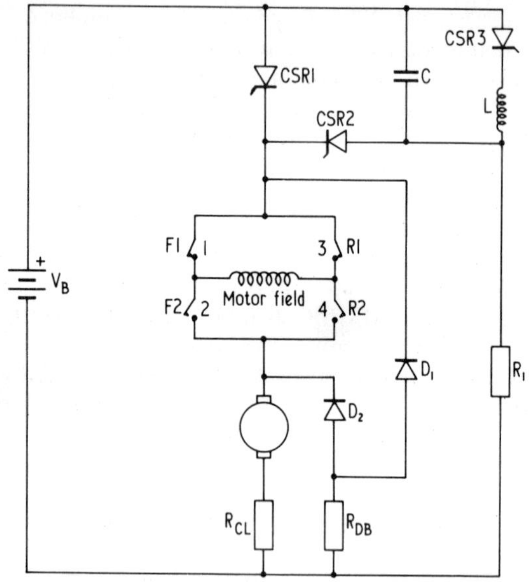

Figure 5.16. Modification to the basic chopper of Figure 5.12

If contactor F or R is energised the state of contacts 1 and 2 or 3 and 4 will be changed so that current can flow through the motor. The current through the field is in the reverse direction when F or R is energised. It is, therefore, possible to run the motor in either direction.

Assuming the truck is running in the forward direction, contactor F is energised and R de-energised so that contacts 3 and 2 are closed. The back e.m.f. of the motor opposes the current, and diode D2 is reverse-biased. If now the motor is thrown into reverse by de-energising contactor F and energising R so as to reverse the field current, the back e.m.f. will reverse since the truck inertia is still keeping the motor armature rotating in the forward direction.

Diode D2 is forward-biased and the kinetic energy of the truck is dissipated as heat in the motor armature and dynamic-braking resistor. During braking, CSR1 can still be switched, as in normal motoring operation, to control the field current. The greater this current the larger the back e.m.f. of the motor and the more severe the braking. After the truck has been brought to a standstill, if the accelerator pedal is still depressed, the truck will move in the reverse direction in a normal motoring mode.

Figure 5.17 shows the control arrangement for the contactors of *Figure 5.16*. Forward F and reverse R contactor coils are energised when CSR5 or CSR6 is fired. SW1 is the forward/reverse selector switch. SW2 is a microswitch on the accelerator pedal. This pedal also controls potentiometer R_2 illustrated in *Figure 5.15*.

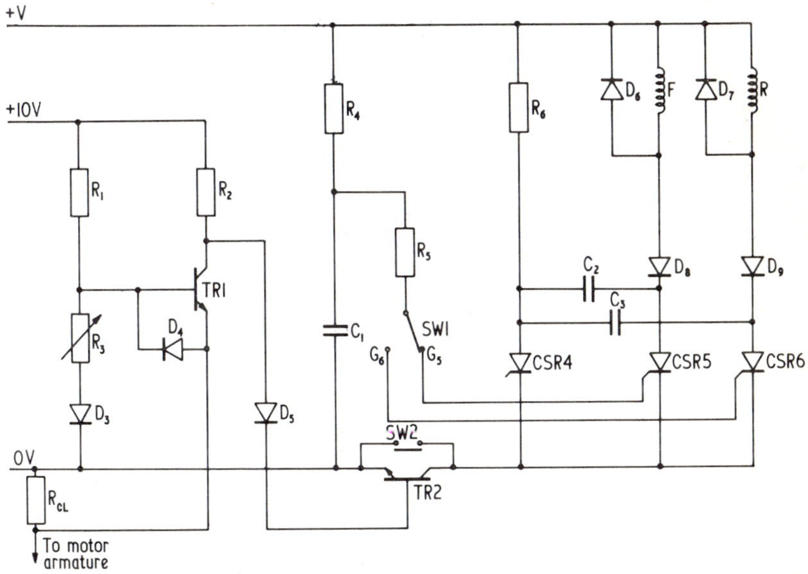

Figure 5.17. Pre-selector reversal and reversal inhibition

Assume SW1 connected as shown. Capacitor C_1 is charged through R_4 to the supply voltage. When the accelerator pedal is depressed SW2 closes and C_1 discharges through R_5 turning CSR5 on. The forward contactor is energised and the vehicle can move in the forward direction. Suppose switch SW1 is now moved to the reverse position. Since the transition time of the switch from the forward to reverse position is short C_1 has not charged sufficiently to fire CSR6 so the vehicle still moves in the forward direction. If the accelerator pedal is now released CSR5 goes off and C_1 can charge via R_4 so that when SW2 again closes CSR6 is fired and the reverse contactor energised. CSR4 and capacitors C_2 and C_3 are additions for fail-safe operation and are described in the next section.

Time delay between forward and reverse operation is incorporated since C_1 must charge to a high enough voltage through R_4 before it can fire the selected thyristor. By incorporating transistor TR2 it is possible to delay the change in contactor states until motor current has fallen below a set value. If load current is small, transistor TR1 is on and TR2 off. However, above a certain value of load current (determined by the setting of potentiometer R_3) TR1 goes off and TR2 is on. This means that even though the accelerator pedal is released and SW2 open, transistor TR2 carries the thyristor currents and prevents a change in the state of the forward/reverse contactors.

PROTECTION

Three aspects of operator protection are considered here:

(1) Slow start, which ensures a gradual build up of speed even though top speed is instantaneously demanded.

(2) Current limit, which prevents excessive battery drain.

(3) Fail safe, which shuts down the system in the advent of a system malfunction.

Figure 5.18 shows the additional components D9, R_{23}, and C_4 necessary to accomplish slow start. TR2, D3, R_2, R_3 and R_6 correspond to *Figure 5.15*.

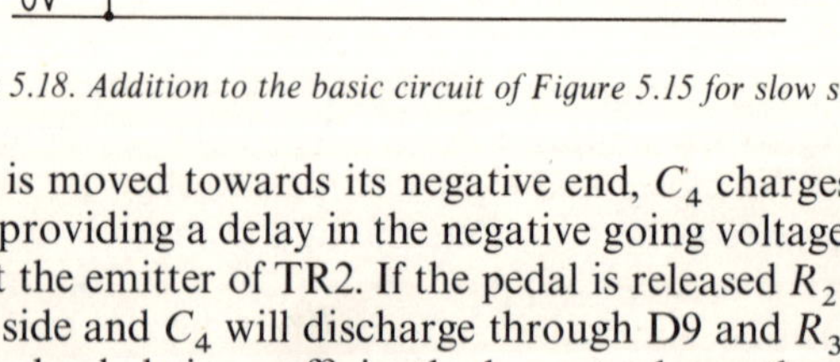

Figure 5.18. Addition to the basic circuit of Figure 5.15 for slow start

When R_2 is moved towards its negative end, C_4 charges through R_3 and R_{23} providing a delay in the negative going voltage (increasing speed) at the emitter of TR2. If the pedal is released R_2 moves to the positive side and C_4 will discharge through D9 and R_3. Resistor R_3 is low valued, being sufficiently large to keep the discharge current to a safe limit. Therefore, although the charging of C_4 occurs slowly, it can discharge rapidly, enabling the speed to be reduced quickly if required.

Figure 5.19 illustrates the current limit circuit. Transistor TR9 is biased by resistor R_{25}, potentiometer R_{26} and diode D11, so that it is normally on. The voltage at the collector of TR9 is at a low value and for any setting of potentiometer R_{26} diode D10 is reverse biased and the current limit has no effect on the operation of the basic circuit. As the motor current increases a point is reached where the emitter voltage of TR9 is raised sufficiently to overcome its base bias and turn it off. Capacitor C_4 discharges through R_{24} and D10. Transistor TR9 does not operate on a snap-action basis. Depending

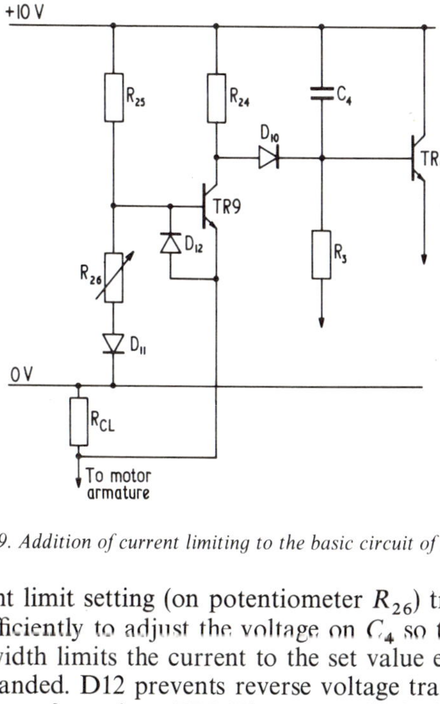

Figure 5.19. Addition of current limiting to the basic circuit of Figure 5.15

on the current limit setting (on potentiometer R_{26}) transistor TR9 is off just sufficiently to adjust the voltage on C_4 so that the maximum pulse width limits the current to the set value even when full speed is demanded. D12 prevents reverse voltage transients across the emitter-base of transistor TR9. The current limit circuit is similar to that used to obtain current delay as described in *Figure 5.17*. By using two separate circuits it is possible to adjust operating currents independently.

A simple fail-safe circuit is illustrated in *Figure 5.20*. Thyristor CSR4, resistor R_6 and switch SW2 correspond to those shown in *Figure 5.17*. Pin A of the fail-safe circuit is connected to the anode of thyristor CSR2 shown in *Figure 5.16*. When capacitor *C* (*Figure*

5.16) is fully charged with the end connected to CSR2 anode at battery negative voltage, transistor TR1 is on and capacitor C_1 is discharged holding TR2 off. (The zero voltage line in *Figure 5.20* is commoned to the battery negative in *Figure 5.16*.) During a resonant reversal of C the anode of CSR2 rises to twice battery voltage. Current is fed via D1 and R_1 turning TR1 off. C_1 begins charging through R_5 towards battery negative voltage. Charging rate is so arranged that at the chopping frequency being used, when operating at maximum pulse width, C_1 does not charge sufficiently to overcome the Zener voltage of D2. When the resonance is completed and CSR2 is fired, capacitor C charges again with the end connected to CSR2 anode at negative, turning TR1 on and discharging C_1.

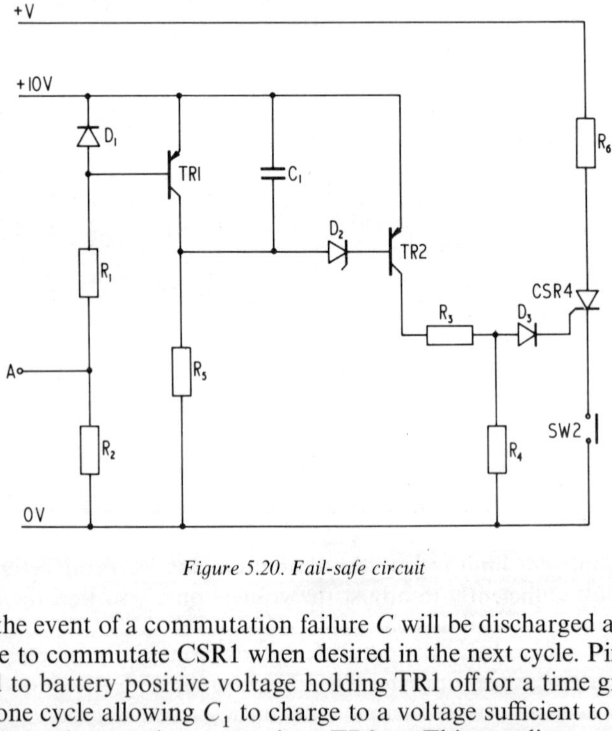

Figure 5.20. Fail-safe circuit

In the event of a commutation failure C will be discharged and so unable to commutate CSR1 when desired in the next cycle. Pin A is raised to battery positive voltage holding TR1 off for a time greater than one cycle allowing C_1 to charge to a voltage sufficient to overcome D2 voltage and turn transistor TR2 on. This supplies gate drive to thyristor CSR4 via diode D3 turning it on. When CSR4 was off, capacitor C_2 or C_3 of *Figure 5.17* was charged to battery voltage via R_6 depending on whether CSR5 or CSR6 was conducting. When CSR4 is turned on the charged capacitor is applied across the conducting thyristor which is turned off isolating the main circuit.

Diode D3 is necessary to prevent voltage being forced via the reverse cathode-gate impedance of CSR4 when SW2 is opened and thyristor CSR5 or CSR6 is fired.

The fail-safe circuit is relatively simple and will protect in certain conditions of commutation failure resulting from the inability of the circuit to commutate CSR1. This can occur:

(1) when the load current exceeds maximum designed value
(2) when thyristor CSR1 goes to short-circuit
(3) when thyristor CSR2 goes to short-circuit
(4) when thyristor CSR3 goes to short-circuit
(5) when thyristor CSR2 goes to open-circuit or gate pulse to it is lost
(6) when thyristor CSR3 goes to open-circuit or gate pulse to it is lost
(7) when the 10V electronic supply fails.

The circuit as described above will protect against failure for conditions (1) to (5) but not (6) or (7). In condition (6) it is difficult to discriminate between the fully-on state when pulses from CSR3 are purposely removed and the failure state when the thyristor is open-circuit or accidentally loses its gate drive. It can be overcome by sensing the control potential. A positive potential (signifying low speed) coupled with the off state of CSR3 for longer than one cycle would then mean that failure had occurred. In condition (7) the loss of the electronic supply would result in failure if thyristor CSR1 was the last thyristor to be fired before the supply was lost. This can be protected against by using a normally energised relay, feeding off the electronic supply, which would be de-energised and break the contactor coil current in the event of a loss of supply.

CONCLUSIONS

When designing electrical vehicles it is important to allow for as many eventualities as possible. For instance the fork lift truck operator should not drive off at top speed if he is carrying a delicately perched load. And of course no one would be foolish enough to run the truck up against a wall and switch the controls to full speed, just to see what happens. Or would they? The wise designer allows for these conditions, in this instance by building in slow start and current limit facilities.

6

Controlled rectification

A conducting thyristor has identical properties to a rectifier, *i.e.* it presents a low impedance to current flow in one direction and a high impedance in the other. But here the similarity ends. A thyristor has a third terminal, called a gate which, when unenergised, causes it to block current in both directions, giving controlled rectification.

Controlled rectifier circuits had been in use long before thyristors were invented. The control element was then the mercury arc rectifier. These are still used for very high power applications where thyristors prove more expensive. As the price of the semiconductor device decreases, however, it is very likely that the mercury arc rectifier will, like the thyratron, be almost entirely replaced.

Rectifier circuits were introduced in Chapter 1. It was shown there that many systems are capable of passing power in either direction, *i.e.* from the supply to the load or back from load to supply. However, not all thyristor controlled rectifier circuits have this property of bidirectional operation. In some applications, where this is not required, other circuits may be used which, although incapable of inversion, do nevertheless possess other advantages. Depending on whether or not rectifier circuits are capable of bidirectional operation they are referred to as bidirectional or unidirectional converters respectively. Each of these types then possesses a variety of circuit configurations which group them further into bridge or push-pull systems. Both these can operate from single- or three-phase supplies.

There are two principal parameters which are of interest in converter circuits. These are the output voltage harmonics and input current waveforms. These will be considered further, and effects of source reactance and discontinuous load current briefly examined.

BIDIRECTIONAL CONVERTER

Perhaps the simplest bidirectional a.c. to d.c. converter is that shown in *Figure 6.1.* The thyristor is capable of supplying power only during half a cycle, when supply line A is positive with respect to B, so that this is a half-wave controller. On resistive loads the supply current follows the shape of the input a.c. voltage when the thyristor is conducting. When the thyristor is off the load current and voltage are zero, if the leakage through the device is neglected. Clearly the delay angle α can be used to regulate the value of the mean d.c. output voltage. The operation of the circuit on inductive loads changes slightly. Now when the thyristor is fired, at t_{01} say, the load current will increase in a finite time through the inductive load. At t_1 the supply voltage reverses but CSR1 is kept conducting while the load

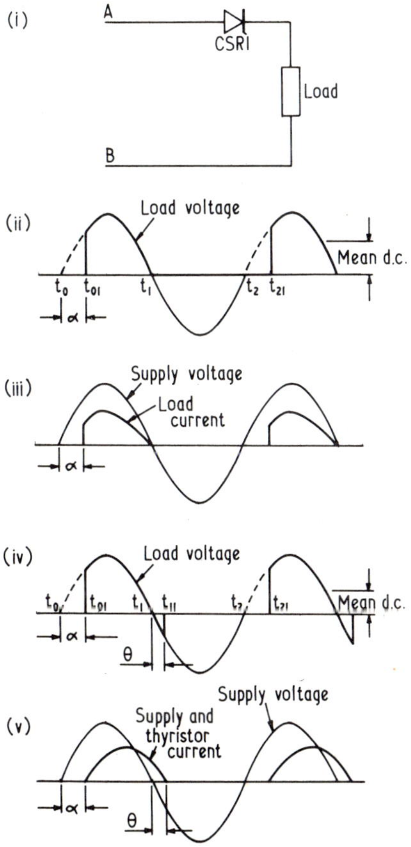

Figure 6.1. Half-wave bidirectional converter. (i) circuit arrangement, (ii), (iii) resistive-load waveforms, (iv), (v) inductive-load waveforms

energy stored during time t_{01} to t_1 is fed back to the supply. The load voltage goes negative, following the reverse half-cycle of the supply voltage. At t_{11} the load current falls to below the holding current of thyristor CSR1 and it goes off.

The half-wave circuit of *Figure 6.1* is not normally used since it produces a large output voltage ripple, and is incapable of providing continuous load current. There are several ways in which it may be extended to full-wave operation. Generally these systems fall into two groups:

(i) push-pull converters, which require a tapped transformer input.

(ii) bridge converters, where an input transformer may be used but is not an essential requirement for correct system operation.

Figure 6.2 shows the circuit of a push-pull bidirectional converter and *Figure 6.3* gives its operating waveforms. It is assumed here that the load current is maintained at a constant d.c. value through the operating cycle. T_1 is a centre-tapped transformer whose turns ratio may be adjusted to give any primary to secondary voltage change. When terminal A of the a.c. supply goes positive CSR1 is forward-biased, and when the supply polarity reverses CSR2 becomes forward-biased. When either thyristor conducts, the load voltage equals the instantaneous a.c. voltage across half the transformer secondary winding.

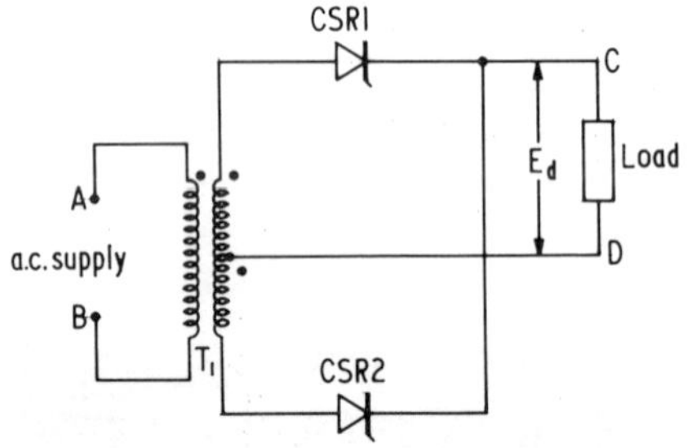

Figure 6.2. Push-pull two-pulse bidirectional converter

Referring to *Figure 6.3*(i), the thyristors are fired at the commencement of the a.c. cycle and, as far as the circuit is concerned, they behave exactly like diodes. The supply current is assumed square (ripple-free load current) and is composed of a fundamental and various higher harmonics. The fundamental of the current is in phase with the input voltage, so that the system behaves like a unity power factor load.

Figure 6.3(ii) shows a delay of α between the start of a positive

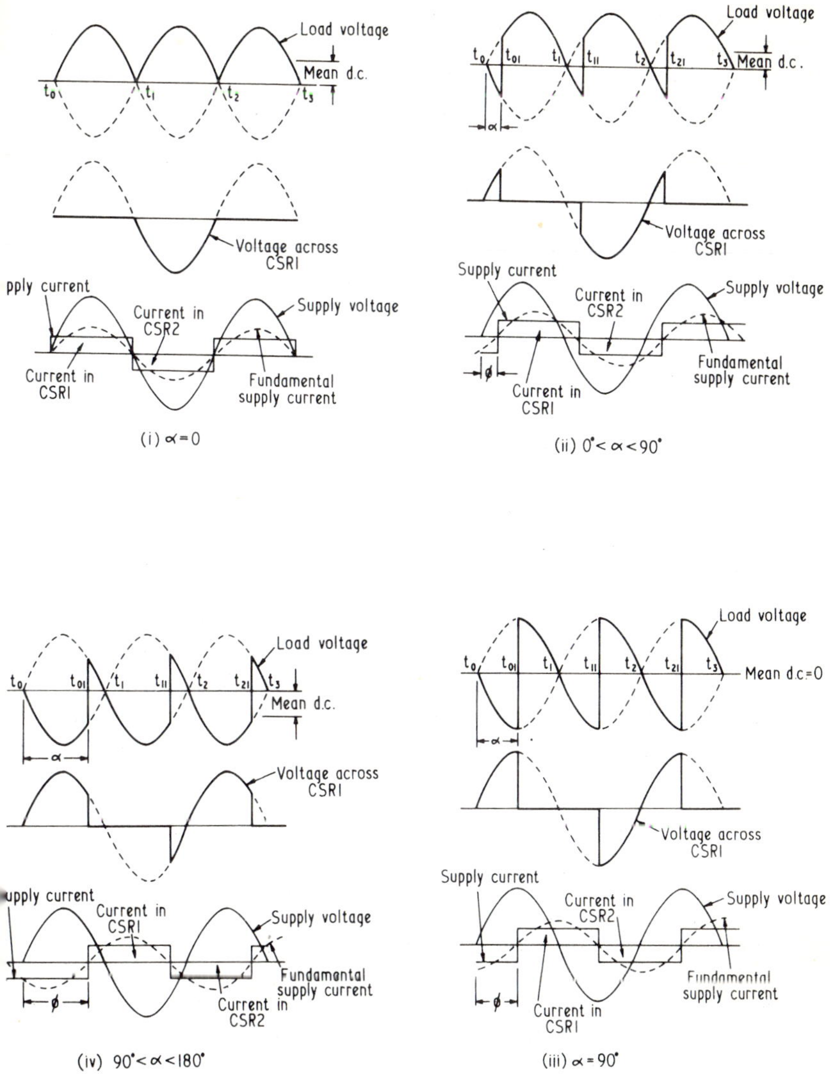

Figure 6.3. Circuit waveforms for a push-pull bidirectional converter

half-cycle and the firing of the corresponding thyristor. Therefore before t_0 thyristor CSR2 was conducting. When the supply voltage reverses at this point it is kept in conduction due to inductive load current, even though the voltage is negative. This is a regenerative period and power flows from the load to the supply. The voltage across CSR1 is positive and, referring to *Figure 6.2*, it is seen to be equal to the combined voltage across both halves of the secondary transformer winding, *i.e.* twice the load voltage. At t_{01} thyristor CSR1 is fired. The voltage across CSR2 now equals that of the two halves of the secondary winding, and since it is negative this thyristor turns off. CSR1 now conducts up to t_{11} when CSR2 is refired. During t_{01} to t_1 power is fed from the supply to the load, and from t_1 to t_{11} it is fed from the load back to the supply.

Figure 6.3 illustrates the circuit waveforms for progressively increasing delay angles. The following can be noted from these curves:

(i) For a delay angle up to 90°, more power is fed into the load than is received back from the supply. Beyond this point there is net transfer from the load to the supply. This cannot be maintained continuously unless the load is provided with a suitable d.c. power source.

(ii) Each thyristor conducts for a 180° period, and the shape of the input current waveform is unchanged. However, the phase angle ϕ between the fundamental component of this current and voltage input, changes such that $\phi = \alpha$. That this is necessarily true can be verified at $\alpha = 90°$. There are now equal periods of power flow in both directions between the load and the supply. As far as the supply is concerned, therefore, the load is purely inductive, and the phase angle ϕ must be 90°. A converter, even though itself consisting of a fixed value of resistance and inductance, will therefore present a varying load power factor as the firing angle is changed. For net rectification, this power factor is lagging although for inversion it changes to leading.

(iii) The mean d.c. load voltage decreases as the firing angle α is increased. Beyond 90° delay the voltage goes negative and at 180° it reaches a peak negative value. Clearly for a.c. to d.c. rectifier systems the negative voltage period is undesirable.

(iv) The value of the d.c. ripple voltage also increases as the firing angle is increased up to 90° delays. Beyond this point the ripple in the negative voltage decreases as α is increased to 180°.

(v) The period for which a thyristor is reverse-biased reduces progressively as the delay angle increases to 180°. A thyristor must of course be reverse-biased for greater than its turn-off time in order to successfully commutate it. Therefore the maximum delay angle can never be raised to 180° and for practical systems is normally limited to about 165° (on 50Hz systems). If a thyristor is not successfully commutated, it will commence conduction the instant its anode voltage goes positive, and so provide a complete half-cycle of power to the load. There will therefore be an abrupt change in the converter operating mode from almost full inversion to full rectification.

Push-pull converter circuits are popularly used in applications which require an input transformer either for isolation purposes or for effective phase number increase. As will be seen later the larger the number of input phases the lower the d.c. voltage ripple and the higher the power which the converter can handle. However, when an input transformer is not essential, a bridge system is often more economical. A single-phase bridge is shown in *Figure 6.4*. Its operation can be followed by the waveforms of *Figure 6.3* where CSR1, CSR4 and CSR2, CSR3 conduct in pairs. There are, however, three points of difference between the bridge and push-pull converters.

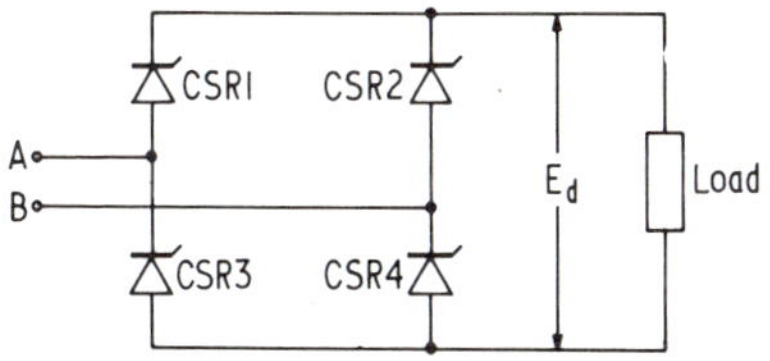

Figure 6.4. Bridge-type two-pulse bidirectional converter

(i) In a bridge system each thyristor must be rated to block the peak voltage across the a.c. inputs of the converter. Therefore the peak load and thyristor voltages are equal. For a push-pull system it was seen that the thyristors are rated for at least twice the peak load voltage.

(ii) A push-pull converter uses two devices compared to the four used for a bridge system. However, their voltage rating is now doubled. For low-power systems the price of a thyristor is usually determined by its voltage rating, so that a bridge converter can prove cheaper; this is not necessarily so when the current ratings of the devices increase. When considering costs it is also necessary to add the price of control systems. Since a push-pull converter uses fewer thyristors, and they have a common cathode, the cost of its drive circuitry should be less than for the bridge converter.

(iii) In a push-pull converter there is only one thyristor in any conduction path between the supply and the load. A bridge system now has two series thyristors. Therefore the efficiency of a bridge converter would be expected to be lower than a comparable push-pull circuit. However, since thyristors normally have a volt drop of the order of 1V, it is clear that this would only have a significant effect on very low voltage supplies.

Although single-phase circuits are relatively simple in construction, they are limited in power handling capabilities, and produce output-voltage ripples which are much greater than those from three-phase systems. The circuits described so far can be termed two-pulse, *i.e.* the ratio of the fundamental d.c. voltage ripple frequency to that of the input a.c. supply is two. The greater the pulse number the lower the smoothing requirements of the circuit.

Figure 6.5 shows a three-pulse push-pull converter and *Figure 6.6* gives its operating waveforms. For zero firing delay the thyristors in the most positive phase conduct to the neutral line. The voltage across the thyristors is zero when they conduct, and equals the line voltage between it and the phase of the conducting device when the thyristor is off. Each thyristor is on for 120° and the supply current and voltage are as shown. The d.c. load current is assumed to be ripple free, and the input a.c. current is seen to have a d.c. component equal to one third of the load current. However, this magnetising current can normally be eliminated by zigzag connection of the input transformer.

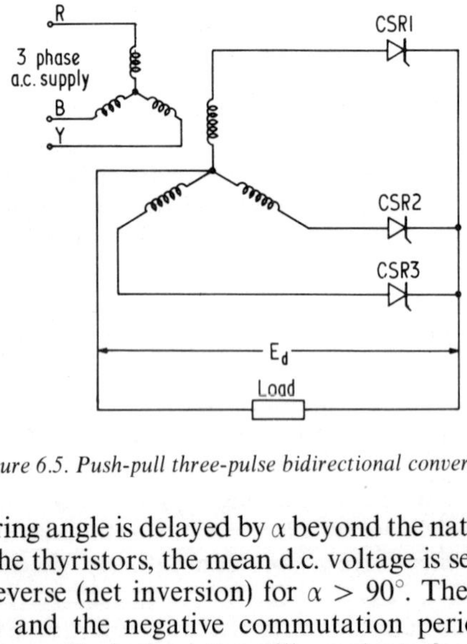

Figure 6.5. Push-pull three-pulse bidirectional converter

When the firing angle is delayed by α beyond the natural commutation point of the thyristors, the mean d.c. voltage is seen to decrease, and then to reverse (net inversion) for $\alpha > 90°$. The ripple voltage also increases and the negative commutation periods across the thyristor decrease. Each thyristor still conducts for a 120° period, although as expected these are shifted relative to the input supply voltage so that phase angle ϕ equals the delay angle α. The d.c. component of the input current is unchanged. It is therefore seen that the waveforms are essentially as for the two-pulse system except that now the d.c. voltage is fabricated from three parts, for each cycle of the input voltage, instead of two.

Three-pulse systems may be extended to six pulses, as shown in

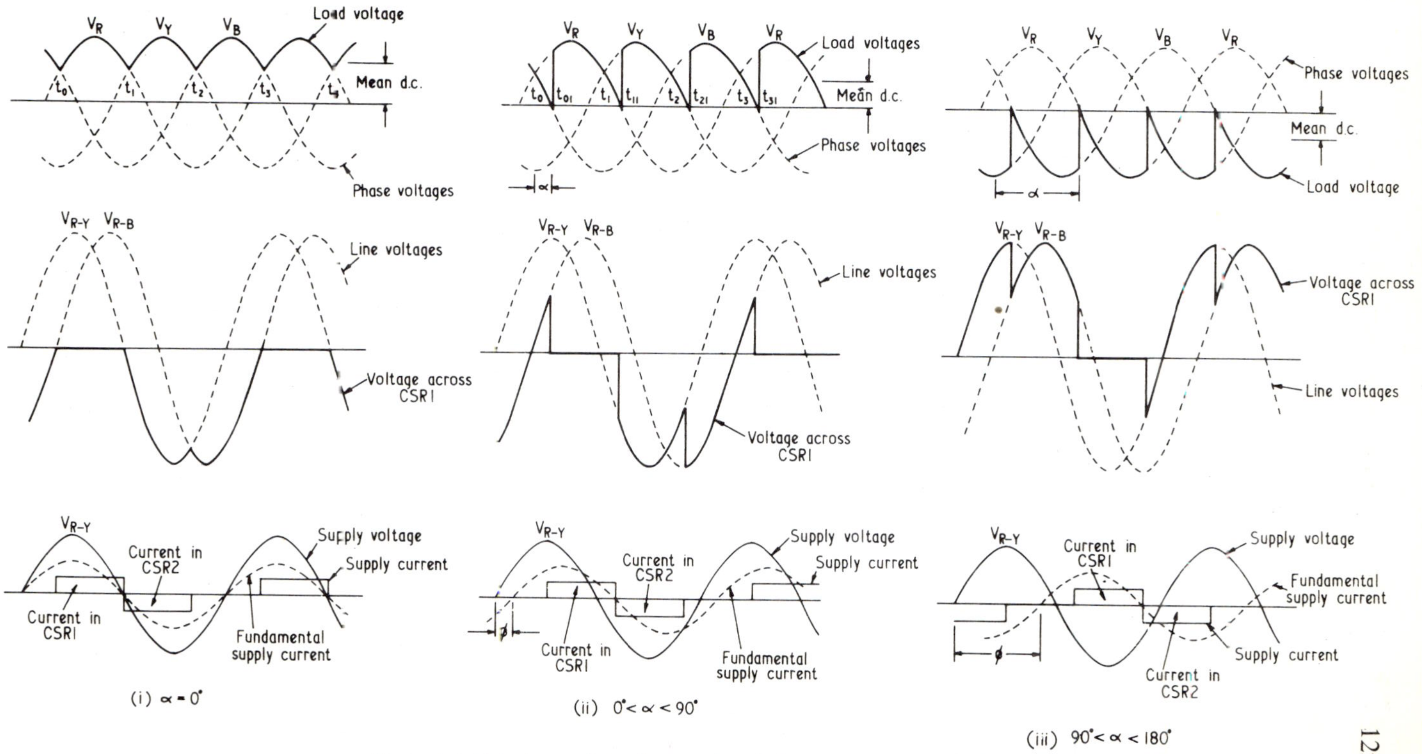

Figure 6.6. Circuit waveforms for a push-pull three-pulse bidirectional converter

Figure 6.7, by creating a six-phase system in which each line conducts to neutral for 60° during a cycle. This, however, results in poor utilisation of the transformer and devices, with a consequent increase in

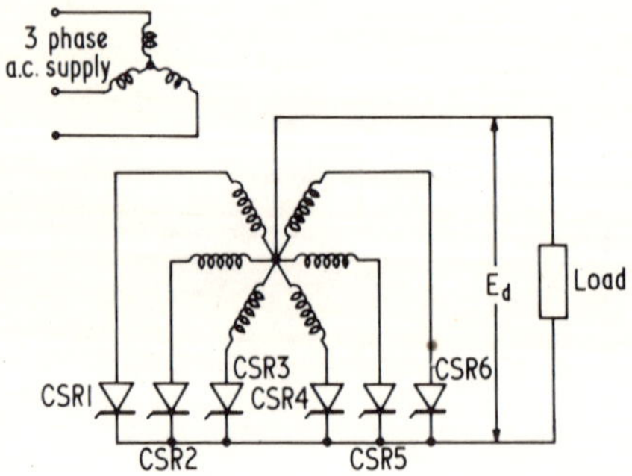

Figure 6.7. Push-pull six-pulse bidirectional converter

their r.m.s. to mean current ratio. A much better solution is given in *Figure 6.8*. In this circuit, two three-pulse systems are operated in parallel through an interphase transformer (also called an absorption coil or a phase equaliser). This is a centre-tapped auto-transformer and its action is such as to cause it to absorb the instantaneous voltage difference across its windings, and to produce a mean potential at its centre point. Therefore the d.c. voltage will contain

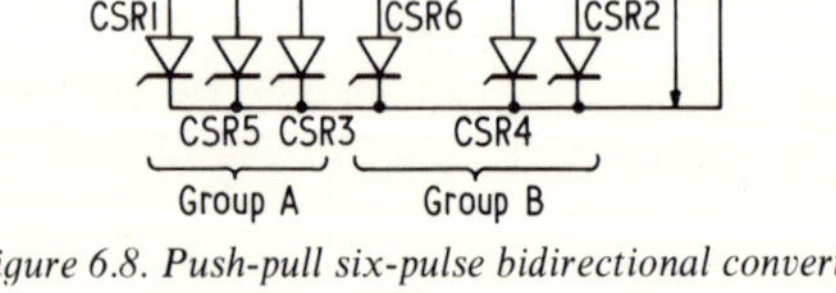

Figure 6.8. Push-pull six-pulse bidirectional converter using an interphase transformer

a six-pulse ripple as shown in *Figure 6.9*. The frequency of the voltage across the interphase transformer is equal to three times that of the a.c. input.

An interphase transformer must be sufficiently fluxed to produce the sharing action between the two three-pulse systems. On light loads this would not occur so that the system then reverts to three-pulse operation, with a consequent shift in mean d.c. voltage. To prevent this change in voltage when the load falls below a critical value (known as the transitant load), the rectifier can be operated with a small permanent bleeder load. As an alternative to this the interphase transformer may be energised by a triple frequency supply current flowing through an auxiliary winding on its core.

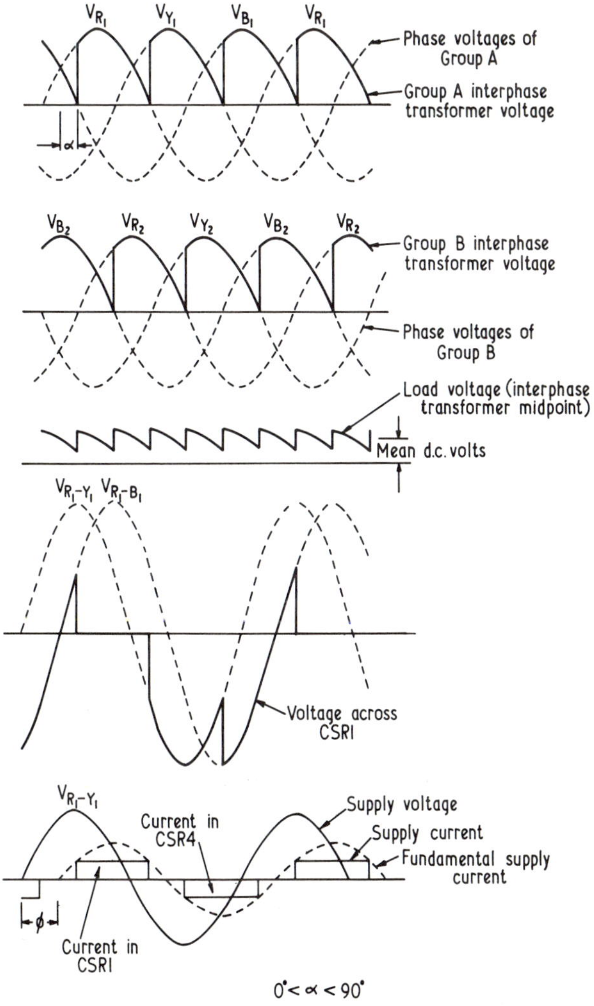

Figure 6.9. Circuit waveforms for a push-pull six-pulse bidirectional converter with interphase transformer

The thyristors in the circuit of *Figure 6.8* are seen to conduct for 120° each, so increasing their utilisation factor. The input load current also resembles a sine wave more clearly than for two- or three-pulse circuits, and contains no d.c. component.

Two six-pulse systems, of the type shown in *Figure 6.8* can readily be connected together, through a third interphase transformer, to form a twelve-pulse converter. Such a circuit is shown in *Figure 6.10.* The principle of operation is essentially unchanged, each thyristor still conducting for a 120° period during a cycle. However, the d.c. voltage has a fundamental ripple frequency twelve times that of the a.c. supply and the a.c. current approaches the mean sine wave even closer than for a six-pulse system.

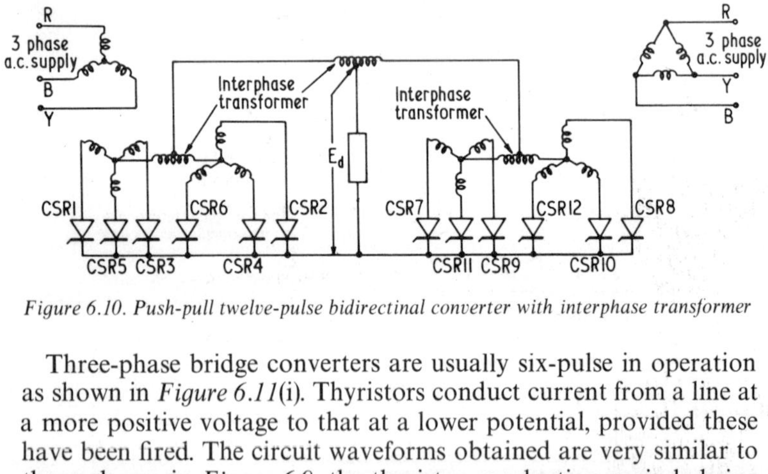

Figure 6.10. Push-pull twelve-pulse bidirectinal converter with interphase transformer

Three-phase bridge converters are usually six-pulse in operation as shown in *Figure 6.11*(i). Thyristors conduct current from a line at a more positive voltage to that at a lower potential, provided these have been fired. The circuit waveforms obtained are very similar to those shown in *Figure 6.9*, the thyristor conduction periods being shown in *Figure 6.11*(ii) and (iii). Thyristors conduct for 120° each and the load fundamental frequency is six times that of the a.c. supply. Similarly two six-pulse circuits can be operated with suitable phase shifts to give a twelve-pulse system as in *Figure 6.12.* The thyristors again conduct for 120° each. The circuit waveforms are shown in *Figure 6.13.* The reduction of d.c. voltage and a.c. current harmonics is self evident.

Unidirectional converters

Unidirectional converters are capable of passing power in one direction only, *i.e.* from the supply to the load. They can consist of a bidirectional circuit with the addition of a free-wheeling diode across

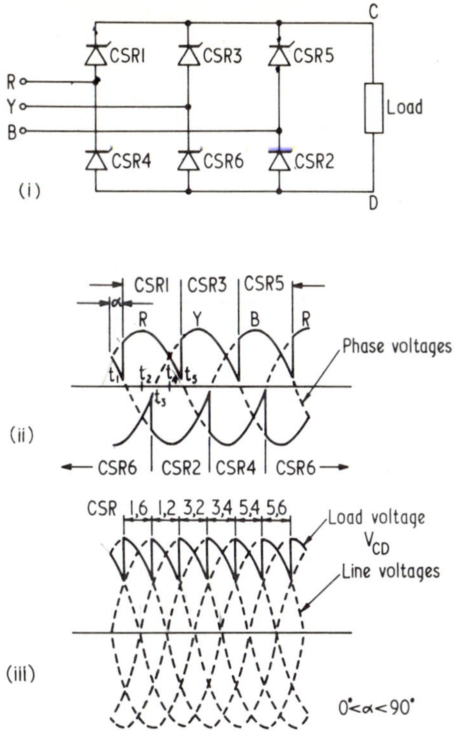

Figure 6.11. Bridge six-pulse bidirectional converter

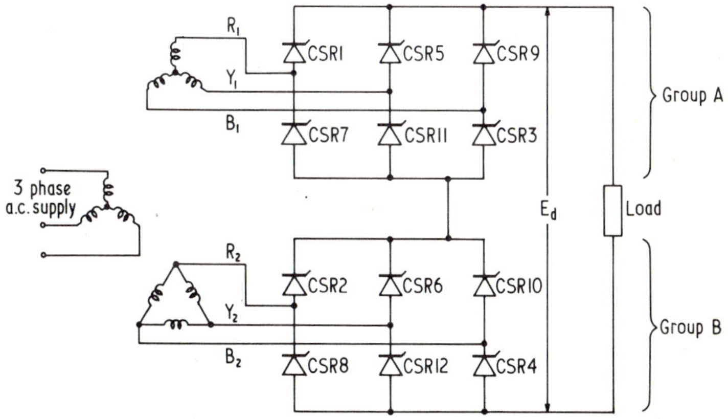

Figure 6.12. Bridge twelve-pulse bidirectional converter

the load, although often other circuit modifications are made. Apart from circuit simplicity unidirectional converters have other advantages, such as lower d.c. voltage ripple and a reduction of quadrature

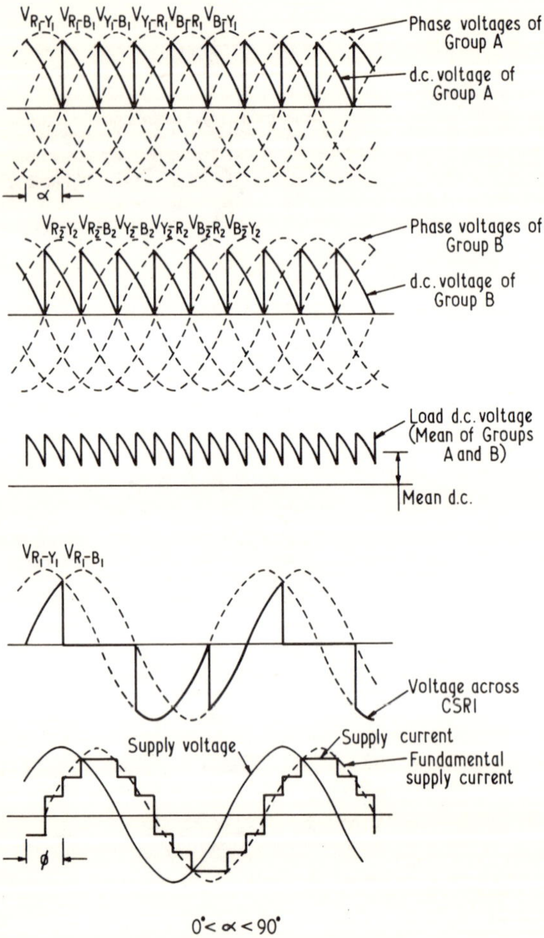

Figure 6.13. Circuit waveforms for the bridge twelve-pulse bidirectional converter of Figure 6.12

phase input current. These will be examined below with the help of typical circuits.

Figure 6.14 illustrates a half-wave circuit which has an additional diode D1 connected across the load. The load waveforms given in

Figure 6.14(ii) can be contrasted with those obtained with the bidirectional circuit of *Figure 6.1*, and are seen to be identical on resistive loads. The operation of the circuit is as follows, assuming the load current is virtually ripple free. At t_{01} thyristor CSR1 is turned on and supplies power to the load. At t_1 the input voltage reverses. The inductive load tends to prevent any decay of load current and this now transfers from CSR1 to D1. Therefore until CSR1 is refired at t_{21} the load current free-wheels in diode D1, hence the term free-wheeling diode is applied to this device. If the volt drop across D1 be neglected the load voltage is zero during the free-wheeling period, so that the converter is now no longer capable of passing power back from the load to the supply.

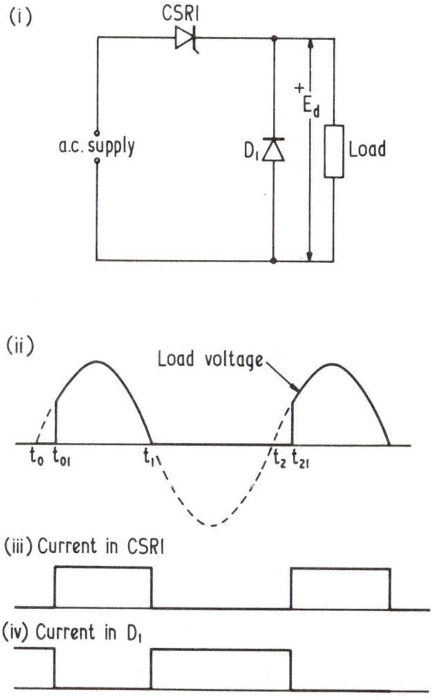

Figure 6.14. Half-wave unidirectional converter

As seen from the above discussion it is possible to modify any bidirectional converter to unidirectional operation simply by connecting a free-wheeling diode across the load terminals. This is illustrated in *Figure 6.15*(i) which shows the modification applied to the circuit of *Figure 6.4*. Thyristors in diagonal arms of the bridge are fired in pairs, as before, at any delay angle α required. However, whereas for a bidirectional converter these would have been maintained in conduction for a 180° period, until commutated by the firing of the opposite pair of thyristors, for unidirectional circuits the thyristors turn off as soon as the supply voltage reverses. Load current is now carried by free-wheeling diode D1.

Several modifications can be made to bidirectional circuits, apart from using free-wheeling diodes, to prevent them returning power to the load. These often simplify the overall circuit arrangement. *Figure 6.15*(ii) shows a system where half the bridge thyristors have been replaced by diodes. Diodes are cheaper than thyristors and do not need any associated gate drive circuitry. Therefore the overall cost of the converter has been reduced. Referring to the load-voltage waveforms the operation of the system can be explained as follows. At t_{01} thyristor CSR2 is fired. Line A is positive to B so that load current flows via D1 and CSR2. At t_1 the input voltage reverses. Load inductance will now cause a free-wheeling current to flow in CSR2 and D2 until t_{11} when CSR1 is turned on and current commutates to this thyristor. It is seen from this description that all devices carry current for 180°, irrespective of the firing delay angle α. This is clearly advantageous for calculating semiconductor ratings. However, the system has a serious limitation. Suppose the converter is working with a low delay angle, CSR2 being fired at t_{01}, and it is desired to turn the bridge off. All thyristor gate pulses are removed. The load current at t_1 would now transfer to CSR2 and D2. If the load inductance is sufficient, this current will not decay to zero during the half-cycle interval, so that at t_2 thyristor CSR2 is still conducting. Since its anode voltage goes positive at this instance the thyristor will continue to conduct, with a zero delay angle, and load current flows via CSR2 and D1 for a complete half-cycle. Depending on the load impedance this state could be maintained continuously, the converter 'half-waving' throughout. The only way to turn the bridge off would now be to refire the thyristors at the appropriate instances and to increase α gradually, to reduce the load current. When this current reaches a value which is insufficient to keep the free-wheeling current on for 180°, the thyristor gate pulses may be removed. This circuit limitation can be overcome by the addition of free-wheeling diode D3, so that each thyristor is commutated at the end of a half-cycle, when the diode conducts. If gate pulses are now removed, say from CSR2, then at t_1 this thyristor will go off. Since none of the thyristors are refired the load current will free-wheel in D3 until it decays to zero.

An alternative system to *Figure 6.15*(iii), which overcomes the 'half-waving' effect and so gives a controller with a greater response speed, is shown in *Figure 6.15*(iv). The number of devices have not

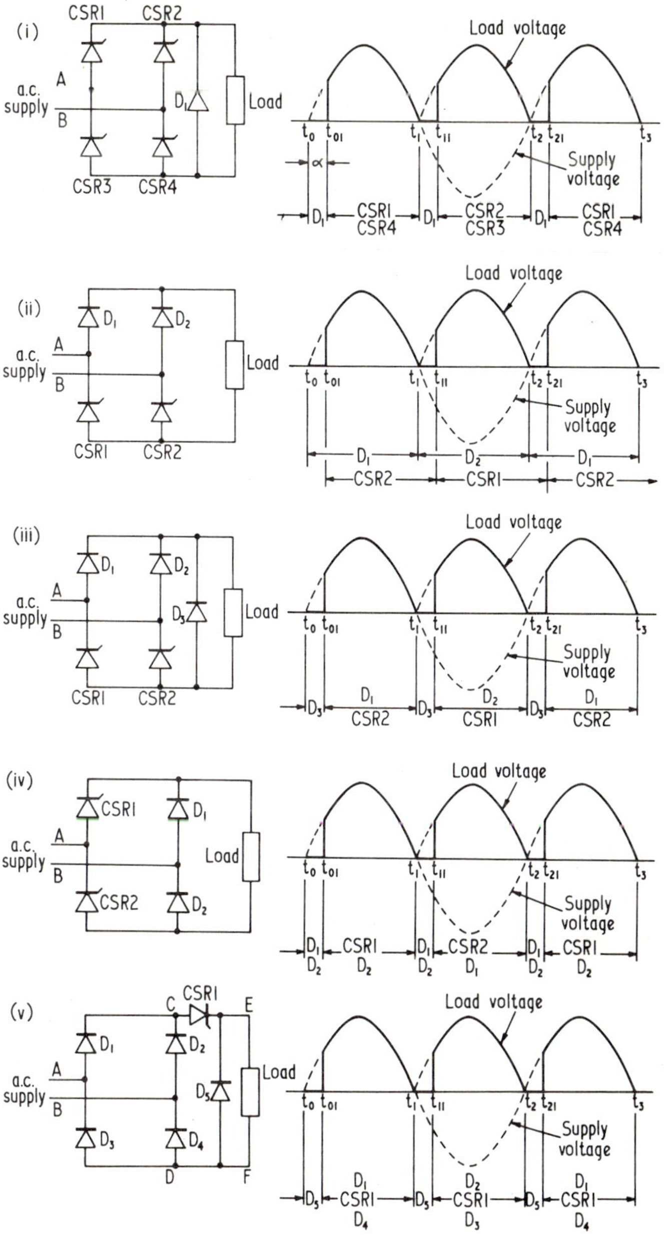

Figure 6.15. Bridge two-pulse unidirectional converters

been increased but they have been re-arranged so that D1 and D2 give a free-wheeling path for load current. At t_{01} thyristor CSR1 is fired and load current flows via CSR1 and D2 as before. At t_1 the current free-wheels through D1 and D2 so that CSR1 goes off. Therefore the system behaves as in *Figure 6.15*(iii) where all conducting thyristors are commutated at the end of half-cycle. However, now the conducting period for the bridge thyristors and diodes are $180^\circ - \alpha$ and $180^\circ + \alpha$ respectively, so that they could be unequally loaded depending on the load duty cycle.

Since thyristors are more expensive than diodes the converter given in *Figure 6.15*(v) sometimes proves economical. Free-wheeling diode D5 is not necessary if the load is purely resistive. As seen the system consists esssentially of providing a fully rectified wave at CD and then regulating this with thyristor CSR1. Since the waveform across the thyristor therefore falls to zero only briefly every half-cycle, a free-wheeling diode is essential, even for slightly inductive loads, to ensure successful device commutation. The operation of the circuit is readily followed by reference to the associated load voltage waveform.

Figure 6.16 shows the load voltage and current waveforms obtained from unidirectional circuits for various delay angles. Several features of these systems are evident from this figure and the above discussions.

(i) The load voltage has a lower ripple due to the absence of negative portions of the waveform.

(ii) The mean load voltage varies from a maximum to zero as the delay angle changes from 0° to 180°.

(iii) The power factor angle ϕ changes proportionally to the delay angle α, as for bidirectional converters.

(iv) Whereas for bidirectional converters the load current wave shape was unchanged as the delay angle varied, for unidirectional converters the load current period decreases with delay angle increase. Therefore for bidirectional converters at $\alpha = 90^\circ$ the load current is unchanged in value from that at zero delay angle. However, since d.c. load voltage is zero there is now no net input power and all the a.c. current is quadrature component or wattless. For a unidirectional converter, on the other hand, the a.c. input current at very low d.c. output voltages is also very small, so that the quadrature component of the current has been reduced.

(v) As mentioned before unidirectional converters are often cheaper than bidirectional converters.

(vi) There are no regenerative periods, so that a unidirectional converter cannot pass power from the d.c. to the a.c. side.

It is seen from items (i), (iv) and (v) that there is an advantage to be gained from using unidirectional converters in two-pulse systems, hence their popularity in applications which do not require regeneration. However, for systems with more than two pulses it will be seen later that the d.c. ripple frequency increases between bidirectional and unidirectional systems by a factor of two. Therefore, depending on the control range, when d.c. voltage filtering requirements are stringent bidirectional converters are sometimes preferred.

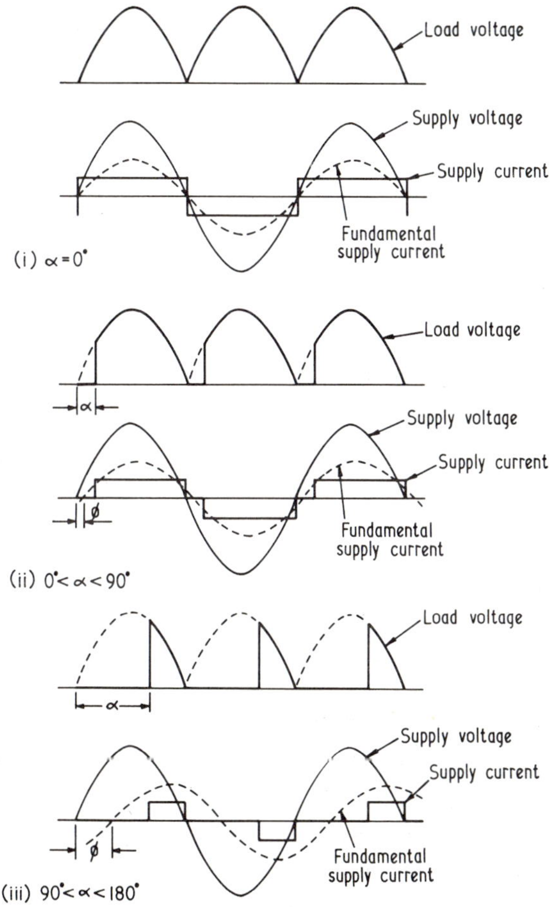

Figure 6.16. Circuit waveforms for a two-pulse unidirectional converter

Bridge circuits can be converted from bidirectional to unidirectional operation by changing half the devices from thyristors to diodes. The same rule does not apply to push-pull converters. The circuit of *Figure 6.2* for instance will not operate correctly if one thyristor is replaced by a diode. All such systems require a free-wheeling diode for unidirectional operation, as shown in *Figure 6.17*.

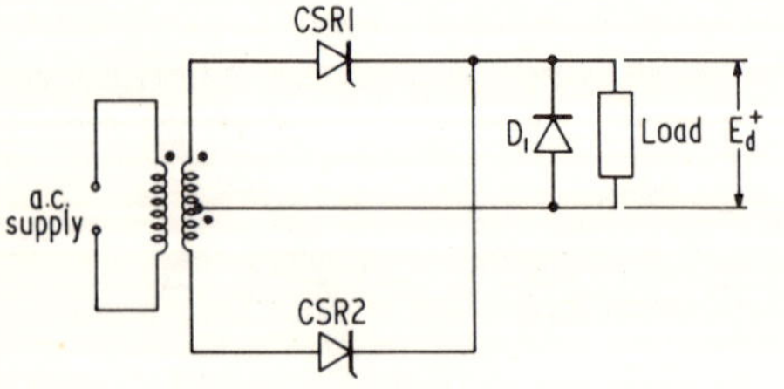

Figure 6.17. Push-pull two-pulse unidirectional converter

Figure 6.18 shows a three-phase unidirectional converter in which half the thyristors of a bidirectional circuit have been replaced by diodes. It suffers from the same disadvantages as that of the single-phase circuit of *Figure 6.15*(ii). However, it is not possible to overcome these by a re-arrangement of components, as was done in *Figure 6.15*(iv). Instead a free-wheeling diode must be used as in *Figure 6.15*(iii). The circuit waveforms for *Figure 6.18* are shown in

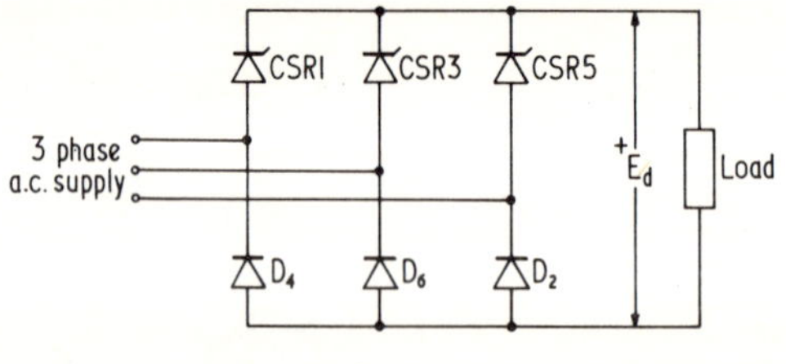

Figure 6.18. Bridge three-pulse unidirectional converter

Figure 6.19. The top half of the bridge is controlled so that conduction will occur from the most positive phase in which a thyristor has been fired. The current in the bottom of the bridge will commutate naturally from one diode to the next depending on which is connected to the most negative phase. In *Figure 6.19* for $\alpha = 0$ the output is a six-pulse waveform such as that obtained for *Figure 6.11*. However, as the delay angle increases, the waveform changes to three-pulse. Therefore in *Figure 6.19*(ii) the fundamental voltage ripple is three times that of the input a.c. At this stage there are as yet no free-wheeling current periods. The a.c. current has two durations of 120° whose phase relation to each other and the supply voltage changes with α. Beyond the 60° delay point free-wheeling times

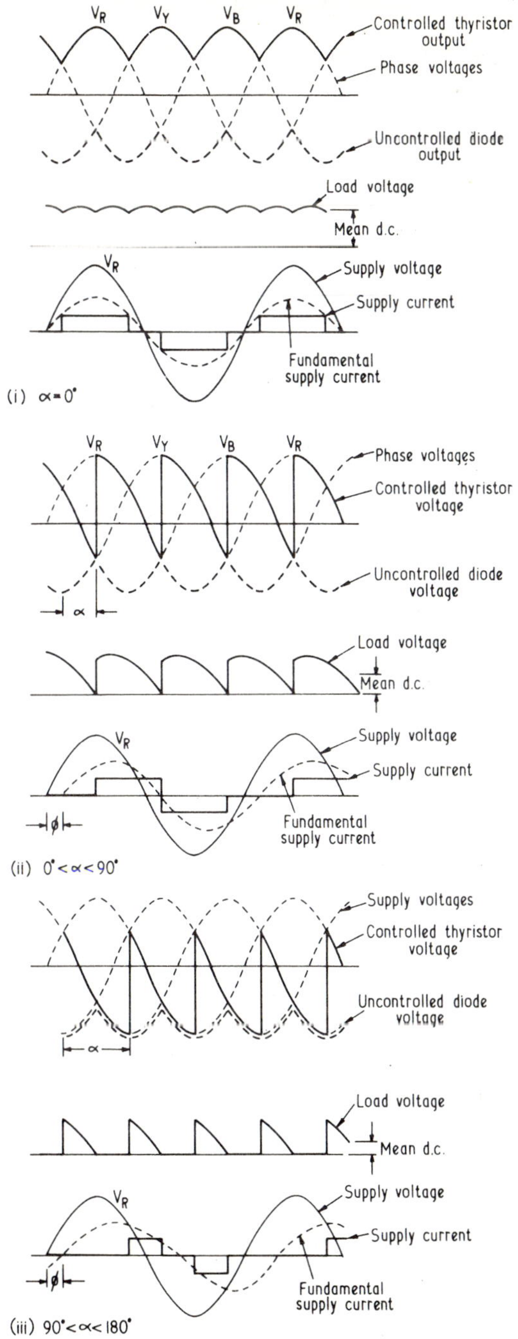

Figure 6.19. Circuit waveforms for a three-pulse unidirectional converter

increase. These lead to periods of zero voltage in the load waveform and a reduction of positive and negative portions of the a.c. current. In effect the a.c. supply current waveform may be considered as being made up from a positive (thyristor) part and a negative (diode) part. As the delay angle increases the diode block remains stationary (since it is uncontrolled) while the thyristor block moves towards it. Beyond 60° delay the two overlap to an increasingly greater extent, resulting in an increase in the free-wheeling period.

It has been stated before that a unidirectional wave has a lower d.c. voltage content owing to the absence of negative portions of the waveform. However, for high pulse numbers it is seen above that there can be reduction of ripple frequency by a factor of 2 from that of bidirectional converters. There is now obviously a compromise situation where, depending on the range of control voltage required, one converter would give lower ripple output than another. Ripple contents from converter circuits are considered again later.

DISCONTINUOUS LOAD CURRENT

In the above discussions it has been assumed that, apart from resistive loads, the load current has been continuous during a cycle of operation. If this is not the case then the load voltage waveform will be modified. Generally these systems are not easy to analyse since the output conditions are dependent on the load.

Figure 6.20 shows the waveforms from a bidirectional converter for various delay angles, where the load Q factor is finite. For delay

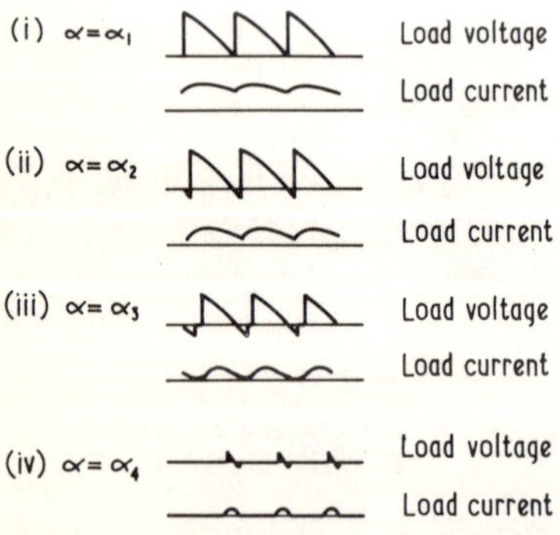

Figure 6.20. Load waveforms for a six-pulse bidirectional converter operating with a finite load inductance and possible discontinuous load current

α_1 the load voltage just dips to zero although the load inductance maintains the current continuous. At α_2 the current is still continuous so that the load voltage follows the contour of the a.c. input voltage and swings negative over certain portions. For larger delays of α_3, however, the load inductance is insufficient to maintain a prolonged regenerative period, and the current decays to zero. For a passive load its voltage will now be zero, as shown. Increasing the delay angle further reduces the d.c. voltage. However, it also reduces the peak load current so that the regeneration period is shortened.

In *Figure 6.20* the load has been assumed to be passive. If in fact it contained a d.c. source, such as a motor back e.m.f., the voltage during zero load current periods would rise to the value of this e.m.f. This is shown in *Figure 6.21*, which illustrates the effect of maintaining the converter firing angle fixed but of changing the load current,

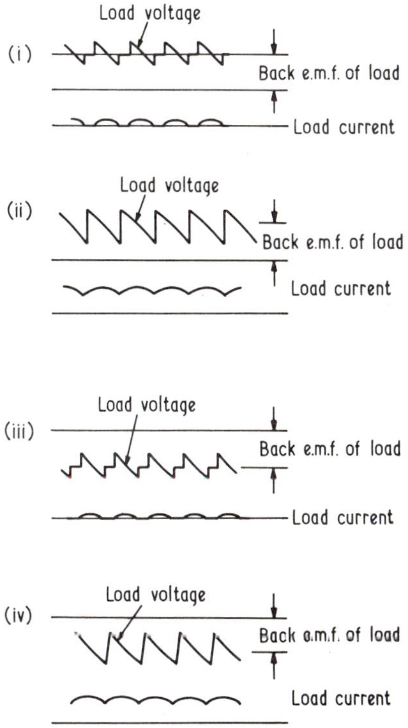

Figure 6.21. Load waveforms for a six-pulse converter operating with a back e.m.f. load, (i), (iii) light loads, (ii), (iv) heavy loads

as would normally happen in a d.c. motor under variable torque conditions. The shape of the load waveform is seen to change. During continuous load current periods it follows that of the a.c. supply, but when the load current becomes discontinuous it rises to the value of the load back e.m.f. Since the mean voltage of the load varies with its waveform the effective load voltage has been changed although the firing angle has not. This is highly undesirable in many applications, and illustrates the advisability of introducing external load inductances so as to maintain continuous load current. *Figure 6.21*(iii) shows the effect of discontinuous load current during the inverting mode of the converter, assuming that the load has an internal back e.m.f. which provides the inverter voltage.

In order to ensure that the load current never becomes discontinuous, it is possible to analyse the converter circuit to determine the minimum load inductance required under various firing angles. Such an analysis would depend on the value of load back e.m.f. and could be involved. It has, however, been done for passive loads. *Figure 6.22* shows load voltage and current waveforms for a single-phase bidirectional converter. The load current reaches a minimum

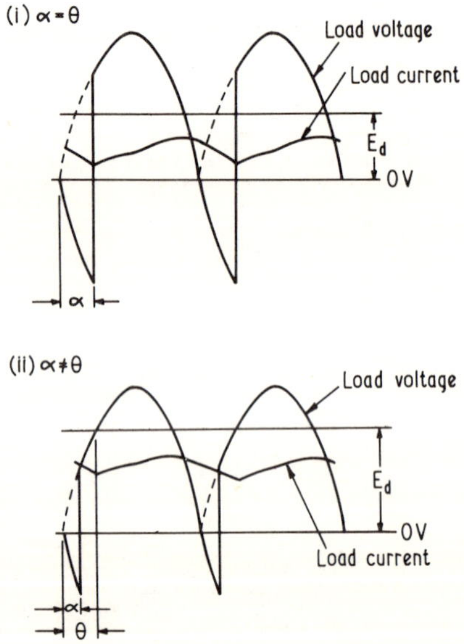

Figure 6.22. Load waveforms for a two-pulse bidirectional converter

value at the intersection points of the instantaneous and mean d.c. waveforms. It can be shown that the value of inductance L_c required to prevent this current from falling below zero is related to the circuit resistance R and angular frequency ω of the supply by

$$\frac{\omega L_c}{R} = \frac{\pi}{2\cos\alpha}\left[\cos\theta + \frac{2}{\pi}\sin\alpha - \frac{2}{\pi}\cos\alpha\left\{\frac{\pi}{2} + \alpha - \theta\right\}\right] \qquad (6.1)$$

For large firing angles, $\alpha = \theta$ so that equation (6.1) reduces to

$$\frac{\omega L_c}{R} = \tan\alpha \qquad (6.2)$$

For unidirectional converters the analysis is complicated further due to zero voltage periods in the load waveform. It is found to be given by

$$\frac{\omega L_c}{R} = \theta - \alpha - \frac{\pi}{2} + \frac{\alpha + \sin\alpha + \pi\cos\theta}{1+\cos\alpha} \qquad (6.3)$$

As before there are two operating conditions according to whether α is smaller or larger than a critical value (35·5°), which gives $\alpha = \theta$. In this case equation (6.3) reduces to

$$\frac{\omega L_c}{R} = -\frac{\pi}{2} + \frac{\alpha + \sin\alpha + \pi\cos\alpha}{1+\cos\alpha} \qquad (6.4)$$

Similar analysis can be made for three-phase converters.

Figures 6.23 and *6.24* show the plot of L_c/R for single- and three-phase converters. The abscissas in both cases have been given in terms of the percentage of maximum d.c. output as well as the delay

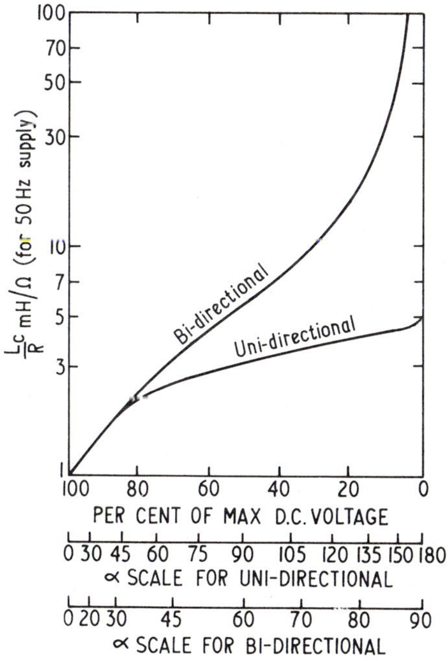

Figure 6.23. Critical inductance requirements for two-pulse converters

angle required for unidirectional and bidirectional converters. From these curves it is seen that the critical inductance requirement for bidirectional circuits tends to infinity at low output voltages. This is understandable since at these delay angles the mean d.c. voltage is low, whereas the a.c. ripple on the voltage is at its peak value. For unidirectional converters the a.c. ripple is also very small at delay angles approaching 180°, so that the critical inductance is a finite value. *Figure 6.23* clearly shows the advantage of unidirectional converters in requiring smaller load inductance. However, *Figure 6.24* illustrates that the unidirectional converter has a higher critical inductance requirement for three-phase circuits except at low load

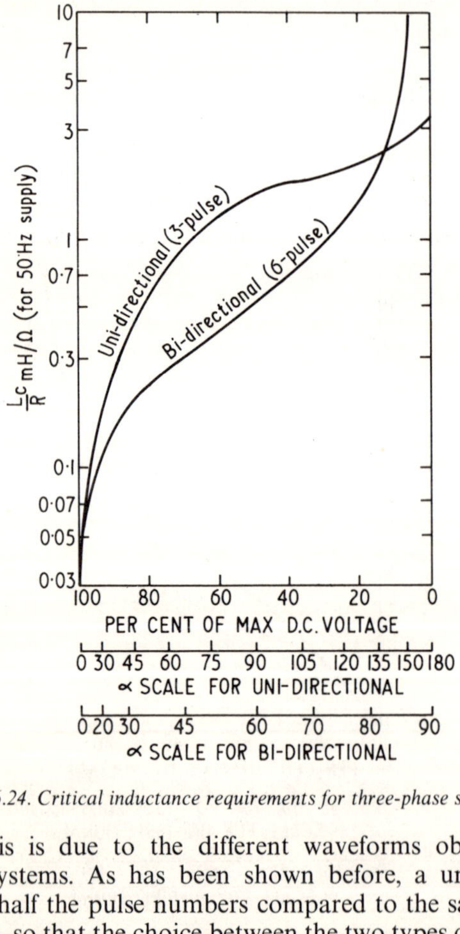

Figure 6.24. Critical inductance requirements for three-phase systems'

voltages. This is due to the different waveforms obtained with multipulse systems. As has been shown before, a unidirectional circuit gives half the pulse numbers compared to the same bidirectional system, so that the choice between the two types of converters would clearly now depend on the output voltage range required. Comparison of *Figures 6.23* and *6.24* clearly shows the less stringent inductance requirements of three-phase circuits so that where continuous load current is required, from a low inductance load, the pulse number of the circuit should be made as large as possible.

SOURCE REACTANCE

In the discussions so far the impedance of the a.c. lines has been omitted for convenience. Generally this would consist of the reactance of the distribution transformer and any associated cabling, and for a reasonably 'stiff' supply it may be neglected. However, there are instances, for example when the converter is operating from its own local generator, or the load current approaches its short-circuit conditions, when this approximation is no longer valid. The effect of source reactance on a converter can be explained with reference to the single-phase bridge circuit of *Figure 6.25*. In the figure shown CSR2 and CSR3 are conducting at t_0 and feeding power back from the supply to the load. Inductors L_1 and L_2 are line impedances. At t_{01}, after a delay α, thyristors CSR1 and CSR4 are fired. In the absence of the line inductors these thyristors would instantaneously carry the full-load current while CSR2 and CSR3 would turn off. However, this would require the reversal of current flow in any line inductors, which cannot occur instantaneously. Therefore for a period μ all bridge thyristors will conduct. The instantaneous supply voltage appears across the line inductors in such a direction as to help build up the current in CSR1 and CSR4, and to decay the current in CSR2 and CSR3. Clearly the overlap period, for any given load current, is minimum when this voltage is at its peak value, *i.e.* for α close to 90°. During overlap the load current free-wheels through the four thyristors so that load voltage is zero. Another way of expressing the same thing is to consider the load voltage to be the instantaneous mean of the overlapping phases, which for a single-phase supply is always zero. With opposite arms of the bridge conducting the voltage across EF is clearly that of the input AB. During overlap, however, it falls to zero.

Comparing the waveforms with and without source reactance, the effect of this reactance can be considered as being four-fold:

(i) The mean d.c. voltage is reduced for any delay angle α. This is clearly so since a portion of duration μ has now been removed from the output waveform.

(ii) The harmonics of the output load have been changed since its waveform has been modified. Since overlap does not affect the output pulse number, the spectrum of frequencies present is not altered. It can be shown, however, that for the same mean d.c. load voltage the amplitude of the harmonics is lower with overlap in the waveform than when overlap is not present. This is primarily due to the fact that with overlap the firing angle must be advanced (α reduced) to maintain an unchanged d.c. voltage.

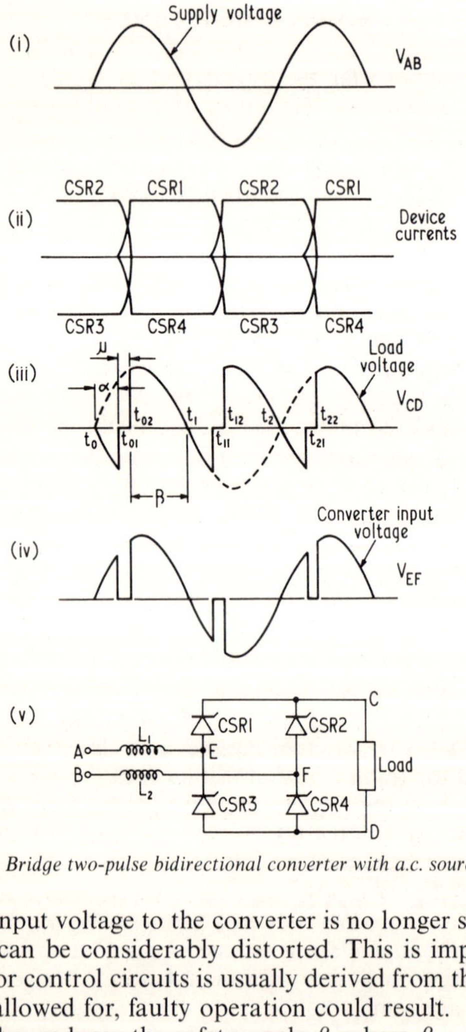

Figure 6.25. Bridge two-pulse bidirectional converter with a.c. source impedance

(iii) The input voltage to the converter is no longer sinusoidal (as at EF) but can be considerably distorted. This is important since the timing for control circuits is usually derived from this wave and, unless it is allowed for, faulty operation could result.

(iv) Overlap reduces the safety angle β, where $\beta = 180^\circ - \alpha - \mu$. This is not important for rectifier operation, but it must be borne in mind that β is also the time that is available to a conducting thyristor for turn off. Since β reduces as α increases commutation failure could occur at large delay angles. *Figure 6.26* shows the circuit waveforms during inversion. The delay angle has been increased so that CSR1

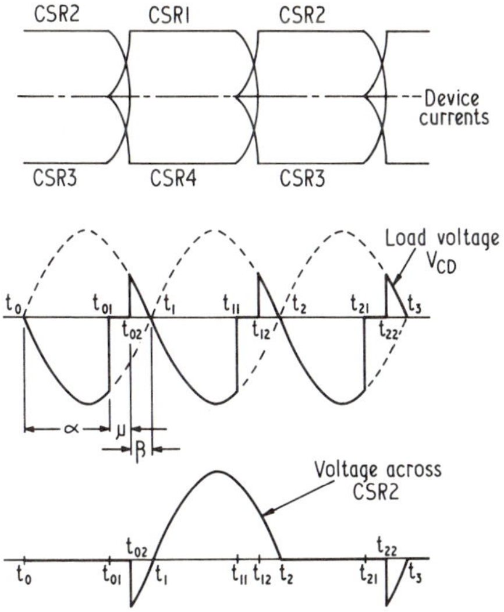

Figure 6.26. Two-pulse bidirectional bridge converter with a.c. source impedance, operating in an inversion mode

and CSR4 are fired at almost the end of the half cycle, at t_{01}. Since the instantaneous supply voltage is low, the overlap angle μ is relatively long. When it is over at t_{02} thyristors CSR2 and CSR3 go off. However, they are reversed-biased for duration β only and they must turn off during this time. If this does not occur, then at t_1 the devices turn on and provide a full half cycle of power from the supply to the load. It will be seen later that μ varies with load current so that the maximum value of α chosen must make allowance for this. Circuits operating from severe line impedance sources, therefore, are often limited in voltage control range.

Overlap effects are not restricted to bidirectional converters. For instance *Figure 6.27* shows the load voltage and device currents for the circuit of *Figure 6.15*(i), assuming that the load current is ripple free and the input lines have series reactances. Prior to t_0 thyristors CSR2 and CSR3 were conducting. At t_0 the supply voltage reverses and the load current begins to transfer from the thyristors to the free-wheeling diode D1. Line reactances prevent this occurring instantaneously, although the load voltage is zero since D1 is conducting. At t_{01} current has transferred completely to D1 and CSR2 and CSR3 go off. When CSR1 and CSR4 are fired at t_{02} there is again an overlap angle μ, due to the finite time for the current to transfer

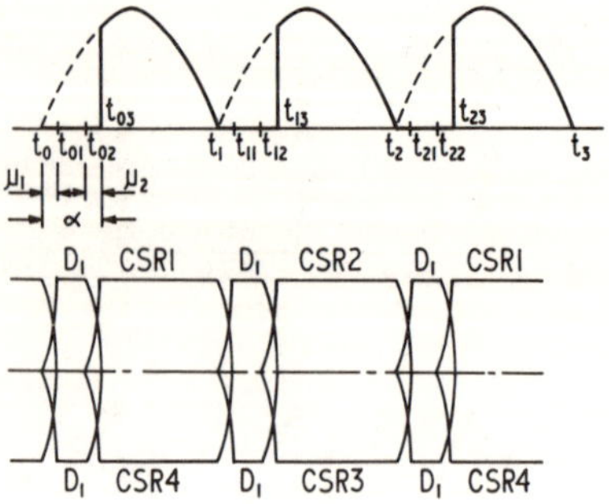

Figure 6.27. Circuit waveforms for the two-pulse unidirectional converter of Figure 6.15(i) operating with a.c. source reactance

from D1 to the thyristors before the load voltage rises to that of the supply. Quite clearly the effect of overlap has once again been to remove a portion of duration μ from the load-voltage waveform during every half cycle. Since the instantaneous line voltages at the overlap periods are different the two overlap angles μ_1 and μ_2 will also differ, with $\mu_1 > \mu_2$.

Figure 6.27 has illustrated a case where $\alpha > \mu_1$. The situation is complicated when the reverse is true. Now the current in D1 never reaches the full d.c. value, since before this can happen the opposite bridge thyristor is fired and the current will be caused to decay in D1. The zero period in the load voltage waveform is equal to the sum of the two overlap periods (partial current transfer from thyristor to diode and diode to thyristor, in parallel with thyristor to thyristor.)

Overlap in the converter has the effect of increasing the rise and fall times of devices currents. This is reflected to the a.c. input current as shown in *Figure 6.28*. The fundamental component of the a.c.

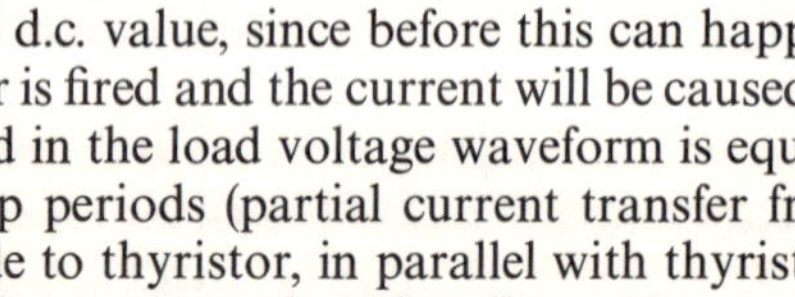

Figure 6.28. Effect of source reactance on the a.c. input current

current is now also seen to be shifted by a lagging angle ϕ_1 from the case for no overlap. This means that the power input to the converter is reduced. This is to be expected since overlap has reduced the mean d.c. voltage and therefore the d.c. power output (for constant load current). It can of course also be explained as being due to the introduction of an inductive impedance in the a.c. lines which would increase the power factor angle.

During the overlap period the two lines between which current is being transferred assume the same potential, which is the mean value of their instantaneous voltages. For a single-phase bridge this has been shown to be zero, but it is not so for three-phase converters. The action of overlap can be illustrated for a three-phase circuit with reference to the diode bridge shown in *Figure 6.29*. *Figure 6.30* illustrates its operation for overlap angles of 30°. The diodes will

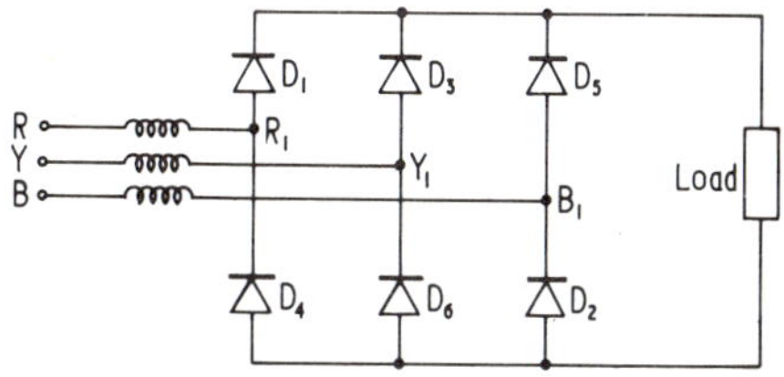

Figure 6.29. Three-phase bridge with source reactance

conduct, as soon as their anode voltages go positive, *i.e.* at t_0, t_1, t_2, etc., since they do not possess gate control. Prior to time t_0 diodes D5 and D6 were conducting. At t_0 the red phase becomes the most positive phase and the load current would normally flow from this phase to the most negative phase, *i.e.* the yellow phase. Therefore the load current should commutate instantaneously from D5 to D1. However, due to line inductances, there is a finite commutation time during which both D5 and D1 pass current to D6. This overlap period lasts up to (depending on load current) say t_{01}. Diode D5 now goes off and diodes D1 and D6 carry the load current. During overlap of the red and blue phases the top junction of the load maintained a potential which was an instantaneous mean between that of the two phases. The load voltage is the difference between this mean phase voltage and that of the yellow phase (which is the potential of the bottom junction of the load). It can therefore be found in *Figure 6.30* as the difference between the heavy lined waveforms.

Also shown in *Figure 6.30* is one line voltage waveform. This is the difference between the instantaneous phase voltages and is clearly zero during phase overlap. The line voltage is the voltage at the input terminals of the bridge (after the series line inductors) and would normally be the waveform which operated any internal control circuits.

In the waveforms of *Figure 6.30* the overlap conditions have been such that there have been alternate periods of conduction by two and three diodes. This is called mode one operation. The overlap period

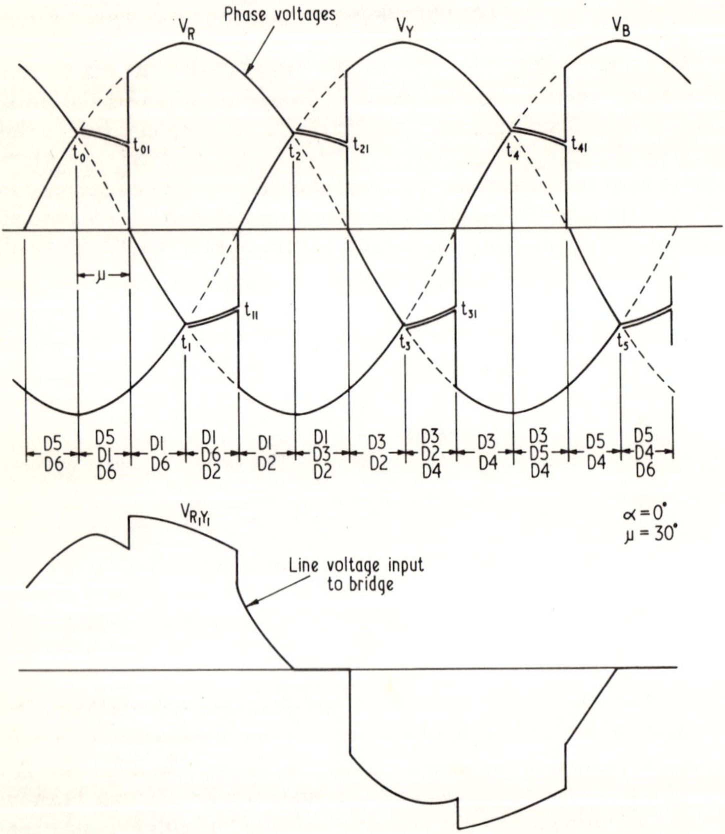

Figure 6.30. Mode one operation in a diode bridge with source reactance

is less than 60°. If the line inductance or d.c. load current is increased a stage is reached where the overlap will be 60°. This is called mode two operation. Further increase in load current will keep the overlap at 60° for some time but the overlap period will be shifted towards the peak of the instantaneous line voltage which would help current transfer. This is illustrated in *Figure 6.31*. In *Figure 6.31*(i), for instance, at t_0 current should transfer to the top diode in red phase. However, the red phase voltage is held below this point due to

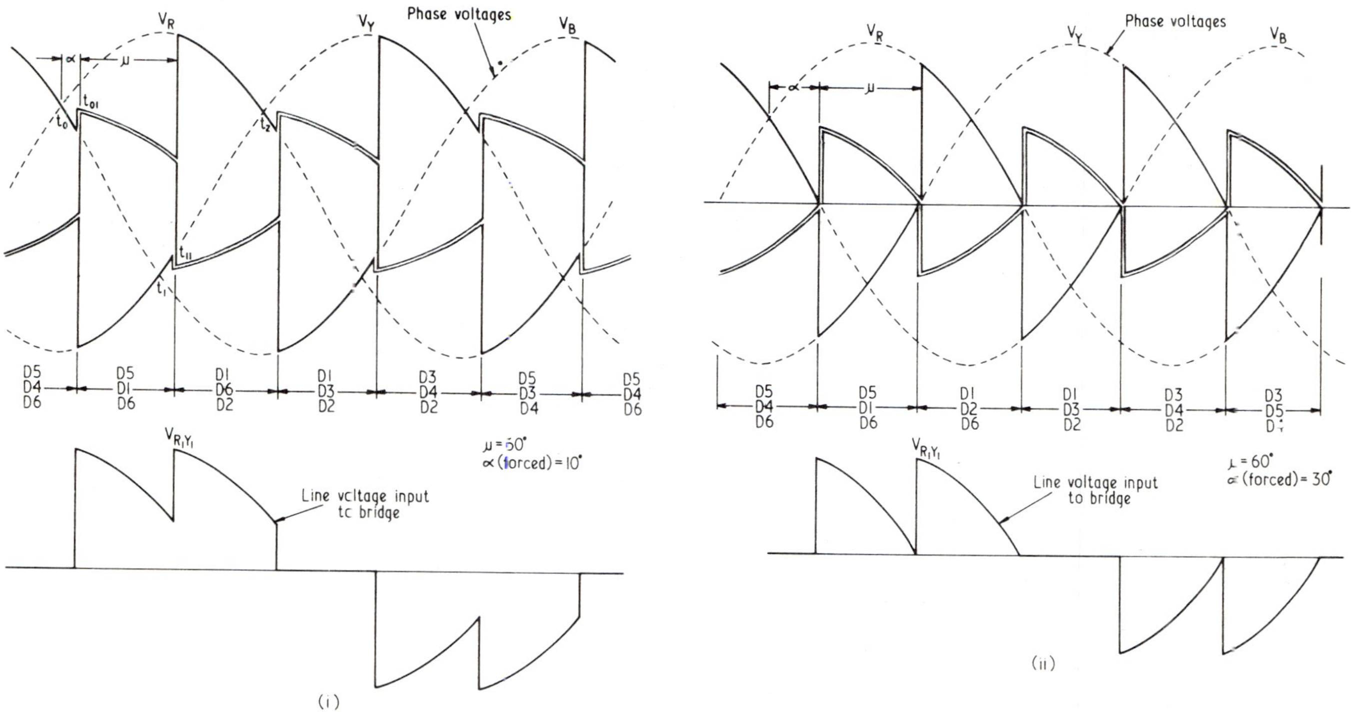

Figure 6.31. Mode two operation in a diode bridge with source reactance

overlap with the yellow phase so that current transfer is delayed until this is completed at t_{01}. As soon as the red phase is released by the yellow phase it goes into overlap with the blue phase as current transfers from D1 to D5 and flows to D6. At t_1 transfer should occur between D6 and D2, but since the blue phase voltage is held above that of the yellow phase owing to its overlap with the red phase, this transfer is delayed till t_{11}. This waveform shows that there has been an artificially created delay angle of about 10° in the changeover point between phases. Similarly in *Figure 6.31*(ii) the increased load current still gives the same value of 60° overlap but increases the delay to 30°. This represents the boundary condition between mode two and mode three operation. It should be noted that, throughout mode two, three diodes are conducting at any one time.

Figure 6.32 shows mode three operation with overlap angles exceeding 60°. Two overlap periods of the top and bottom halves of the bridge now run into each other so that four thyristors begin to

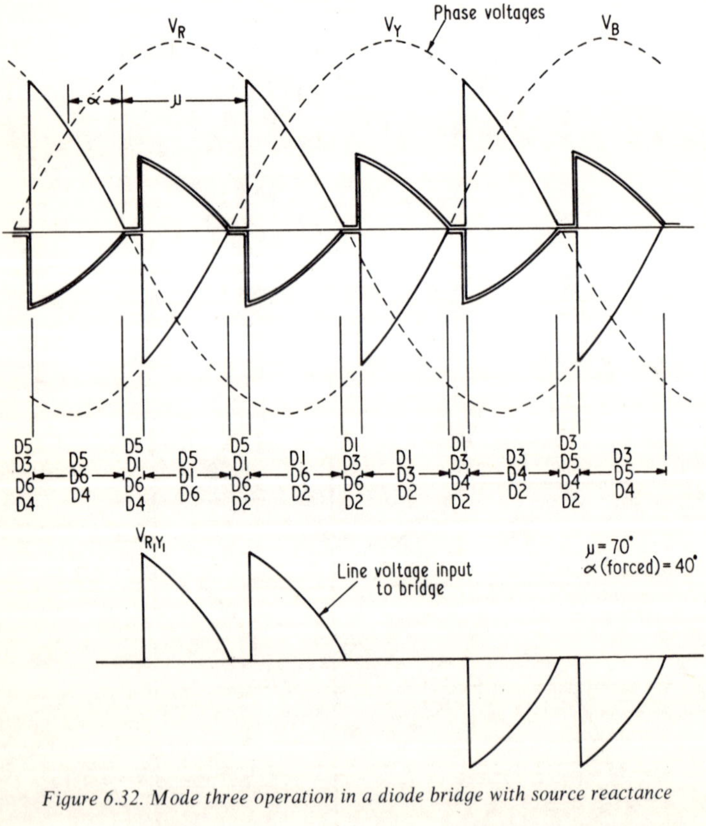

Figure 6.32. Mode three operation in a diode bridge with source reactance

conduct simultaneously over certain parts of the waveform. The load voltage is now zero since the load current free-wheels through devices in the top and bottom half of the bridge. The effect on the input line voltage is also seen to be such as to introduce further instances of zero voltage.

Overlap in thyristor converters can result in three operating modes as well. However, the occurrence of mode two, which is so important in diode circuits, is rare and is often considered as a special case of mode one. *Figure 6.33* gives the waveforms for the bidirectional

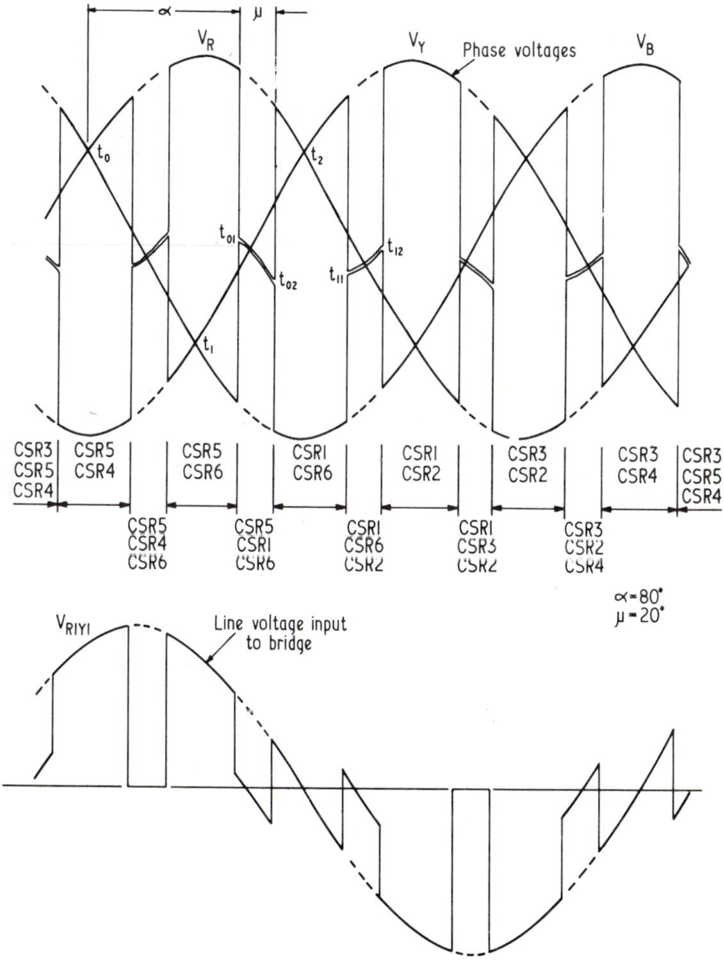

Figure 6.33. Mode one operation in a thyristor bidirectional bridge with all fully controlled arms and source reactance

bridge shown in *Figure 6.11*. The delay angle α is taken as 80° and the overlap angle μ as 20°. At time t_0 the red phase becomes the most positive line. The firing of CSR1 is, however, delayed to t_{01}. When this is done current transfer occurs between CSR5 and CSR1. This overlap terminates at t_{02} and CSR5 goes off. Similarly the commutation between the negative half of the yellow and blue phases is delayed from t_1 to t_{11}, and so on. The load voltage is given by the difference between the two instantaneous waveforms as before, and is now seen to have negative (regenerative) periods. The line voltage into the converter is also distorted. As a further example *Figure 6.34* gives the voltage waveforms for the unidirectional converter

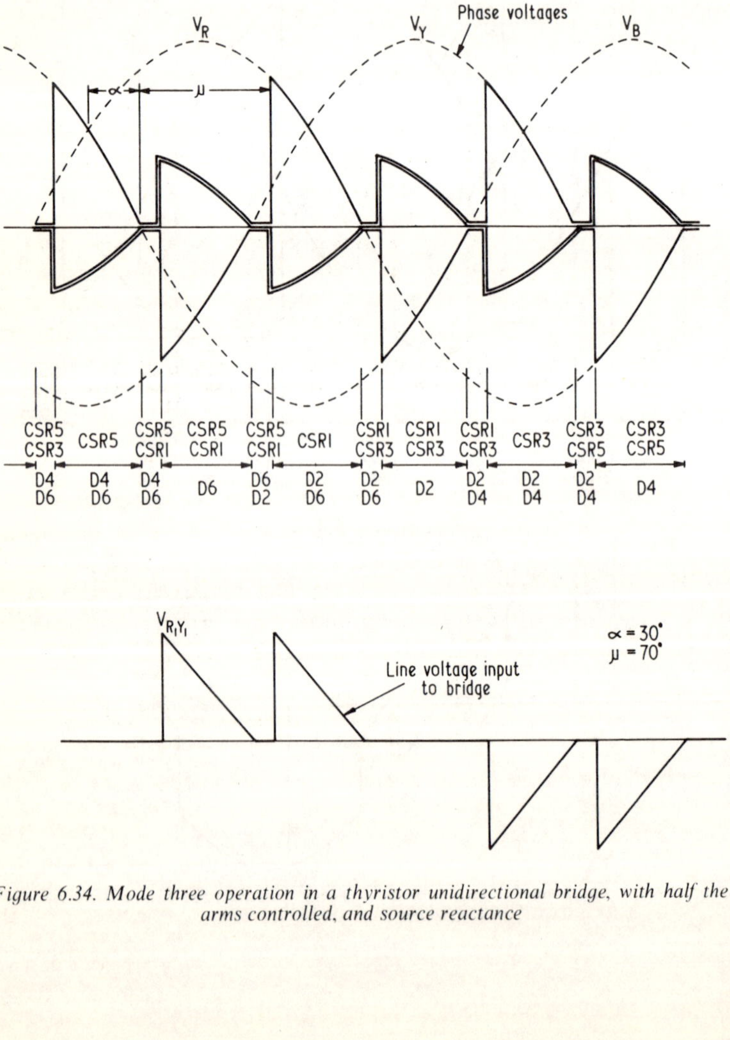

Figure 6.34. Mode three operation in a thyristor unidirectional bridge, with half the arms controlled, and source reactance

of *Figure 6.18* which is operating with a delay angle α of 30° and an overlap angle μ of 70°. There are now periods when three and four devices conduct simultaneously so that this is mode three operation. As before the load voltage dips to zero when four devices conduct.

Performance factors

The performance of converter circuits can be compared by means of several factors. The commonest in use are the d.c. voltage ratio, the input displacement factor, the input power factor and the input current distortion factor. The d.c. voltage ratio gives the relation between the converter output at any delay angle α and that obtained when the same converter is operating free of delay angles or overlap. The input displacement factor is the cosine of the angle between the fundamental component of a.c. current into the bridge and the supply voltage. It has been shown, when comparing the various circuits, that this angle ϕ increases (lagging) as the phase control angle α or overlap angle μ increase. For bidirectional converters $\phi = \alpha$. It is not proposed to consider this further here. The input power factor is the ratio of the total mean input power to the total r.m.s. input volt amperes. The input current distortion factor is the ratio of the r.m.s. value of the fundamental to the total input current. Both these factors define the relation between the in-phase and quadrature components of a.c. current. This has been discussed in previous sections and it was shown then that unidirectional converter circuits often have the advantage over bidirectional converters in that they have a lower quadrature (or wattless) current input.

In this section the value of the mean d.c. voltage from a converter will be found and the harmonic content of this waveform discussed.

The output voltage waveform from a converter with zero delay angle or phase overlap is composed of a series of cosine curves which are linked together as in *Figure 6.35*(i). The duration of each section is equal to $2\pi/p$ where p is the pulse number of the converter. The average value of this waveform is clearly given by

$$E_{\text{do}} = \frac{p}{2\pi}\int_{-\pi/p}^{+\pi/p} \sqrt{2}E\cos\theta\, d\theta$$

where E is the r.m.s. input voltage.

Therefore
$$E_{\text{do}} = \sqrt{2}E\frac{p}{\pi}\sin\frac{\pi}{p} \tag{6.5}$$

If a delay angle α is now introduced into the converter, as in *Figure 6.35*(ii) then, since the waveform still follows a cosine wave over its control range,

$$E_d = \frac{p}{2\pi}\int_{-\pi/p+\alpha}^{+\pi/p+\alpha} \sqrt{2}E\cos\theta\, d\theta$$

$$= \sqrt{2}E\frac{p}{\pi}\sin\frac{\pi}{p}\cos\alpha$$

$$= E_{do}\cos\alpha \qquad (6.6)$$

The d.c. voltage ratio is equal to E_d/E_{do} and is given by $\cos\alpha$.

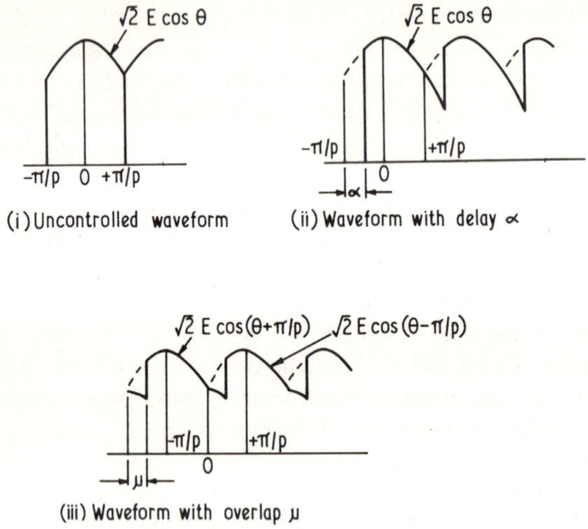

Figure 6.35. Waveform used in calculating load d.c. voltage for a p-pulse converter

To consider the effect of the overlap angle suppose the delay angle is zero as in *Figure 6.35*(iii). The datum for the overlapping waveforms has now been moved to the commencement point for overlap, so that the equations for the two waveforms are given by

$$e_1 = \sqrt{2}E\cos\left(\theta+\frac{\pi}{p}\right)$$

$$e_2 = \sqrt{2}E\cos\left(\theta-\frac{\pi}{p}\right)$$

During the overlap period the load voltage assumes a value intermediate between e_1 and e_2 so that the d.c. voltage is given by

$$E_d = \frac{p}{2\pi}\left[\int_0^{\mu}\frac{e_1+e_2}{2}d\theta + \int_{\mu}^{2\pi/p} e_2\, d\theta\right]$$

$$= \sqrt{2}E\frac{p}{\pi}\left[\int_0^{\mu}\frac{\cos(\theta+\pi/p)+\cos(\theta-\pi/p)}{2}d\theta+\int_{\mu}^{2\pi/p}\cos\left(\theta-\frac{\pi}{p}\right)d\theta\right]$$

$$= \sqrt{2}E\frac{p}{\pi}\sin\frac{\pi}{p}(1+\cos\mu)$$

$$= \sqrt{2}E\frac{p}{\pi}\sin\frac{\pi}{p}\cos^2\frac{\mu}{2}$$

$$= E_{do}\cos^2\frac{\mu}{2} \tag{6.7}$$

Therefore the effect of overlap is similar to that of firing-angle delay in that the mean d.c. output voltage is reduced, although the extent of variation in the two cases is different. For a circuit operating with delay α and overlap μ, the value of mean d.c. voltage is clearly

$$E_d = E_{do}\cos\alpha\cos^2\frac{\mu}{2} \tag{6.8}$$

In order to estimate the overlap angle μ it is necessary to know the value of d.c. current I_d and reactance X in each phase. *Figure 6.36* shows the equivalent circuit at the moment of overlap between two

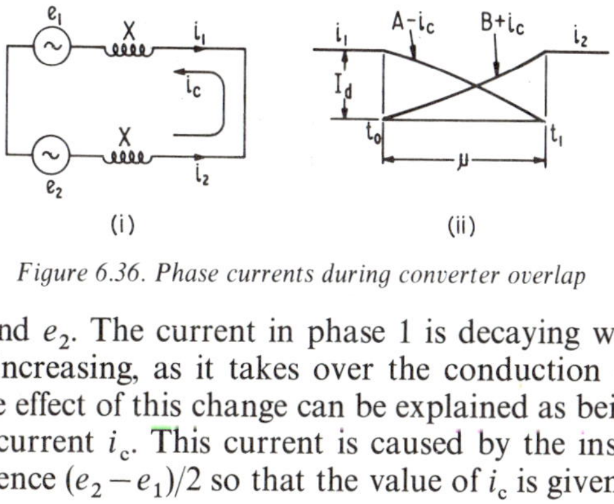

Figure 6.36. Phase currents during converter overlap

phases e_1 and e_2. The current in phase 1 is decaying while that in phase 2 is increasing, as it takes over the conduction of the load current. The effect of this change can be explained as being due to a circulating current i_c. This current is caused by the instantaneous phase difference $(e_2-e_1)/2$ so that the value of i_c is given by

$$i_c = \frac{e_2-e_1}{2X}$$

$$= -\frac{\sqrt{2}E\sin\pi/p}{X}\cos\theta$$

$$= -I_c\cos\theta$$

where I_c is the peak value of the circulating current.

The current delivered by the phases during overlap can be considered to be composed of a constant part and a variable circulating part, as shown in *Figure 6.36*(ii). When these two components of current are equal the overlap period is terminated. Considering the instant t_0 at the commencement of overlap,

$$i_1 = I_d = A - i_c = A + I_c$$

since at t_0 the value of $\theta = 0$ and $i_c = -I_c$.

Therefore $$A = I_d - I_c$$

Also at this point $$i_2 = 0 = B - I_c$$

Therefore $$I_c = B$$

At t_1 the value of $\theta = \mu$ and

$$i_2 = I_d = B + i_c = I_c(1 - \cos\mu)$$

Therefore $$\cos\mu = 1 - \frac{I_d}{I_c}$$

Substituting for I_c this gives

$$\cos\mu = 1 - \frac{I_d X}{\sqrt{2}E \sin \pi/p} \tag{6.9}$$

where X is the equivalent reactance per phase in the a.c. lines and E is the r.m.s. voltage per phase of the input.

Equation (6.6) for d.c. voltage ratio is a cosine curve for all bidirectional converters. It is shown plotted in *Figure 6.37*. Unidirectional converters have a different characteristic due to the discontinuity in the cosine curve and the introduction of zero voltage

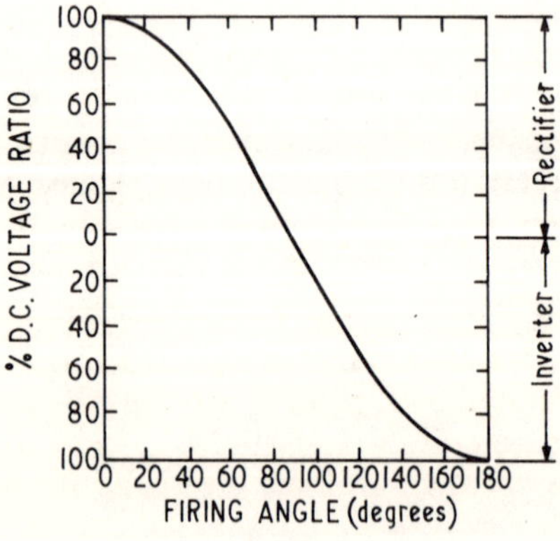

Figure 6.37. Variation of d.c. voltage ratio with firing angle for a bidirectional converter

periods during free-wheeling load current. The equation for the d.c. voltage ratio now depends on the circuit used, since the incidence of this point will change. It can be shown for half-controlled converters (in which half the thyristors in the converter are replaced by diodes) that the mean d.c. voltage is given by

$$E_{\mathrm{d}} = E_{\mathrm{do}}(1+\cos\alpha) \tag{6.10}$$

so that the d.c. voltage ratio is $(1+\cos\alpha)$. This is plotted in *Figure 6.38*. Also illustrated here are ratios for various other systems which use free-wheeling diodes in order to give unidirectional converters.

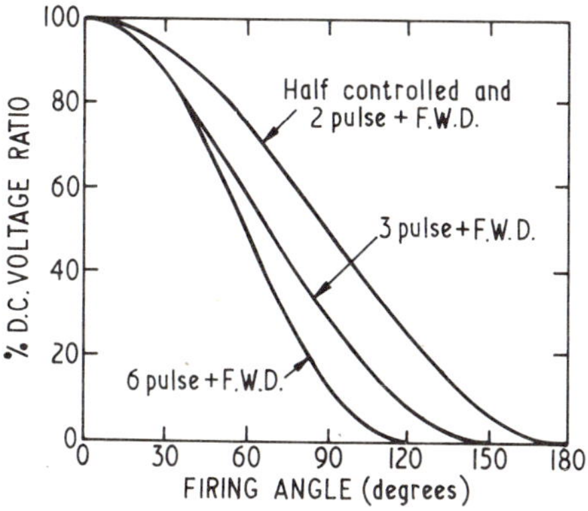

Figure 6.38. Variation of d.c. voltage ratio with firing angle for various unidirectional converters

The harmonics present in the d.c. voltage from the converter have been shown to be related to its pulse number. For a bidirectional system the ratio of the amplitude of the nth harmonic to the peak d.c. voltage is given by

$$\left[\frac{1}{(n-1)^2}+\frac{1}{(n+1)^2}-\frac{2\cos 2\alpha}{(n+1)(n-1)}\right]^{1/2} \tag{6.11}$$

This is plotted in *Figure 6.39* up to the 24th harmonic. All the harmonics are not present in every system. Generally for a p-pulse converter harmonic numbers present are kp where k is an integer 1, 2, 3, etc. The harmonic plot clearly shows the peaking at 90° firing angles when the mean d.c. voltage is zero, all the output being a.c. ripple.

For unidirectional converters the situation is again complicated due to the variation of the instance of current free-wheeling with

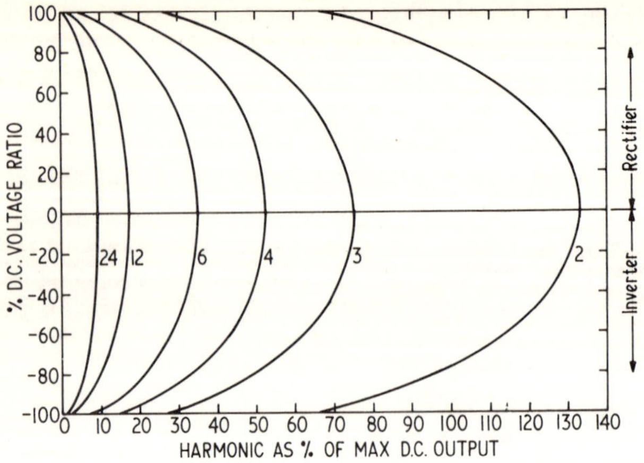

Figure 6.39. Predominant harmonics in the output of bidirectional converters

circuit type. *Figures 6.40* and *6.41* give the lower harmonic content for single- and three-phase half-controlled converters. Comparing this with the bidirectional converter it is seen that harmonics of the same number are lower in unidirectional circuits.

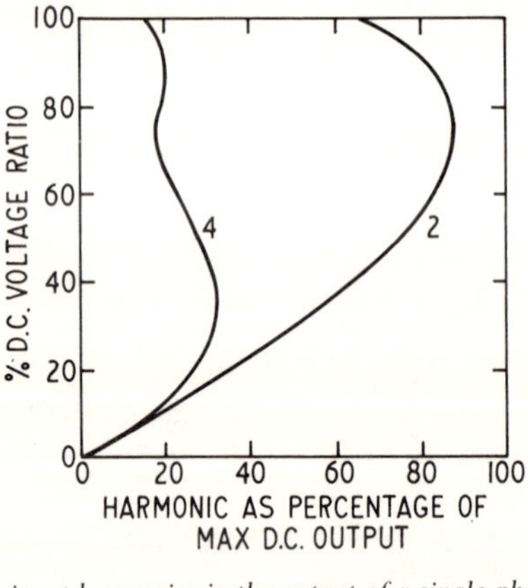

Figure 6.40. Predominant harmonics in the output of a single-phase half-controlled bridge converter

A useful way of comparing harmonics from several control rectifier systems is to determine their harmonic factors. The r.m.s. voltage output E_R is composed of a d.c. voltage E_d plus a spectrum of harmonic voltages of magnitude E_1, E_2, E_3, etc.

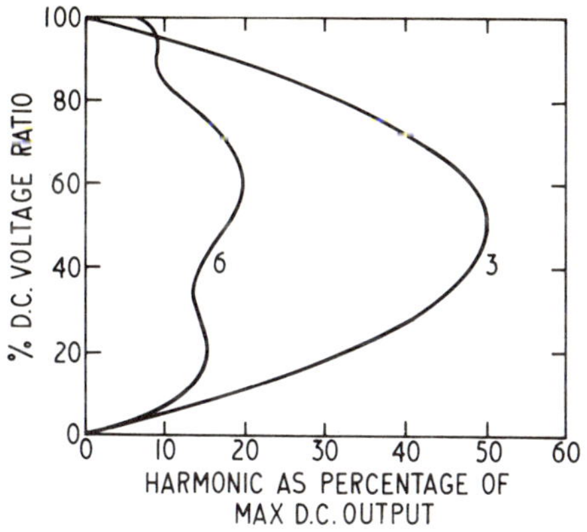

Figure 6.41. Predominant harmonics in the output of a three-phase half-controlled bridge converter

If
$$E_h = \sum_{n=1,2,3...} E_n$$
then
$$E_R^2 = E_d^2 + E_h^2$$
or
$$E_h = [E_R^2 - E_d^2]^{1/2}$$
The harmonic factor is defined as the ratio:
$$\frac{E_h}{E_{do}} = H = \left[\frac{E_R^2 - E_d^2}{E_{do}^2}\right]^{1/2}$$
The ratio can be evaluated for several different circuits. For example considering the bidirectional bridge rectifier circuit of *Figure 6.4*,
$$E_R = \left[\frac{1}{\pi}\int_{\alpha}^{\pi+\alpha} (\sqrt{2}E\sin\theta)^2 d\theta\right]^{1/2}$$
where E is the r.m.s. supply voltage,

since
$$E_{do} = \sqrt{2}E\frac{2}{\pi}$$
and
$$E_d = E_{do}\cos\alpha$$
the harmonic factor is given by
$$H = \left[\frac{\pi^2}{8} - \cos^2\alpha\right]^{1/2} \tag{6.12}$$
Table 6.1 gives the harmonic factors for some of the circuits discussed in the previous section, which are representative of the various pulse numbers.

Table 6.1

Circuit reference	*Harmonic factor H*
Figures 6.2 and *6.4* (two-pulse)	$\left[\frac{\pi^2}{8} - \cos^2\alpha\right]^{1/2}$
Figure 6.5 (three-pulse)	$\left[\frac{2\pi^2}{27} + \frac{\sqrt{3}\pi}{18}\cos 2\alpha - \cos^2\alpha\right]^{1/2}$
Figure 6.8 (six-pulse)	$\left[\frac{\pi^2}{18} + \frac{\sqrt{3}\pi}{12}\cos 2\alpha - \cos^2\alpha\right]^{1/2}$

CONTROL CIRCUITS

The firing circuits for phase-controlled rectifiers are essentially similar to that used for a.c. phase control, as shown in *Figure 4.10*. The input detector determines the incidence of zero supply voltage, the delay section provides a variable delay from this point, and this signal is then used to energise the gate drive circuits which fire the thyristors. There are two further considerations. On inductive loads the firing pulses must be of sufficient width to overcome any overlap angle. This was illustrated in *Figure 6.27* where it was shown that a pulse width of at least μ_1 is required to ensure that the thyristor reaches its full conduction on inductive load. This problem of extended firing pulses is not as serious as for a.c. line control, where for inductive loads a pulse duration of 90° is normally required. However, controlled rectifier circuits have an added feature that the input waveform could be distorted if line reactances are severe. Generally the firing circuits should be able to start the timing pulse when this voltage rises from the zero voltage point.

Figure 6.42 shows an inexpensive half-controlled bridge firing circuit. When line A goes positive capacitor C_1 charges through R_1 and R_2. After a delay, determined by the value of R_1, its voltage is sufficiently large to break over trigger diode D6. A pulse of current is applied via D3 and R_3 to CSR2. Load current can now flow via D1 and this thyristor. When the supply reverses CSR2 goes off. C_1 now charges with reverse polarity applying a pulse through T_1 to CSR1 when D6 breaks over. Load current flows through D2 and CSR1.

The circuit of *Figure 6.42* is capable of providing a short duration firing pulse only. *Figure 6.43* shows an alternative scheme which gives a high-frequency gate drive for the complete duration of the thyristor conduction period. Transistor TR1 operates as the a.c. zero voltage detector. When the supply voltage falls to zero this transistor comes on due to current flow through R_2. Capacitor C_1 is discharged. When the supply is non-zero TR1 goes off, its base-emitter being

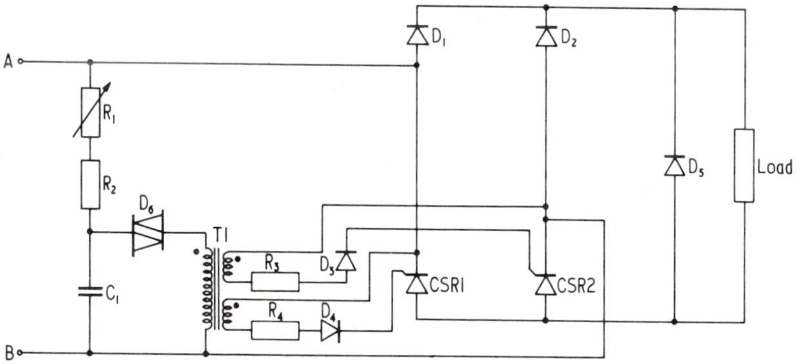

Figure 6.42. Inexpensive firing circuit for bridge converter

reverse-biased due to current flowing in D1. C_1 charges through R_{18}, R_{19} and turns TR2 on when its voltage exceeds that of Zener diode D17. Transistors TR4 and TR6 form two free-running blocking oscillators whose outputs are half-wave rectified and then fed to the thyristors in opposite arms of the bridge. TR3 and TR5

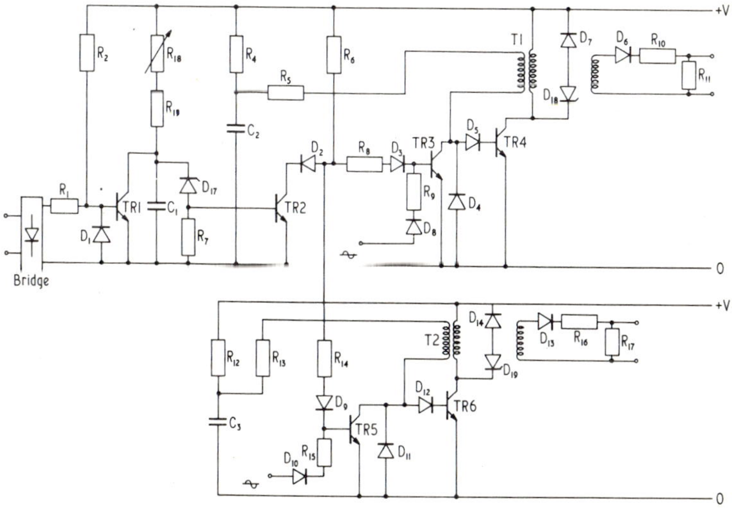

Figure 6.43. Alternative thyristor firing circuit

control these oscillators so that when these transistors are on the oscillators are held off. When TR2 comes on, both TR3 and TR5 are able to go off. However, a.c. signals, in anti-phase to each other, are fed to the bases of TR3 and TR5. Therefore when TR2 is on, either TR3 or TR5 is allowed to go off in any half cycle. This ensures that gate pulses are not applied to a thyristor which is reverse-biased, so increasing its dissipation. With TR2 off, as would be the case at the start of a half cycle, both TR3 and TR5 are on, preventing any of the thyristors from being fired. This firing circuit is extendable to three-phase operation by turning TR1 on at the phase intersection points.

CONCLUSIONS

Although only typical circuits have been discussed in this chapter, it should be clear that there are a bewildering number of different combinations for controlled rectification. However, they can all be divided as follows:

(1) Bidirectional or unidirectional converters.

(2) Each of these can then be push-pull or bridge in configuration.

(3) Both these converters can have any number of output pulses. The higher the pulse number the lower the ripple content.

Bidirectional converters are capable of inversion in addition to rectification, and this is often a desirable feature. Where it is not required unidirectional converters can be used. The resulting circuit is usually cheaper and has a smaller d.c. voltage ripple and input 'wattless' current content.

Push-pull circuits have the advantage that there is only one conducting device in series with the load. This was an important consideration with mercury arc rectifiers which could have arc drops of 30V or more. However, modern thyristors lose only a volt or two so that this is important only when working from abnormally low supply voltages. Generally bridge circuits have better transformer utilisation and are more frequently used.

7

Frequency converters

An a.c. output of a given frequency and voltage can be produced either from a d.c. source, or directly from another a.c. source of a different frequency and voltage. Frequency converters can therefore be classified broadly into two types, d.c. to a.c., called inverters, and a.c. to a.c. known as cycloconverters.

Cycloconverters and inverters were introduced in Chapter 1. In that chapter a simplified operating mode, *i.e.* single-phase resistive load, was used in order to illustrate the circuit principles. In this chapter, cycloconverters will be analysed in greater detail. In particular it will be shown that they form an extension to controlled rectifier circuits, as described in Chapter 6, and can be treated similarly. It is important to keep this closeness in mind since, although controlled rectification forms one of the most frequent uses of thyristors, the cycloconverter is often regarded as something mysterious, and complicated, by the average engineer. This reputation has no doubt been gained by the cycloconverter owing to its rather formidable looking circuit connections. However, it will be seen that each system can be broken down into basic controlled rectifier circuits which are interconnected.

Inverter-type frequency changers are not covered in detail. Only sufficient information is given here to allow a comparison to be made with cycloconverters. Chapter 9 covers inverter systems in greater detail.

D.C. TO A.C. CONVERTERS

An inverter converts from direct to alternating voltage. If operating from an alternating-voltage source, it is necessary first to rectify and smooth this to a direct voltage before it can be fed into the inverter.

A basic inverter circuit is shown in *Figure 7.1.* The arrows on the thyristors indicate that they are turned off, when required, by some external method, which is not shown. Diodes D1 to D4 are necessary for inductive loads, but not if the load is resistive. The operation of

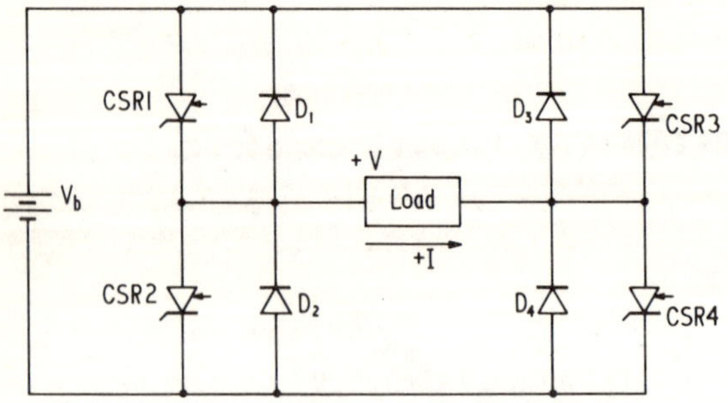

Figure 7.1. Basic inverter circuit (excluding commutation components)

the inverter can be explained with reference to *Figure 7.2*. This shows the instantaneous load voltage, the mean load voltage and the load current after filtering.

At t_0 thyristors CSR1 and CSR4 are fired. Load current is negative and flows through D1 and D4 preventing the thyristors from conducting. However, since these diodes conduct the load voltage is still positive.

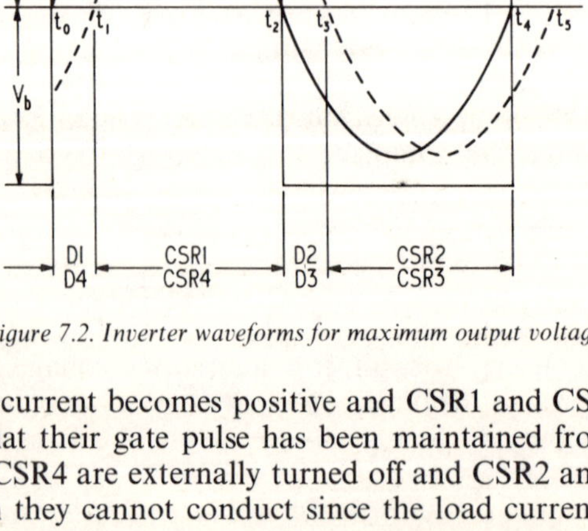

Figure 7.2. Inverter waveforms for maximum output voltage

At t_1 the current becomes positive and CSR1 and CSR4 turn on, provided that their gate pulse has been maintained from t_0. At t_2, CSR1 and CSR4 are externally turned off and CSR2 and CSR3 are fired. Again they cannot conduct since the load current is positive and flows through D2 and D3 until t_3 when the current becomes negative and the thyristors turn on. The frequency of the output is clearly controlled by the thyristor firing circuit.

Normally the frequency changer operates into an iron-cored load (such as a motor) which is to function at constant flux. Since:

$$\text{flux} \propto \frac{\text{voltage}}{\text{frequency}}$$

the mean voltage must be varied in proportion to frequency. One method is illustrated in *Figure 7.3*. At t_0 thyristors CSR1 and CSR4

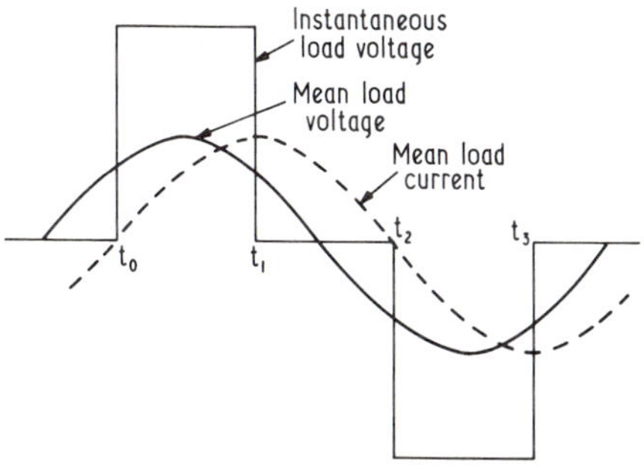

Figure 7.3. Inverter waveforms with reduced output

are fired and conduct since the current is starting positive. At t_1 thyristor CSR4 is turned off. The current continues to circulate through thyristor CSR1 and diode D3 effectively short circuiting the load. At t_2 thyristor CSR1 is turned off and CSR2 and CSR3 are turned on. The current reverses. At t_3 thyristor CSR3 is turned off and load current circulates through CSR2 and D4. The load voltage is again zero. As seen the mean voltage is much less than in *Figure 7.2*.

In the waveforms shown in *Figures 7.2* and *7.3* it has been assumed that the load current is filtered and sinusoidal. In all probability this will not be the case, as in inverters used for motor drives. Current and voltage waveforms can then be represented by *Figure 7.4*. The output is rich in harmonics which causes excessive load losses. To reduce the harmonic content of the load voltage alternative control techniques can be used. One such system is shown in *Figure 7.5*. The load voltage is made up from several sections of varying duration, each block switching between zero and d.c. supply voltage. The placing and width of these voltage periods are such as to give a mean a.c. voltage which approximates closely to a sine wave. The load current waveform is also shown in *Figure 7.5* and is seen to have a smaller low-frequency harmonic content than in the traditional voltage control system of *Figure 7.4*. The various inverter voltage control systems are covered in Chapter 9. The introduction here has been made to show that methods exist for giving an inverter output which is as good as that obtained from traditional cycloconverters.

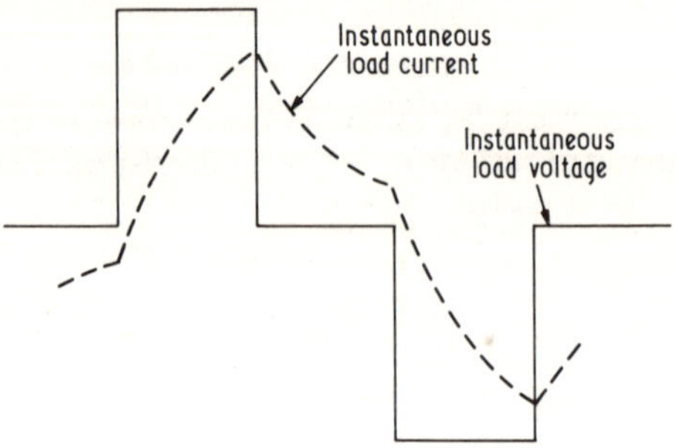

Figure 7.4. Inverter waveforms with unfiltered load current

A frequent requirement in frequency changer drives is that of regeneration. For instance a locomotive, driven from a d.c. source through an inverter and a.c. motor, is travelling uphill. When it reaches the top, and starts to move down again, the inverter may be used to regenerate back the kinetic energy of the motor into its battery supply. Regeneration can be obtained in inverters. It is, however, not a natural phenomenon of most inverter systems, as for cycloconverters, and special circuit modifications are necessary to accomplish it successfully. For instance unless the d.c. supply is 'soft' the

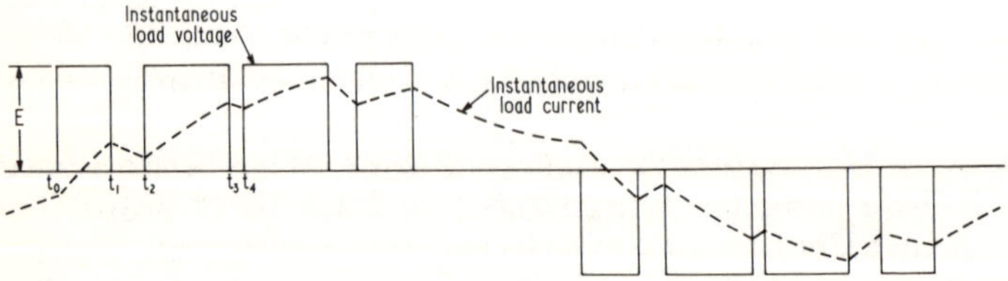

Figure 7.5. Multipulse voltage control for an inverter

regenerated energy would cause excessive voltage increases in the system. For inverters operating from a rectified a.c. source, it is necessary to connect an inverting bridge to feed this energy from the d.c. to the a.c. supply, as shown in *Figure 7.6*. D1 to D4 and CSR1 to CSR4 form the bridge circuit of *Figure 7.1*. The d.c. supply for the inverter is derived from D5 to D8 and smoothed by L_5 and C. CSR5 to CSR8 form the usual arrangement of an a.c. line commutated inverter feeding energy from the d.c. side back to the a.c. supply. It is used during regeneration to prevent excessive voltage build up across the capacitor. During motoring, chokes L_1 and L_3 are saturated by load current and L_2 and L_4 absorb the instantaneous voltage difference between rectifier and inverter. During regeneration energy flow is from motor to a.c. supply, and the functions of the chokes are reversed.

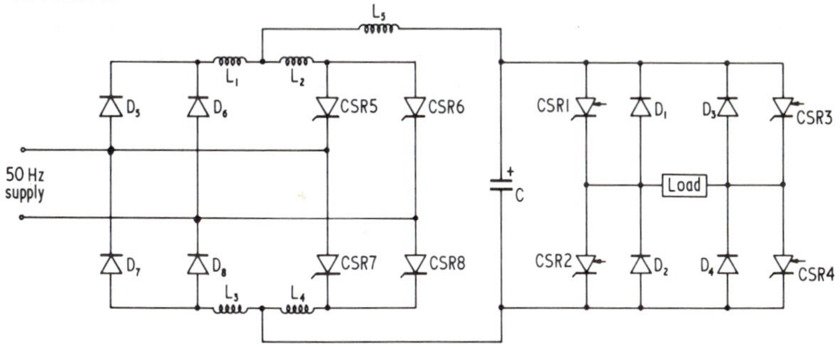

Figure 7.6. Regeneration in an a.c. rectified inverter

The control circuit required for an inverter system is relatively simple. It is shown in *Figure 7.7* as a functional block schematic. The reference input fixes the frequency at which the thyristors are fired and determines the control sequence and hence the output voltage level. The 'distribute' block guides the gate pulses to the correct

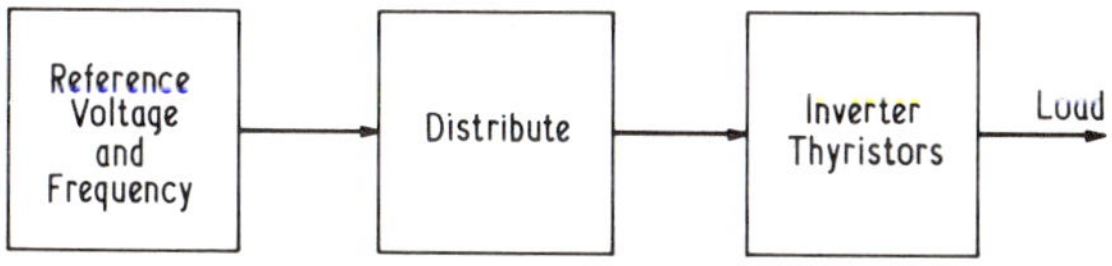

Figure 7.7. Basic inverter control circuit

inverter thyristor. As seen it is not necessary to have a feedback signal from the load voltage or current unless a stabilised output or current limit is required. The reference determines the firing point of the thyristors and therefore the load current and voltage.

A.C. TO A.C. CONVERTERS

Cycloconverters can operate in one of two modes, envelope or phase-control. In both systems the circuit arrangement is identical, it is the operating mode which determines the type of converter. In Chapter 1 the phase-control system was introduced. This is the commonest method in use and will be used here to illustrate the fundamental properties of cycloconverters. In the following sections several cycloconverter arrangements are described and this is followed by considerations of envelope type cycloconverter, and a performance analysis of the phase-controlled converter.

A basic single-phase cycloconverter circuit was shown in *Figure 1.14*. This is redrawn here in *Figure 7.8* in a form which illustrates clearly that the system consists essentially of two bridge-type two-pulse bidirectional converters, connected in opposite directions across the load. The converter can be considered to consist of a positive and a negative group, the load voltage and current polarity being as indicated.

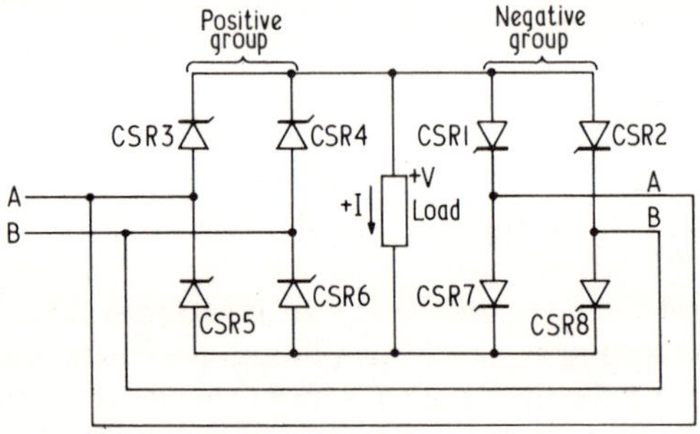

Figure 7.8. Modified arrangement of a single-phase bridge cycloconverter

The operation of the phase-controlled cycloconverter can be explained by reference to the waveforms given in *Figure 7.9*. The load current is assumed here to be filtered and therefore sinusoidal. At t_0 line A is positive to B and the load current is negative (*i.e.* opposite to the direction shown in *Figure 7.8*) so that CSR7 and CSR2 are

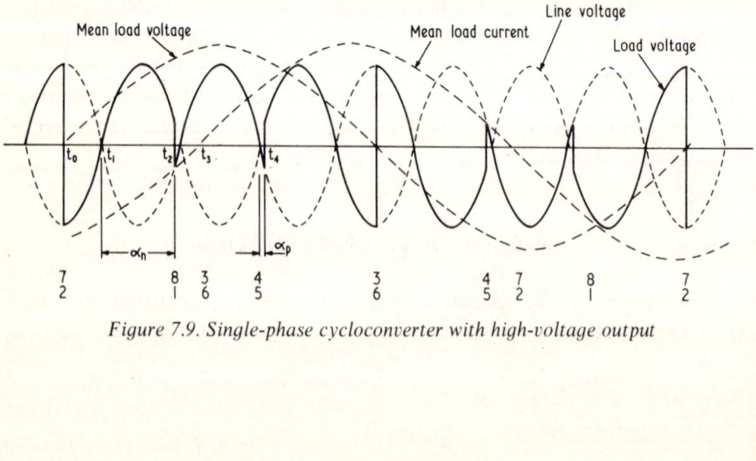

Figure 7.9. Single-phase cycloconverter with high-voltage output

fired. These thyristors are maintained in conduction, even when the voltage across them reverses at t_1, due to the energy stored in the inductive load.

At t_2, CSR8 and CSR1 are fired turning CSR7 and CSR2 off and driving the instantaneous load voltage negative. When load current reverses at t_3, CSR8 and CSR1 turn off and in order to maintain the load waveform as shown, CSR6 and CSR3 are fired. These conduct until t_4 when CSR4 and CSR5 are fired turning them off; and so on through the cycle.

The instant of firing the thyristors can clearly be varied as desired. In *Figures 7.9* and *7.10* α_p and α_n represent the minimum and maximum delay angles for the positive and negative group of converters

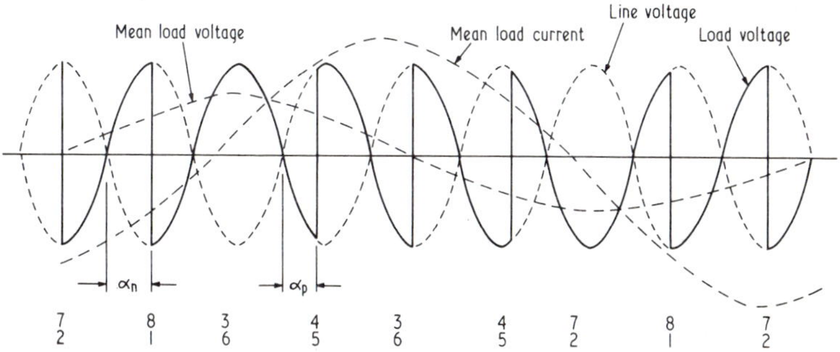

Figure 7.10. Single-phase cycloconverter with low-voltage output

respectively. When the delay is 90° the mean output is clearly zero. The variation of the delay about this point determines the amplitude of the output, as shown in *Figure 7.10* where the mean load voltage is lower than that in *Figure 7.9*. Referring to these two figures the following can be deduced regarding the cycloconverter:

(i) The positive group of thyristors, 3, 6, 4 and 5, conduct in pairs when the load current is positive. The negative group, 7, 2, 8 and 1, conduct when it is negative.

(ii) A firing delay of 90° in any phase corresponds to zero load voltage. The firing points of the thyristors oscillate about this delay at a frequency determined by the required output frequency. Clearly the greater the swing of oscillations about 90° (maximum total excursion 180°), the greater the mean output voltage. The maximum delay angle is limited by the requirement of ensuring that the negative voltage available to turn off a thyristor during a commutation period is of sufficient duration.

Figures 7.9 and *7.10* show a sinusoidal current waveform, although in practice this will not be so unless load filters are used. However, it will be shown later that cycloconverters are more frequently used in three-phase systems, where the output current and voltage can be made to approximate closely to a sine wave. During periods when the load voltage and current are in the same sense, either positive or negative, there is a flow of power from the supply to the load. When voltage and current are in an opposite sense there is regeneration from the load to the supply. Careful examination of *Figures 7.9* and *7.10* will show that there are greater periods of input power than there are of regeneration. Net regeneration can be obtained by a re-arrangement of firing pulses as shown in *Figure 7.11.* The problem

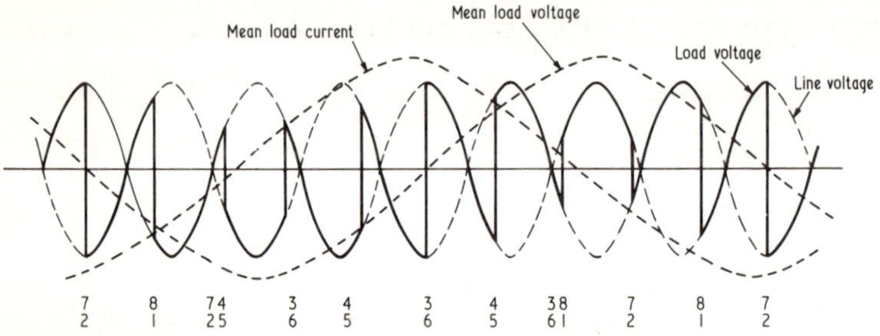

Figure 7.11. Regeneration with a cycloconverter

of regeneration is identical to that encountered in usual phase-control circuits, as discussed in Chapter 6. Therefore a cycloconverter can be readily made to control the flow of power in either direction. This is one of its greatest assets.

The operation of the cycloconverter can be explained more clearly with reference to its group diagram as shown in *Figure 7.12*(i). The load-voltage waveforms are shown for converter delay angles varying from 0° to 180°. The function of the two converters is seen to change from full rectification to full inversion, in varying stages. It is clear that each converter must therefore be able to rectify and regenerate within any half cycle. Since only one converter carries the load current at any instant, it is possible to fire only this system when required. However, it will be seen later that there are often advantages to firing both converters simultaneously, but with their delay angles such that their sum always equals 180°. *Figure 7.13* now illustrates the operating mode. From this it is seen that at any setting one group is in the rectification mode while the second is in inversion, with such a delay angle that the mean output from the two groups are equal. This therefore prevents the transfer of mean power between the two converter groups. This system will be referred to again later in this chapter.

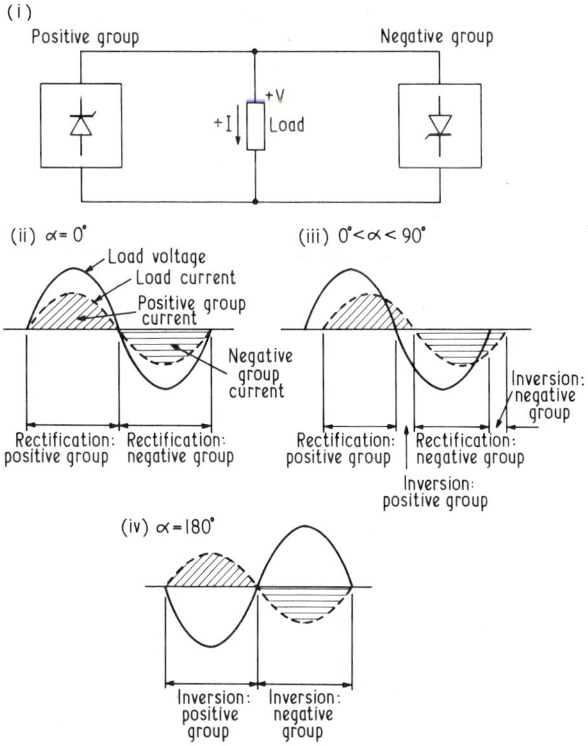

Figure 7.12. Group representation of a cycloconverter

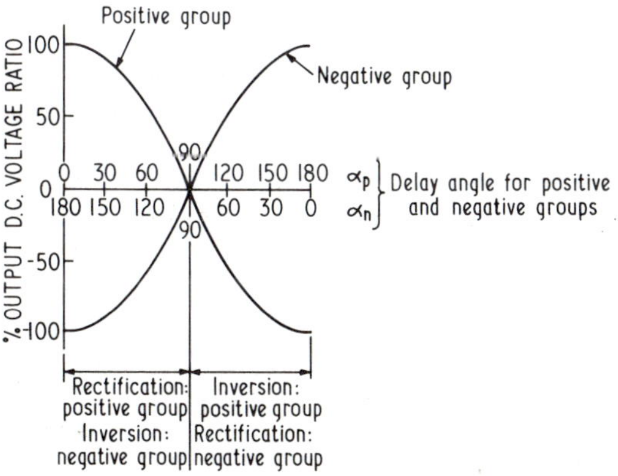

Figure 7.13. Control method for the positive and negative groups in a cycloconverter

CYCLOCONVERTER ARRANGEMENTS

Figure 7.12(i) illustrates very clearly that a cycloconverter is basically a combination of various groups of thyristor converters, of the type described in Chapter 6. *Figure 7.8* illustrated the use of two-pulse bridge converters and *Figure 7.14* shows a functionally identical circuit made up from two push-pull converters. The operation of this system is represented by the previous load waveforms if CSR3 and CSR6 for the bridge are replaced by CSR1 for push-pull, CSR4 and CSR5 for bridge are replaced by CSR2 for push-pull, CSR7 and

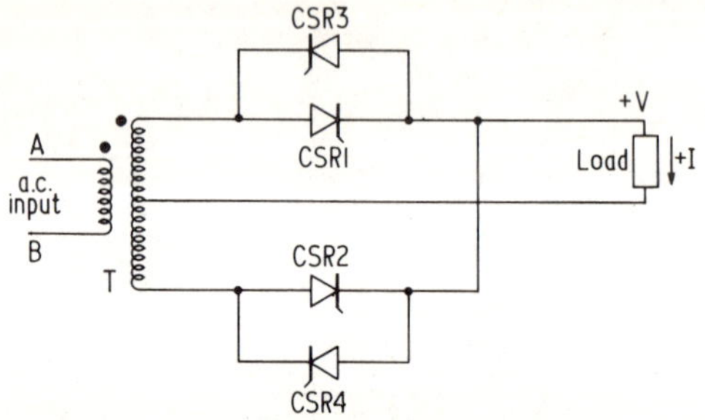

Figure 7.14. Push-pull two-pulse cycloconverter

CSR2 for bridge are replaced by CSR4 for push-pull, and CSR1 and CSR8 for bridge are replaced by CSR3 in push-pull. CSR1 and CSR2 therefore correspond to the positive group and CSR3, CSR4 to the negative group. In common with rectifier converters it can be seen that although push-pull systems use half as many thyristors as a bridge circuit, these have now to be rated at twice the load voltage.

The higher the pulse number of the converter the lower the load ripple voltage. This too was found to be the case for a.c. to d.c. converters. Therefore single-phase cycloconverters are rarely used in practice. *Figure 7.15* shows a three-pulse push-pull cycloconverter circuit which is supplying a balanced three-phase load. To obtain a single-phase output only one converter may be used and the load returned to the neutral point of the transformer secondary. The three-pulse circuit can be extended to six pulses either by using a double-star transformer secondary as in *Figure 6.7* or by an interphase transformer connection. This latter system is given in *Figure 7.16*, where only one phase of the converter is shown in detail. If a single-phase output is required, only one converter is used and the load output is returned to the transformer neutral point. Interphase connections are preferable to double star since, as discussed when considering the equivalent a.c. to d.c. converters, the utilisation of the system is much higher. The six-pulse push-pull system can be extended to twelve pulses by using the circuit of *Figure 6.10*.

Push-pull circuits have a single output point so that there exists a common load terminal to which three-phase systems can be connected. A bridge circuit on the other hand requires both ends of the

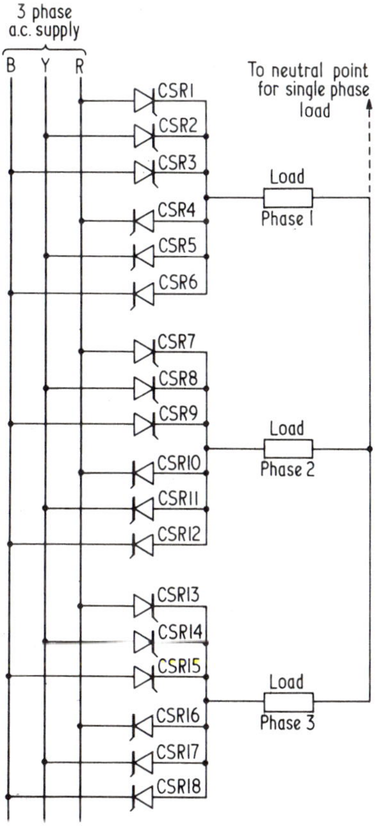

Figure 7.15. Push-pull three-pulse cycloconverter

load to be connected to the converter terminals. Therefore for three-phase loads either the different load phases or the input to the bridges must be isolated. *Figure 7.17* shows a six-pulse bridge converter with isolated load and *Figure 7.18* gives a system in which the load has a common terminal so that the individual bridge inputs need to be supplied from isolating transformers. Isolated load circuits are preferred owing to the complete absence of any input transformer.

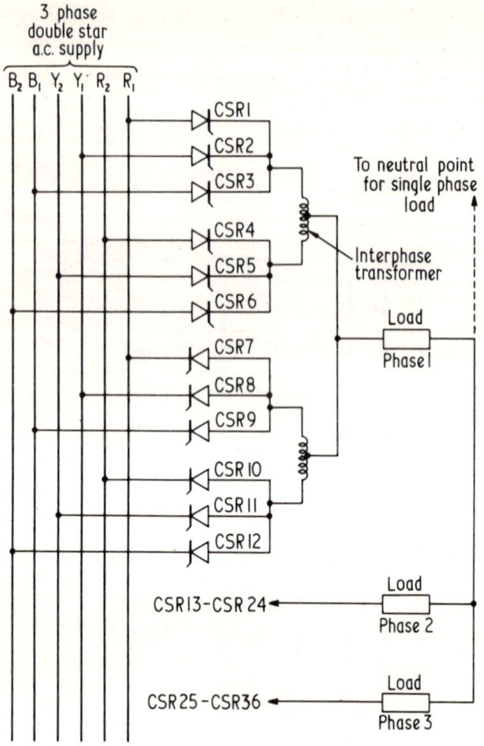

Figure 7.16. Push-pull six-pulse cycloconverter

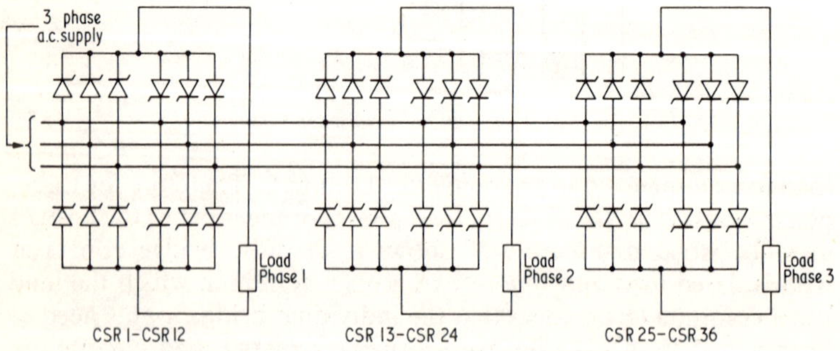

Figure 7.17. Bridge six-pulse isolated load cycloconverter

In the isolated supply system each secondary winding in effect feeds a separate single-phase load, so that the total secondary rating exceeds that of the primary by over 20%. Extension of the six-pulse

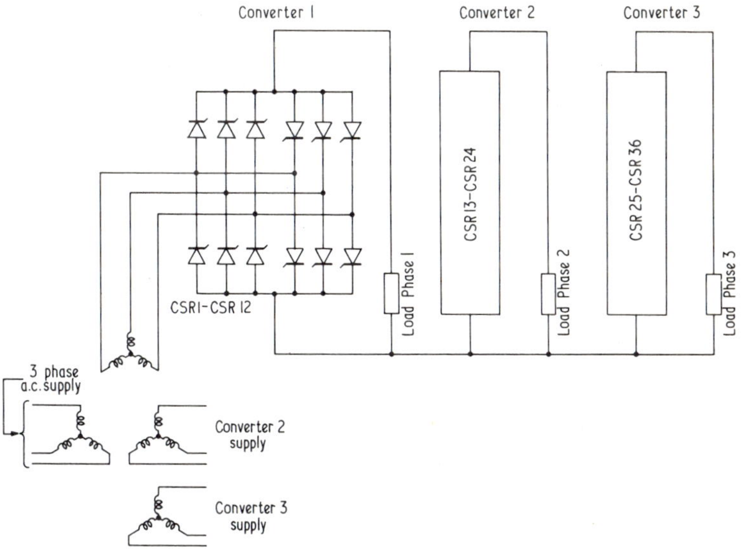

Figure 7.18. Bridge six-pulse cycloconverter with non-isolated load

bridge to twelve pulses by the use of series connected converters is shown in *Figure 7.19*.

The circuits described so far can be referred to as symmetrical cycloconverters since each phase of a three-phase output is made up from a single-phase unit. This is not always necessary. For instance, *Figure 7.20* shows a three-phase load supplied from an asymmetrical cycloconverter which has only two single-phase modules. Although the component content of this converter has been reduced by one third compared to comparable symmetrical circuits, the utilisation of the thyristors and transformer has also been appreciably reduced. Furthermore it is now difficult to maintain a balanced supply to the three-phase load. In spite of this, however, this system is often economical to use in low-power circuits where good performance is not a prime criterion. *Figure 7.21* shows an alternative form of asymmetrical six-pulse converter. It consists of a ring arrangement of bidirectional a.c. to d.c. converters and is therefore not made up from basic single-phase cycloconverter modules. As is seen the

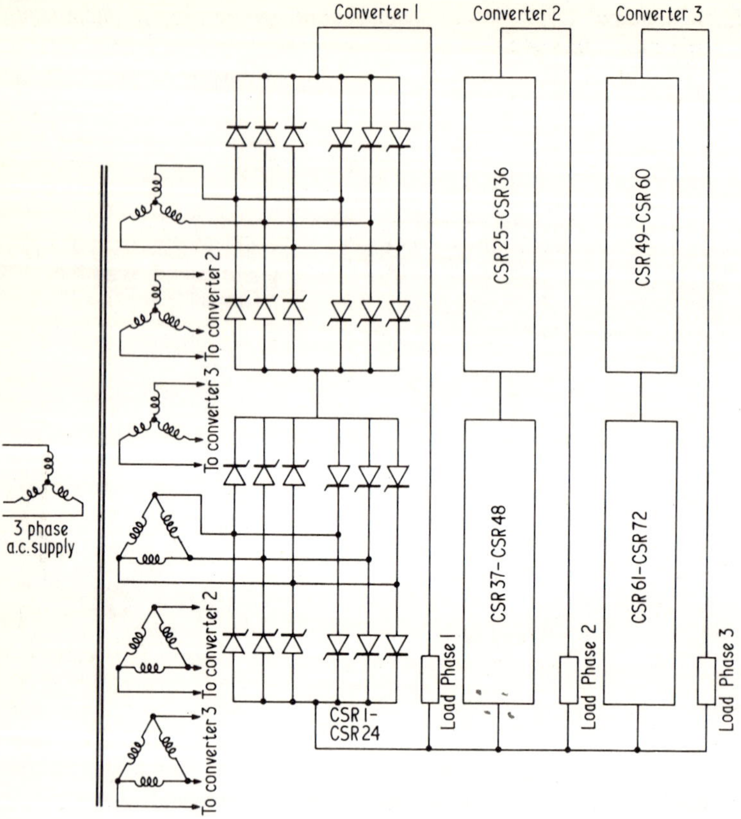

Figure 7.19. Bridge twelve-pulse cycloconverter

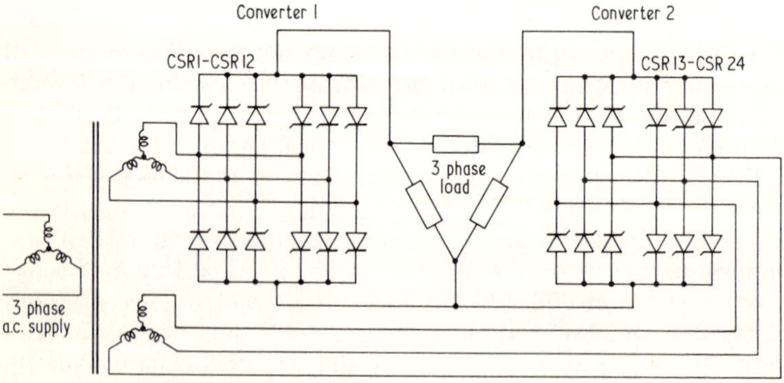

Figure 7.20. Bridge six-pulse two-converter cycloconverter

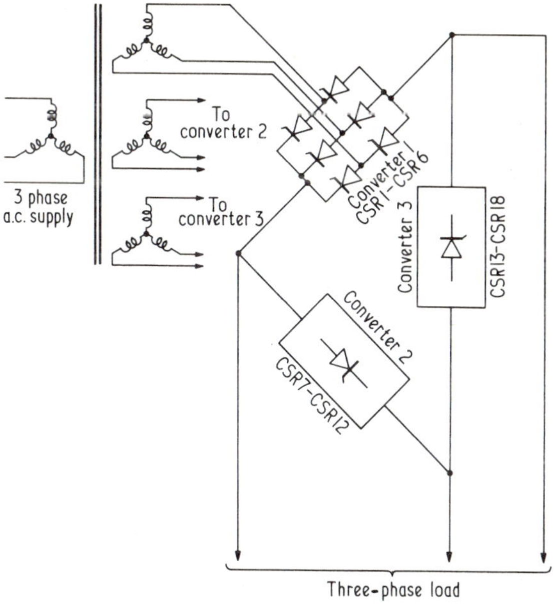

Figure 7.21. Bridge six-pulse ring-type cycloconverter

component content has been reduced by 50% over comparable symmetrical cycloconverters, although as expected the transformer and thyristor utilisation has now also been drastically reduced. Once again, however, this circuit provides a low-cost medium performance converter which is economical in many applications.

ENVELOPE CYCLOCONVERTERS

In the discussions so far the operation of a cycloconverter has been illustrated with reference to its phase-control mode. As mentioned before, this is the most common technique and gives a system in which the load voltage and frequency can be readily controlled. However, in certain applications, especially where the ratio of input to output frequency is fixed, envelope converters can prove simpler and more economical.

Figure 7.22(i) shows the output from a single-phase supply in which the ratio of input to output frequency is three. The converter thyristors are operated at the beginning of every half cycle, so that they follow the envelope of the a.c. supply waveform. Voltage control is now obtained by a correct choice between primary and secondary ratios on the input transformer. Clearly the load voltage has a very high harmonic content. This can be reduced as in *Figure 7.22*(ii) by using a six-pulse cycloconverter. Once again the load voltage follows the envelope of the various phase voltages, and the output voltage is almost rectangular in shape.

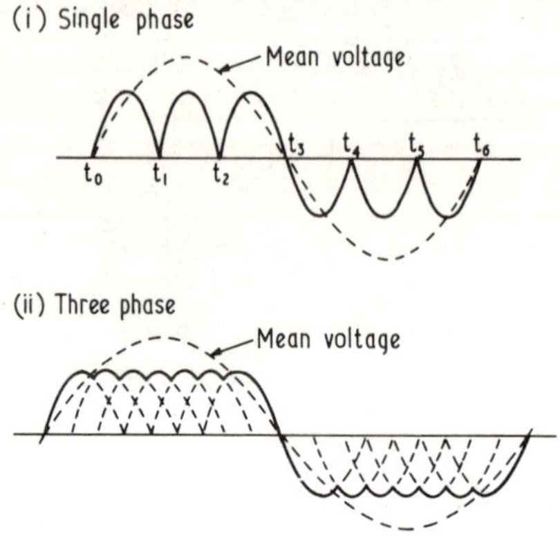

Figure 7.22. Load-voltage waveforms from synchronous envelope cycloconverters

Since most cycloconverters have an input transformer, it is possible to vary the ratio of their secondaries with respect to one another. For instance, *Figure 7.23* shows the output voltage obtained from a system in which phase 1 has 33%, phases 2 and 6 have 73%, and phases 3 and 5 have 93% of the voltage of that of phase 4. The operation is once again of the envelope type where the load voltage

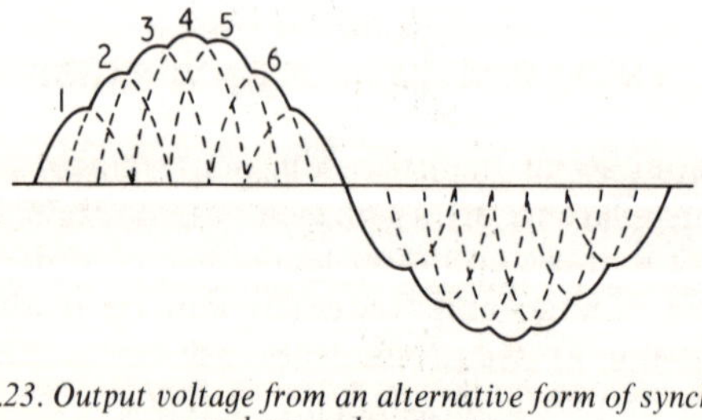

Figure 7.23. Output voltage from an alternative form of synchronous envelope cycloconverter

commutates naturally from phase to phase, so that it is always equal to that of the most positive phase. It can in fact be seen from the above discussions that the half cycle output waveform from envelope cycloconverters is identical to that obtained if the thyristors were all replaced by diodes, so that conduction begins at the commencement of the supply cycle. The function of the thyristor is now purely to ensure that one of the converter groups is switched off when the other group is conducting, so that there is no short-circuit path across the supply.

Figure 7.24 illustrates an alternative form of envelope cycloconverter, which is referred to as asynchronous to distinguish it from the synchronous type of envelope converter described above. *Figure 7.24*(ii) shows that the circuit is essentially the usual arrangement of a cycloconverter with the addition of an induction regulator, which is fed from the three-phase supply, in series with the load. The output voltage from the six-pulse converter is as illustrated in *Figure 7.22*(ii), and to this is added the induction regulator output so as to give the resultant load voltage of the form shown in *Figure 7.24*(i). Once again the approximation to a sine wave is very close.

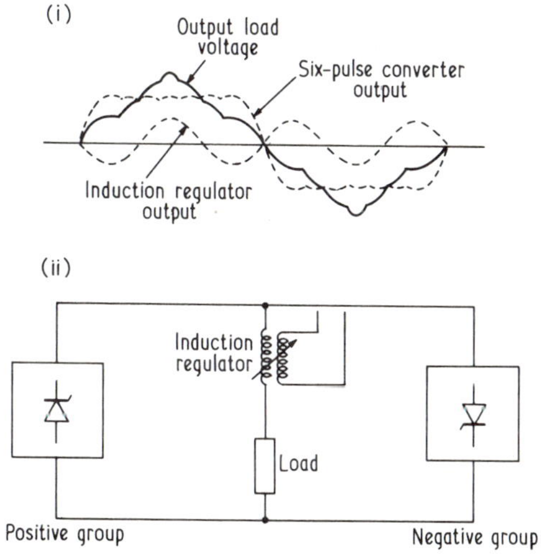

Figure 7.24. Load voltage and circuit connections for one type of asynchronous envelope cycloconverter

Envelope converters are of course incapable of variable-frequency operation. Furthermore, since the converter thyristors operate like diodes during any half cycle, they cannot regenerate back load current to the supply. For this to occur the firing angle on the thyristors would need to be delayed beyond the start of a cycle, as in normal a.c. to d.c. converters. Therefore envelope cycloconverters are incapable of handling inductive loads since they cannot absorb its reactive power. For stable operation it is now necessary to connect a capacitor in parallel across the load, so as to raise its overall power factor. This disadvantage is, of course, not met with phase-controlled converters, in which the firing angle can be shifted readily to meet any required direction of load-current flow.

PHASE-CONTROLLED CYCLOCONVERTERS

The operation of a phase-controlled cycloconverter has already been described with reference to a single-phase system as in *Figures 7.9–7.11*. It is seen that the load frequency can be controlled by the frequency of the oscillations of the firing point about 90°. The load voltage amplitude is governed by the extent of this oscillation about the mean firing point, and the converter can be readily switched from rectification to inversion by regulating the firing angle. The same considerations apply for a three-phase converter, and the load waveforms for the three phases of a typical six-pulse system are shown in *Figure 7.25*. Each phase of the cycloconverter is made up of basic single-phase converter blocks, as in *Figure 7.12*(i) and by

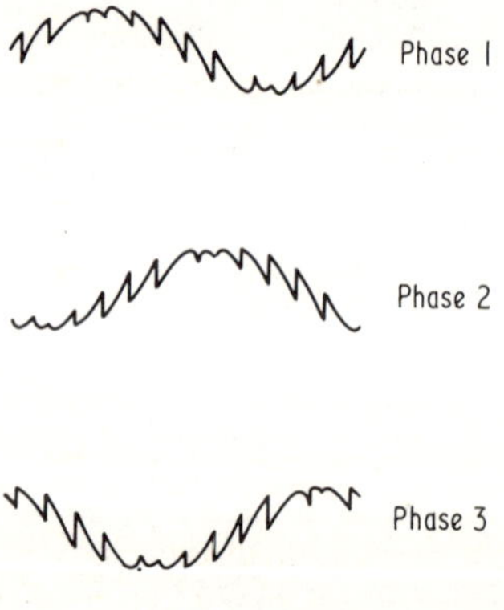

Figure 7.25. Three-phase cycloconverter load voltage for a six-pulse bridge converter

adjusting the firing angles of positive (α_p) and negative (α_n) systems so that $\alpha_p + \alpha_n$ is always equal to 180°, the mean output voltage from the two groups is equal so that there is no net transfer of power round the two converters. However, *Figure 7.26* shows that even though the mean outputs from the two converters are equal, there are instantaneous voltage differences which can give rise to a large circulating current around the loop, unless this is limited by series

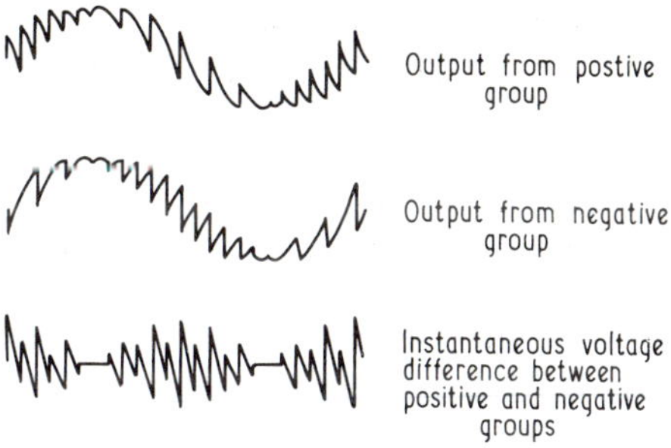

Figure 7.26. Instantaneous voltage difference between positive and negative groups of a cycloconverter

reactors. *Figure 7.27* illustrates the most usual position for such a reactor. This can be added to all converter circuits described in the previous sections, *Figure 7.28* showing its use with one phase of the converter illustrated in *Figure 7.16*. In this position the reactor clearly affects both the load and circulating current, but since only half the turns are in series with the load, the inductance to load current is one quarter of that presented to the circulating current.

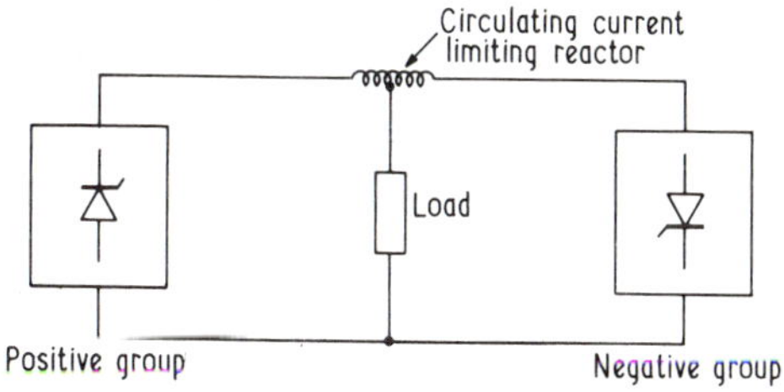

Figure 7.27. Position of a reactor to limit the circulating current between cycloconverter groups

Therefore for a reactance of X, at supply frequency f_S, the reactance presented to the load at frequency f_L is given by $(f_L/f_S)(x/4)$. An alternative position for the circulating-current reactor is clearly in the supply lines between the two converters, although then it affects the load current to a greater extent. If p is the pulse number of the converter system, E the r.m.s. supply voltage to the cycloconverter, and X the effective source reactance, the circulating current is limited to

$$i_c = \frac{E\sqrt{2}}{X}\left(1 - \cos\frac{\pi}{p}\right) \tag{7.1}$$

The flow of circulating current can of course be suppressed

totally by ensuring that only one converter conducts at any time. This is by far the most satisfactory arrangement since the circulating currents can often become extremely large. It does, however, result in a much more complex thyristor control system in which the load current must be sensed and used to block one or other group of converter, depending on the current direction. Another disadvantage of the non-circulating current mode is that the load voltage distortions can become much higher on light loads when the current

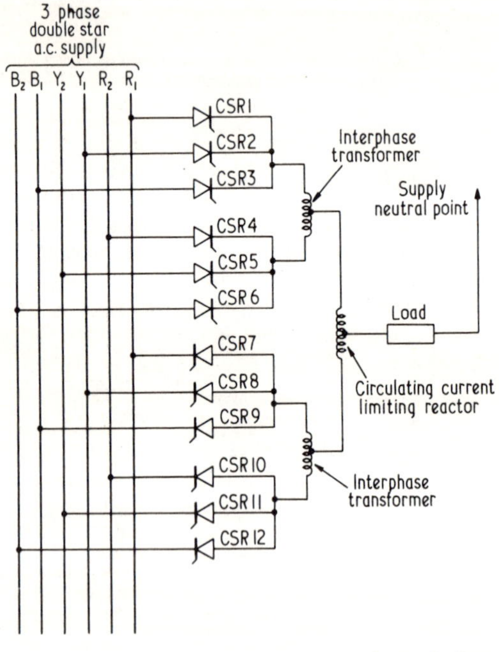

Figure 7.28. Push-pull six-pulse cycloconverter with circulating-current limiting reactor

becomes discontinuous. Since discontinuous operation is only serious on light loads, it is feasible to run a cycloconverter in a circulating-current mode when the load current falls below a certain critical level, and to switch to the group blocking mode when the current increases. The circulating-current reactor can now be made small since it is not required to be effective on heavy load currents, and can therefore saturate. This method of control obviously requires a more complex thyristor firing circuit. Generally, therefore, circulating-current systems are a natural extension from ordinary phase-controlled rectifiers, and ensure the correct thyristor conduction when required for rectification or inversion. It gives the simplest and smoothest control method. The one disadvantage of circulating-current operation is the rather large current flow between the two converter groups, which results in increased device rating and reduced efficiency. Non-circulating current converters prevent this, by blocking one of the converters at all times, so that there can be no interchange of power between the two groups. The price to be paid is greater circuit complexity and higher load voltage harmonics, although this is only important when the ratio of input to output frequency is low.

Other factors affect the load harmonics as well. These are directly dependent on the pulse number of the converter, the ratio of input to output frequency, the level of the output voltage, the load power factor and the technique used to control the load voltage. For instance the load voltage is varied by adjusting the oscillation of the thyristor delay angle by about 90°. This oscillation should follow a cosine law to give a load voltage closely resembling a sine wave. However, imperfections in the control electronics often mean that this perfect law cannot be followed, and the load voltage is distorted. If α_m is the minimum delay angle (maximum delay being $180-\alpha_m$) then the r.m.s. load voltage V_L is given by

$$V_L = \sqrt{2}E\frac{p}{\pi}\sin\frac{\pi}{p}\cos\alpha_m \tag{7.2}$$

This is directly comparable to the equation obtained with phase-controlled rectifiers described in Chapter 6. In common with these converters the load voltage waveform from a cycloconverter is affected by phase overlap caused by source reactance. This is illustrated in *Figure 7.29* and shows the shift in the load voltage harmonic

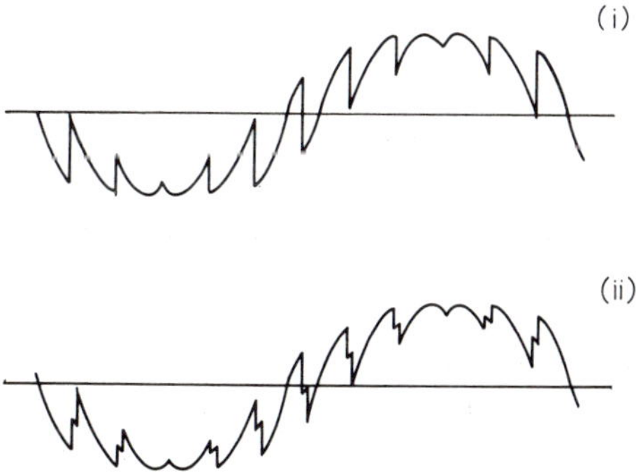

Figure 7.29. Load-voltage waveforms for a cycloconverter: (i) with no source reactance; (ii) with source reactance

content. Source reactance effects are specially important when the cycloconverter is used in variable-speed constant-frequency (v.s.c.f.) systems. This is frequently met in aircraft where the cycloconverter is fed from a local alternator which has a high-frequency output. Since the alternator speed is variable, so also is its frequency. The cycloconverter is now used to ensure that the load frequency is maintained at a fixed value. The source reactance is that of the alternator, and since the cycloconverter represents its biggest load it would create considerable distortion, and this would need to be taken into account in any system design. If μ is the overlap angle caused by instantaneous output current I through source reactance X, α is the firing delay angle and E the r.m.s. supply voltage, then

$$\cos(\alpha+\mu)-\cos\alpha = \sqrt{\left(\frac{2}{3}\right)}\frac{XI}{E} \tag{7.3}$$

The input current conditions to the cycloconverter can be defined by three factors:

(i) The power factor P, which is the ratio of the system watts to the volt amperes.

(ii) The distortion factor D, which is the ratio of the r.m.s. fundamental input current to r.m.s. total input current.

(iii) The displacement factor L, which is the cosine of the angle between the fundamental supply voltage and the fundamental component of the input current.

The relationship between these three factors is

$$P = DL \tag{7.4}$$

Therefore for a sinusoidal input current $D = 1$ and the power factor equals the displacement factor, as expected.

The performance criterion for a cycloconverter is dependent on many different conditions. Exact analysis is difficult and is complicated further by the oscillation of the firing angle throughout the output cycle. The equations given above are based on the assumption of a relatively high input to output frequency ratio.

CYCLOCONVERTER CONTROL SYSTEMS

The block diagram of *Figure 7.30* illustrates functionally the basic requirements for the thyristor firing circuits of a cycloconverter. The reference voltage is ideally a steady d.c. level, which produces a firing angle delay of 90°, on which is superimposed a cosine voltage. The frequency of this voltage will determine the load frequency, and the amplitude of the cosine curve relative to the d.c. level fixes the modulation depth of the load voltage, and hence its magnitude. The distribute circuit feeds thyristors in both the positive and negative converter group. However, a feedback signal from the load current is now used to blank pulses to the group which is not conducting load current, so that circulating currents are prevented.

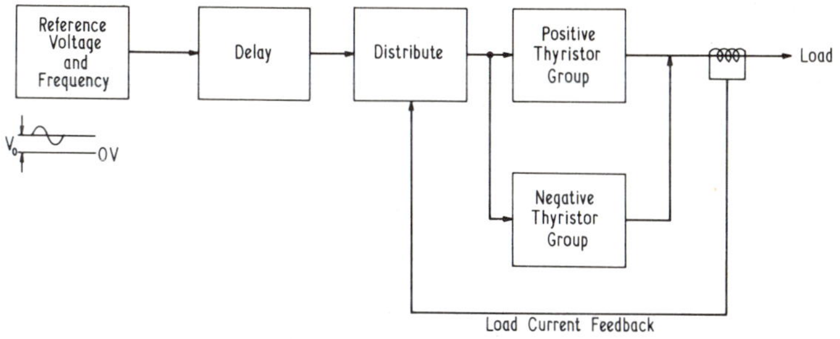

Figure 7.30. Block schematic of cycloconverter control electronics

The control circuits for cycloconverters can become very complex owing to the relatively involved logic functions that have to be performed, and the large number of thyristors that require gate signals. To construct such a drive using discrete components would result in a physically large and expensive system. Modern integrated-circuit techniques, and especially those using hybrid methods, have gone a long way towards reducing this cost, and account for the recent increase in popularity of cycloconverter frequency changers.

Monolithic integrated-circuit techniques are well known. They enable extremely complex functions to be produced which occupy a very small volume. However, these systems are restricted in power output, and to certain capacitor and resistor values. Hybrid integrated circuits may then be used, as a complement to the monolithic system, in order to overcome its shortcomings. A hybrid circuit is essentially a small inorganic 'printed circuit board' on which the interconnection tracks are produced by special techniques, and to which have been added discrete or monolithic components in the form of chips or blocks. Since the packaging from the individual discrete units may be absent, the overall circuit size can be made very small. There are basically two hybrid systems, which differ in the methods used to produce the interconnection paths between the components and some of the components themselves. These are called thick- and thin-film hybrid circuits.

In a thick-film circuit the interconnection patterns are usually produced by means of screen printing on an alumina substrate. The alumina is held on a flat support and is covered by a fine wire mesh which has all the areas blocked except those representing the interconnection pattern. Thick-film conducting ink is squeezed through the mesh of the mask onto the substrate to form the required pattern which is fired on at a high temperature. Crossovers between tracks are made by printing and firing a low-permittivity insulating glaze between conductor layers at the junction. This allows two conductors to cross over without introducing appreciable capacitance.

Thick-film techniques may also be used to produce resistors and capacitors. The resistors are screen printed between the required conductor tracks using resistor ink and the correct mask. Capacitors require a three-step process to sandwich the correct amount of high-permittivity ink between two conductor areas and are made in a similar way to crossovers. Although thick-film resistors can be made over a fairly wide range (typically 10Ω to 1MΩ per square), stable capacitors are limited to very low values. For large capacitances it is common to add small unencapsulated ceramic or tantalum capacitor chips which can be soldered onto the conducting tracks. Chip resistors too can be mounted in this manner if required.

Semiconductor (including monolithic integrated circuit) dice can be added to a hybrid system by several techniques. Dice may first be attached to the alumina by various different types of glues or by die bonding techniques. In die bonding, gold is first screen printed onto an area of the alumina substrate to which the die is to be attached. The die is then placed on the gold, and by using heat and a gentle elliptic movement of the die an alloy junction is formed between the gold-silicon interface. Once the die is fixed onto the substrate its connection pads can be connected to the required interconnection tracks with fine gold or aluminium wires, either by thermal compression or ultrasonic bonding techniques. For thermal compression methods intense heat and pressure is used to make the joint, whereas for ultrasonic bonding the wire is rubbed at a frequency of about 50 kHz against the chip or track, to break down any thin residual oxide layers present and allow fusion of the two surfaces.

Instead of using plain silicon dice, these can also be obtained from several semiconductor manufacturers already mounted in small carriers or with auxiliary bumps or flying beam leads. These alternatives may aid their use in hybrid systems. In the leadless inverted device (l.i.d.) the chip is mounted inside a small ceramic cradle structure and connections are made from the chip to the various 'legs' of this device. When inverted onto the hybrid circuit, these 'legs' can now be soldered directly onto the interconnection tracks.

Beam lead devices are now becoming more commonly used in hybrid circuits. In these the die terminals are brought out by means of flexible thin conducting metal leads which can then be connected to the tracks by thermal compression or ultrasonic methods. Beam lead resistors and capacitors are also becoming available. In some high volume applications special solder bumps are fused direct to chip or film circuit and the die is inverted (flipped) and reflow soldered.

Thin-film circuits differ from their thick-film counterparts in that the basic process of producing interconnection tracks, or resistors and capacitors, is by means of vacuum evaporation onto a glass substrate. For conductors, gold is used, resistors are usually of nickel-chromium alloy or tantalum, and the capacitor dielectric is silicon dioxide. The material to be deposited is subjected to a high temperature in vacuum so that it evaporates. The glass substrate is placed in front of this source and the correct pattern produced either by means of a mask between the two, with a cut out for the required deposition area, or by first coating the substrate all over with evaporated metal and subsequently etching patterns in a manner similar to that used in the manufacture of printed-circuit boards. Although capacitors can be produced by the same techniques as with thick-film circuits, they are limited to low values, and high capacitances, when required, are usually added on in the form of chips.

Generally thin-film circuits are more expensive to produce, require more elaborate equipment and more robust encapsulation. They do, however, give better quality components with much superior accuracy, stability and temperature performance. Usually thick films are used in high-volume applications where lower cost and ruggedness are important.

STEP-UP CYCLOCONVERTERS

The usual form of operation for a cycloconverter is in the step-down mode, where the output frequency is less than that of the input. Generally this frequency is limited to a maximum of one third the supply frequency since at lower ratios the voltage distortion becomes appreciable.

When the cycloconverter is running in its step-down mode it is naturally commutated. The leading kVA required to turn off conducting thyristors is derived from the higher-frequency side, which in this case is the a.c. supply. There is, of course, no reason why a cycloconverter cannot run with an output frequency greater than that of the input. It is now called a step-up cycloconverter, or a cycloinverter. In this instance the turn-off energy required by the conducting thyristors must again be derived from the high-frequency side, *i.e.* the load. If the load has a leading power factor then this requirement will be met. If not then a capacitor must be connected across it to reproduce this condition artificially.

Figure 7.31 shows one form of commutation which may be used in a step-up cycloconverter. Comparison with the chopper circuits of Chapter 5 will show that this is a parallel-capacitor commutation system. It is not the only method that may be used, and Chapter 9

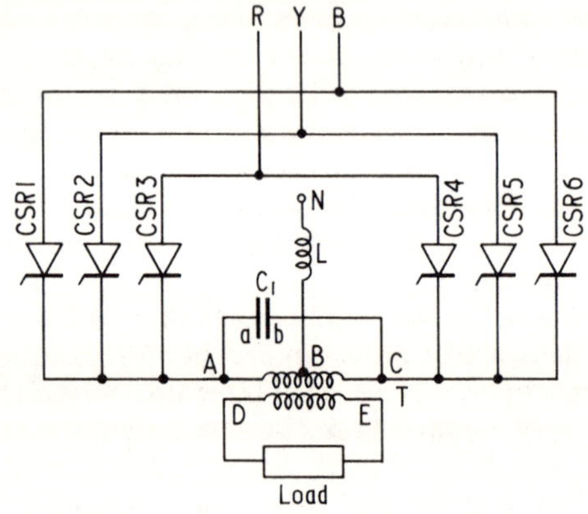

Figure 7.31. Three-pulse push-pull step-up cycloconverter

shows various other inverter circuits, most of which can be used here. The principle of the system is that the thyristors treat the instantaneous value of the a.c. supply voltage as a d.c. base from which the commutation voltage is derived. The overall system is push-pull where the load is supplied from a centre-tapped transformer T. When CSR1, CSR2 or CSR3 conduct, load current flows from one of the input lines to the neutral (assuming that their instantaneous value is positive) and the supply voltage is impressed across AB. This makes the secondary side D positive to E. When CSR4, CSR5 or CSR6 conduct the voltage across the primary and secondary are reversed. Therefore the load sees an alternating voltage, of a magnitude equal to that of the supply modified by the turns ratio between half the primary and the secondary of transformer T. As an example suppose CSR3 is conducting. The load voltage is now proportional to the instantaneous value of supply phase R. At the same time side A of transformer T is raised positive by twice this value with respect to side C since B is the centre point.

Therefore C_1 charges to twice the instantaneous supply voltage with plate *a* positive to *b*. To reverse the load voltage it is necessary to turn off CSR3 and fire, say, CSR4. CSR3 will not go off naturally till the end of the half cycle. However, if CSR4 is fired then plate *b* of C_1 is raised to the same voltage as that of the anode of CSR3. This places a reverse voltage (that on C_1), equal to twice that of the instantaneous a.c. supply at this instant, directly across CSR3, and it turns off. Load current is now supplied by CSR4 and the load voltage has been reversed. Clearly C_1 need not be connected across the transformer primary. The same effect would have been obtained with this on the secondary side. Generally step-up cycloconverters are not frequently used since they do not give any significant advantage over forced-commutated inverter systems.

CONCLUSIONS

The main differences between a cycloconverter and an inverter can be summarised here.

(i) An inverter uses fewer thyristors than a cycloconverter and is usually simpler in construction, even when the commutation components are considered. This usually results in a cheaper system. However, in very large current drives the thyristors used for a cycloconverter would be smaller than in a comparable inverter circuit, since the devices now have a lower duty cycle. Therefore the cost advantage would be slightly offset. In low-power circuits the voltage rating principally determines the cost of thyristors, so that this no longer applies. It must, however, be remembered that inverter thyristors need to be specially selected for turn-off time and dv/dt which results in an approximate 10% increase in their price. The thyristor firing circuits for inverters are also simpler than for cycloconverters, and this results in a further reduction in system cost.

(ii) An inverter can work from a d.c. battery or from an a.c. supply which is first rectified and smoothed by relatively cheap components. A cycloconverter, on the other hand, can only work from an a.c. input. Therefore if a d.c. supply is the only available source it is necessary first to convert this to a.c. before it can feed the cycloconverter. Such a d.c. to a.c. inverter can add appreciably to the original cycloconverter cost even when rotating converters are used. An inverter is therefore much more versatile than a cycloconverter and is the system most frequently used.

(iii) Both inverters and cycloconverters can be used to regenerate power from the load to the supply. However, whereas an inverter requires a fairly complex control system to do this, regeneration occurs as part of the natural process in a cycloconverter. Regeneration is frequently required in high-power motor drive systems with frequent stops and starts. In this application a cycloconverter has the obvious edge over an inverter.

(iv) Both inverters and cycloconverters (three-phase) can provide a stepless output frequency variation. For an inverter this can vary from any value below to any value above the base frequency. For cycloconverters, on the other hand, the output frequency is usually limited to about one third that of the input, and at any rate to below that of the supply unless forced commutation is used. The load-current waveforms from inverters can be made to resemble those of cycloconverters by using pulse-width voltage control techniques.

(v) The inverter has an intermediate d.c. store so that when operating from an a.c. supply the power factor imposed on this source is high, irrespective of that of the load. A cycloconverter does not have any equivalent storage capability so that the load power factor is reflected directly to the supply. However, by the very nature of the phase control involved the power factor seen by the supply is always lagging. Therefore a 60° leading or 60° lagging load angle would both produce an approximate 60° lagging supply power factor angle.

Generally an inverter is used where a wide frequency variation is required, and cycloconverters are preferred when most of the output requirements are at low frequencies.

8

The design of choppers and inverters

Choppers and inverters have already been introduced in previous chapters. It was seen that these devices are very similar since both operate from a d.c. source and need forced commutation for the thyristors. They differ in one important aspect: the output of a chopper is d.c. but that from an inverter is a.c. It is usual, therefore, to think of output and component ratings for a chopper in terms of mean values, while employing r.m.s. quantities for an inverter.

This chapter gives the analysis of chopper and inverter circuits to enable their design characteristics to be readily obtained. Initially, the commutation components will be ignored, so that the results are equally applicable to any form of switching control, for example those using transistors. Then the effects of commutation, on a typical chopper and inverter, are considered.

CHOPPERS WITHOUT COMMUTATION

Figure 8.1 shows an equivalent circuit of a chopper operating into a load of voltage V_F. L, R and C form filter components and D1 is a free-wheeling diode, which carries the inductive load current during the off period of thyristor CSR1. *Figure 8.2* gives this steady-state current waveform. Time t_c corresponds to the period for which CSR1 is on and t_0 for that when it is off. The following assumptions are made:

(1) The thyristors and diodes have zero voltage drop when conducting.

(2) The devices have infinite resistance when non-conducting.

(3) Turn-on time of the thyristors is short compared to the switching period. Therefore switching losses can be neglected.

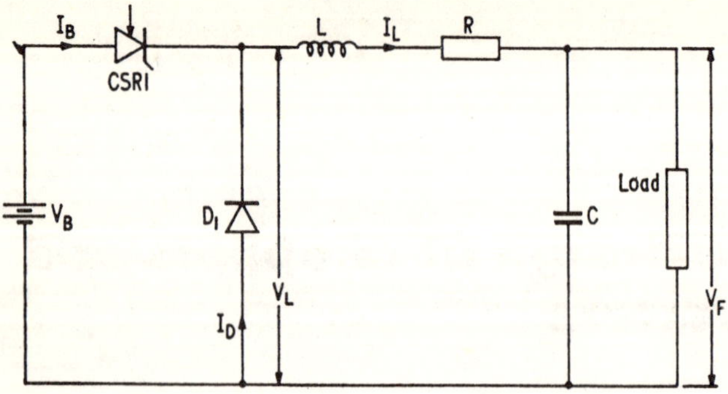

Figure 8.1. Equivalent circuit of a chopper

(4) The d.c. source impedance is negligible so that energy can flow in either direction through it without affecting terminal voltage.

(5) V_F is constant during a cycle of operation.

(6) Load current is continuous.

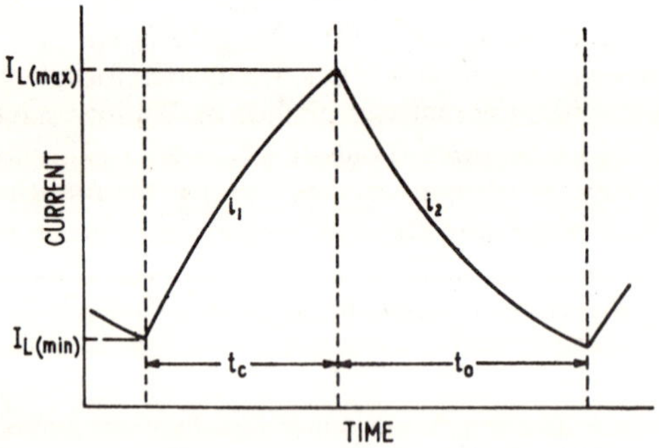

Figure 8.2. Steady-state load-current waveform of a chopper

With these assumptions the values of $I_{L(max)}$ and $I_{L(min)}$ can be written down, from *Figure 8.2*, as:

$$\frac{I_{L(min)}}{V_B/R}+\frac{V_F}{V_B} = \exp\left(-\frac{R}{L}t_0\right)\left[1-\exp\left(-\frac{R}{L}t_c\right)\right]\left[1-\exp\left(-\frac{R}{L}(t_c+t_0)\right)\right]^{-1} \tag{8.1}$$

$$\frac{I_{L(max)}}{V_B/R}+\frac{V_F}{V_B} = \left[1-\exp\left(-\frac{R}{L}t_c\right)\right]\left[1-\exp\left(-\frac{R}{L}(t_c+t_0)\right)\right]^{-1} \tag{8.2}$$

The mean current I_B through thyristor CSR1 and I_D through diode D1 can be found as the mean of i_1 over time t_c and i_2 over t_0, respectively.

$$\frac{I_B}{V_B/R} = \frac{t_c}{t_c+t_0} - \frac{L/R}{t_c+t_0}\left[1-\exp\left(-\frac{R}{L}t_c\right)\right]\left[1-\frac{I_{L(min)}}{V_B/R}\right]$$

$$-\frac{V_F}{V_B}\left[\frac{t_c}{t_c+t_0} - \frac{L/R}{t_c+t_0}\left(1-\exp\left(-\frac{R}{L}t_c\right)\right)\right] \qquad (8.3)$$

$$\frac{I_D}{V_B/R} = \frac{L/R}{t_c+t_0}\left[1-\exp\left(-\frac{R}{L}t_0\right)\right]\left[\frac{I_{L(max)}}{V_B/R}+\frac{V_F}{V_B}\right] - \frac{V_F t_0}{V_B(t_c+t_0)} \qquad (8.4)$$

Mean load current I_L is given by:

$$I_L = \frac{I_B+I_D}{V_B/R}$$

or

$$\frac{V_L-V_F}{R}$$

Therefore

$$\frac{I_L}{V_B/R}+\frac{V_F}{V_B} = \frac{t_c}{t_c+t_0} \qquad (8.5)$$

Equations (8.2) to (8.5) are shown plotted in *Figures 8.3* to *8.6*. The ripple current, derived from equations (8.1) and (8.2) is plotted in *Figure 8.7*. These curves allow device ratings, at any operating frequency, determined by the ratio of $(t_c+t_0)/(L/R)$ to be obtained. The peak load current, $I_{L(max)}$, is that value which has to be commutated in thyristor CSR1. This varies with operating frequency and pulse-width, reaching a maximum value of $(V_B-V_F)/R$ when the thyristor is on all the time. Peak current should be as low as possible for any pulse width. Therefore it is clearly advantageous to operate at higher frequencies. *Figure 8.7* shows that the current ripple is then also minimal and from *Figure 8.4* the thyristor mean current rating is reduced. Although a penalty is paid in terms of increased diode current, a diode is cheaper than an equivalent rated thyristor so that this is acceptable. It is interesting to note that ripple current is independent of the value of V_F and reaches a peak at half control width.

INVERTERS WITHOUT COMMUTATION

A bridge inverter circuit is shown in *Figure 8.8*. The operation of this circuit has been briefly explained in Chapter 7, and may be readily

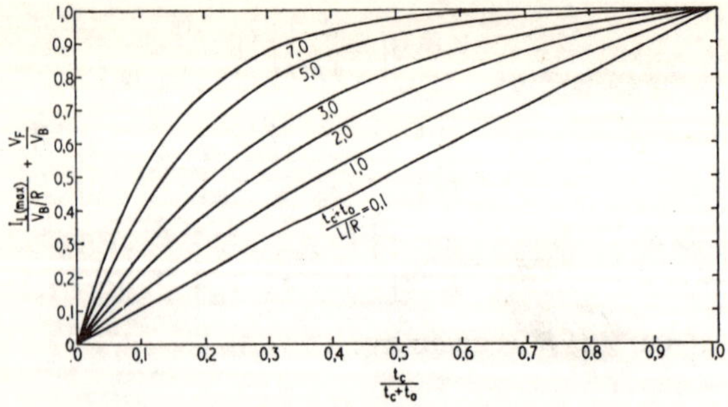

Figure 8.3. Variation of $I_{L(max)}/(V_B/R)+V_F/V_B$ with $t_c/(t_c+t_0)$

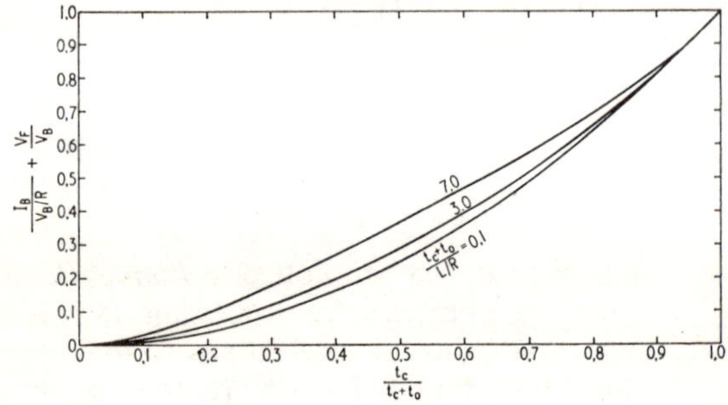

Figure 8.4. Variation of $I_B/(V_B/R)+V_F/V_B$ with $t_c/(t_c+t_0)$

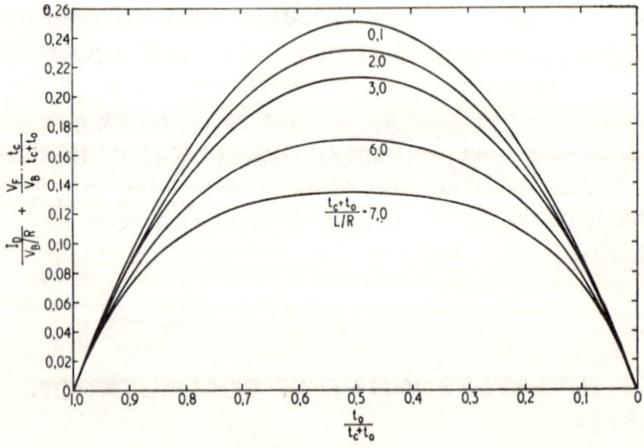

Figure 8.5. Variation of $I_D/(V_B/R)+(V_F/V_B)[t_c/(t_c+t_0)]$ with $t_0/(t_c+t_0)$

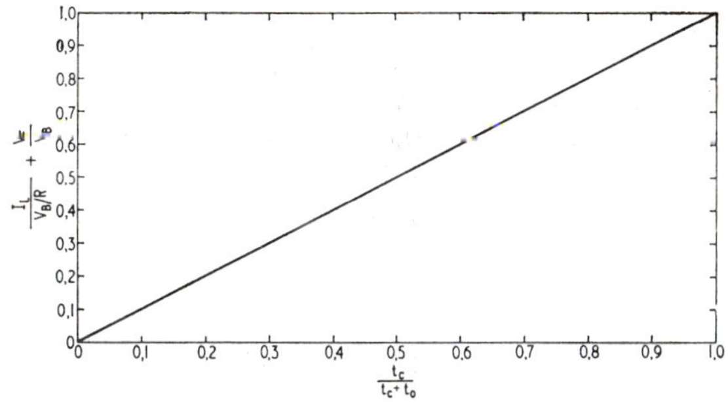

Figure 8.6. Variation of $I_L/(V_B/R) + V_F/V_B$ *with* $t_c/(t_c + t_0)$

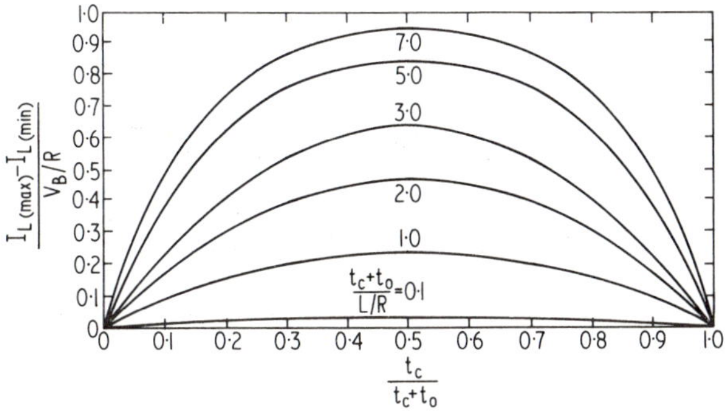

Figure 8.7. Variation of $(I_{L(max)} - I_{L(min)})/(V_B/R)$ *with* $t_c/(t_c + t_0)$

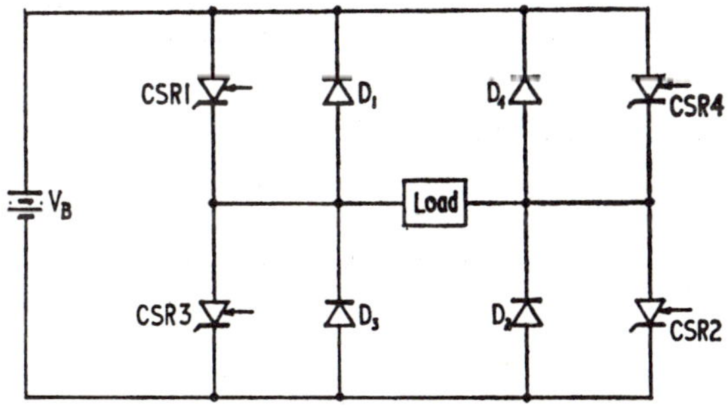

Figure 8.8. Basic bridge inverter circuit without commutation components

followed by reference to the circuit waveforms shown in *Figure 8.9*. As expected the battery current shows periods of regeneration, which is a maximum when there is no zero dwell period in the load-voltage waveform. It also increases as the power factor of the load decreases, and for a purely inductive load the current shows equal areas above and below the zero line, indicating that no net power is taken from the supply.

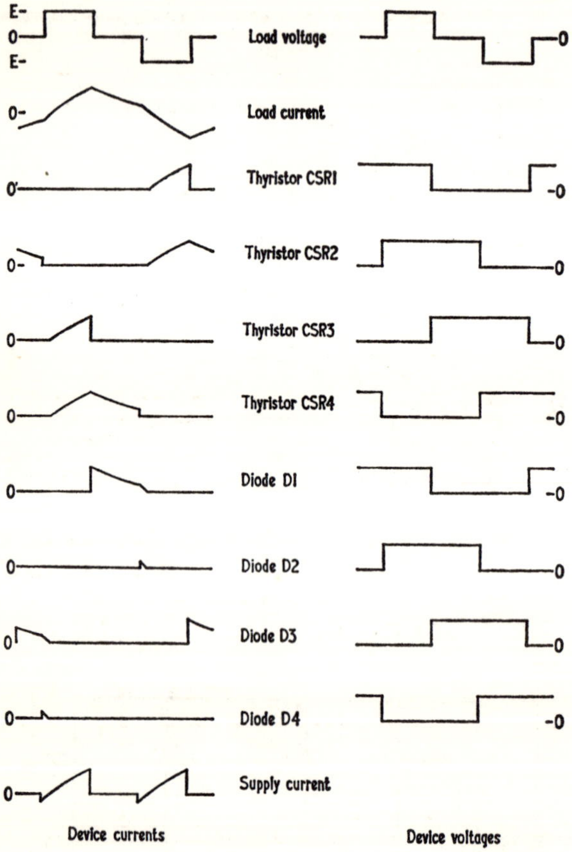

Figure 8.9. Circuit waveforms of the inverter shown in Figure 8.8

The load current and voltage are drawn to an enlarged scale in *Figure 8.10*. This also indicates the device conducting periods. Using the same assumptions as made for the chopper, the characteristics of the inverter can be determined as before. If $T = t_c + t_0$ is the chopping period, then:

$$\frac{I_{L(max)}}{V_B/R} = \frac{1-\exp\left[-\frac{R}{L}\left(\frac{T}{2}-2D\right)\right]}{1+\exp\left[-\frac{R}{L}\frac{T}{2}\right]} \tag{8.6}$$

The r.m.s. current rating of the devices is obtained by considering the current and voltages indicated in *Figure 8.10*. It is important to note that one arm of the bridge carries a larger current than the other,

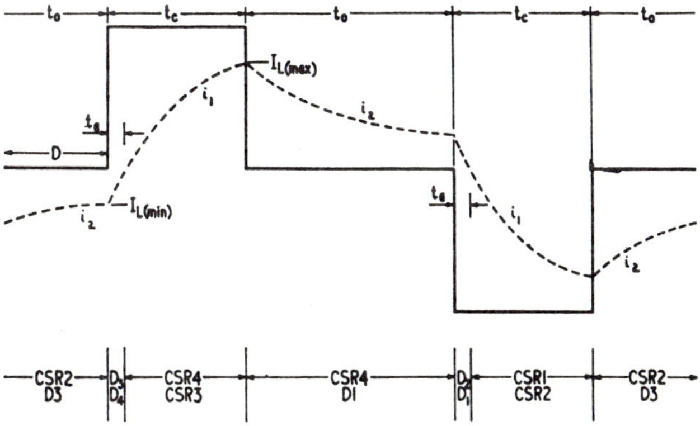

Figure 8.10. Inverter load voltage and current per cycle

and the 'worst case' should be considered. Therefore the rating of the thyristors can be taken as i_1 over $t_c - t_a$ or as the sum of i_1 over $t_c - t_a$ and i_2 over t_2. The latter gives the 'worst case'.

The values of thyristor (I_T), diode (I_D) and load (I_{LR}) r.m.s. currents are given by

$$\frac{I_T}{V_B/R} = \left[\left(\frac{I_{L(max)}}{V_B/R}\right)^2 \frac{L/R}{2T}\left\{1-\exp\left[-\frac{T}{L/R}\cdot\frac{4D}{T}\right]\right\}+\left(\frac{I_{TO}}{V_B/R}\right)^2\right]^{1/2} \tag{8.7}$$

where

$$\frac{I_{TO}}{V_B/R} = \left[\frac{L/R}{T}\left\{\frac{T/2-2D}{L/R} - \log P + 2P\exp\left(-\frac{T/2-2D}{L/R}\right) - \frac{P^2}{2}\exp\left(-\frac{T-4D}{L/R}\right)-\frac{3}{2}\right\}\right]^{1/2}$$

and

$$P = 1+\frac{I_{L(max)}}{V_B/R}\exp\left(-\frac{Rt_0}{L}\right)$$

$$\frac{I_D}{V_B/R} = \left[\left(\frac{I_T}{V_B/R}\right)^2 - \left(\frac{I_{TO}}{V_B/R}\right)^2 + \frac{L/R}{T}\left\{\log P - 2P + \frac{P^2}{2} + \frac{3}{2}\right\}\right]^{1/2} \quad (8.8)$$

$$\frac{I_{LR}}{V_B/R} = \left[2\left\{\left(\frac{I_D}{V_B/R}\right)^2 + \left(\frac{I_{TO}}{V_B/R}\right)^2\right\}\right]^{1/2} \quad (8.9)$$

Although equations (8.7) and (8.9) could be plotted, it would be more informative to show the current variations, with fundamental output voltage, as a ratio of the load current. The quasi-square waveform of *Figure 8.10* is rich in harmonics. *Figure 8.11* gives the

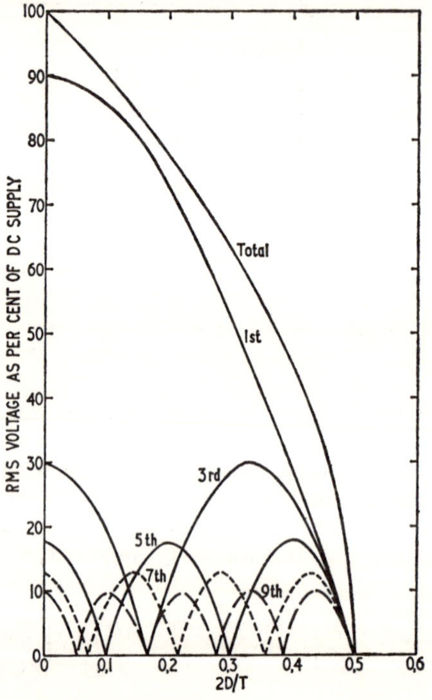

Figure 8.11. Harmonic content of quasi-square wave

variation of the first nine harmonics with the mark-space ratio. Of special interest is the fundamental waveform which is seen to reach 90% of the d.c. value when the waveform has no zero dwell periods. Using this curve the graphs shown in *Figures 8.12, 8.13* and *8.14* can be derived. These are extremely useful plots, and allow the peak current and r.m.s. thyristor or diode current to be determined, at any load voltage and r.m.s. current, for a given operating frequency.

Figure 8.12 shows the large increase in commutation current requirements at low frequencies and voltages, illustrating the unsuitability of this control method for low mark-space operation. The thyristor r.m.s. current is almost constant at $1/\sqrt{2}$ times the load current indicating that the devices conduct for approximately half the load current period. The diode ratings also tend to this value at low voltages.

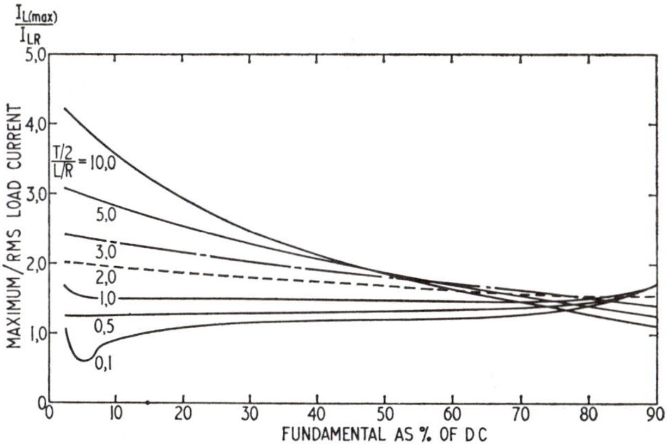

Figure 8.12. Variation of (peak load/r.m.s. load) current with fundamental load voltage

The equations derived so far have been on the assumption of zero inverter loss. This being so there is power balance at every stage. If $\cos\phi$ be the power-factor of the load, I_{BM} the mean battery current,

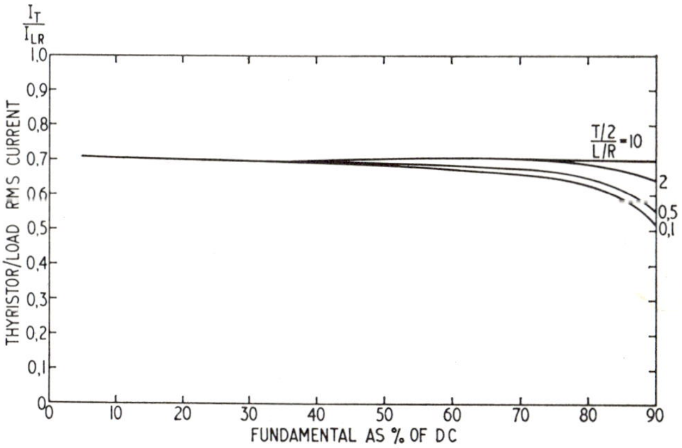

Figure 8.13. Variation of (r.m.s. thyristor/r.m.s. load) current with fundamental load voltage

and V_{LR} the total r.m.s. load voltage including harmonics, then

$$I_{LR} V_{LR} \cos\phi = I_{BM} V_B$$
$$\therefore \cos\phi = \frac{I_{BM}}{I_{LR}} \frac{V_B}{V_{LR}} \tag{8.10}$$

Equation (8.10) is plotted in *Figure 8.15*. It may seem odd at first to notice that power factor at any frequency is not fixed although both L and R are assumed to be so. But it must be remembered that the applied voltage is not sinusoidal so that the load reactance is not $2\pi f L$. Each harmonic must be treated individually and the harmonic composition of the wave changes with pulse width control. As

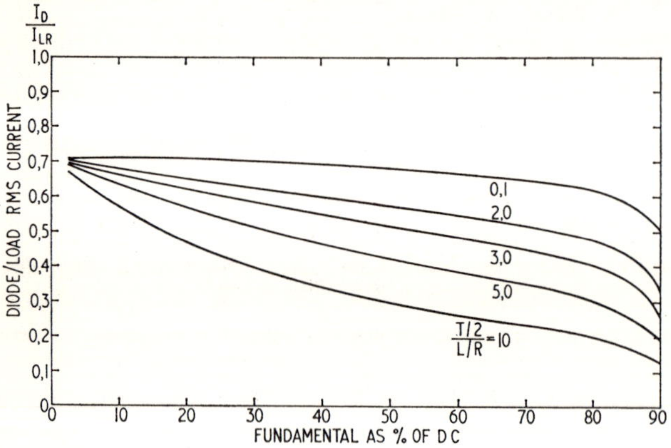

Figure 8.14. Variation of (r.m.s. diode/r.m.s. load) current with fundamental load voltage

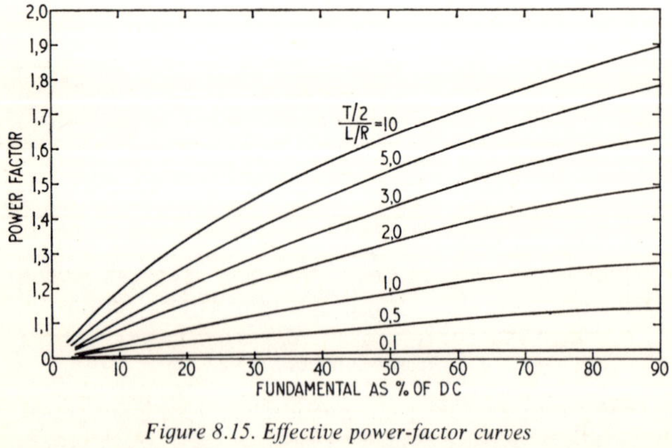

Figure 8.15. Effective power-factor curves

expected there is an increase in power factor as frequency is reduced and load resistance has a greater effect. The curves converge at low voltages where the harmonics become a greater percentage of the fundamental. This also accounts for the decrease, with reduction in pulse width, at any frequency.

THE EFFECT OF COMMUTATION

In this section a typical chopper and inverter are considered along with their commutation components. A method of specifying the device ratings and the value of the commutation inductor and capacitor will also be studied. Although these will differ according to the circuit adopted, the calculation procedure remains unchanged.

Figure 8.16 shows a popular chopper circuit configuration. CSR2 is fired initially. This charges C to V_B with plate b positive. CSR1 now conducts at the start of the load cycle. Simultaneously CSR3 is fired. This allows C to resonate with L_1 and recharge to $L_F V_B$ with plate a positive, where L_F is the loss factor of the resonant circuit. To turn CSR1 off, CSR2 is fired. C now discharges through the load (at constant current if the load inductance be assumed large enough). Until the capacitor voltage falls to zero CSR1 is reverse-biased, and

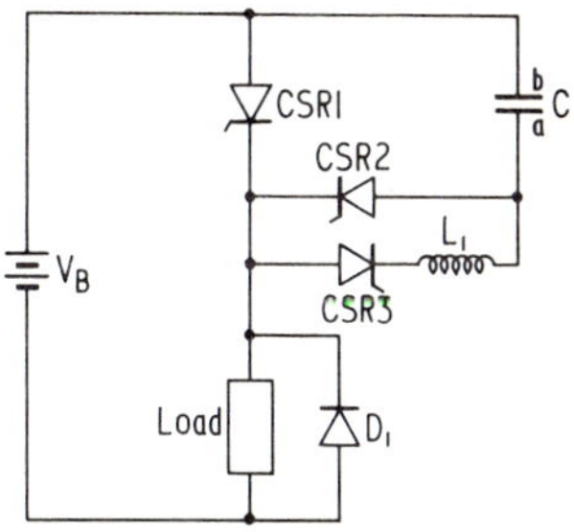

Figure 8.16. Typical parallel-capacitor commutated thyristor chopper circuit

the time from the instant of firing CSR2 is known as the commutation interval. It should exceed the rated turn-off time for CSR1. Thereafter C continues charging via CSR2 and the load until it reaches voltage V_B with plate b positive. CSR2 now turns off, and load current free-wheels in D1 until CSR1 is again fired to commence the next cycle. The circuit action is in fact similar to that of *Figure 5.6*, except that the rating of CSR1 is increased by the resonant capacitor current.

The mean current rating of diode D1 is clearly still given by equation (8.4), its peak rating being $I_{L(max)}$, the current just before

CSR1 turns off (assuming the commutation interval to be short relative to the chopper operating period). The diode voltage rating should exceed $2V_B$ which is obtained when $L_F = 1$. Mean current rating of thyristor CSR1 is I_B, as given by equation (8.3), plus the resonant current due to C and L_1. This gives a total current

$$I_{CSR1(mean)} = I_B + \frac{2CV_B}{t_c + t_0} \tag{8.11}$$

The peak current is either $I_{L(max)}$ or the resonant value of $V_B\sqrt{(C/L_1)}$ whichever is greater. The voltage rating must exceed V_B. Assuming C to discharge at constant load current, CSR1 must have a turn-off time shorter than $CV_B/I_{L(max)}$.

Thyristor CSR2 carries a current of $I_{L(max)}$ during the charge period of C only. Its mean current rating is $2CV_B/(t_c+t_0)$ although its size is normally determined by its peak current capability of $I_{L(max)}$. Similarly thyristor CSR3 has a mean rating of $2CV_B/(t_c+t_0)$ but a peak resonant current of $V_B\sqrt{(C/L_1)}$ which can be high. The value of C is fixed by the commutation requirements and L_1 is chosen such that the resonant time $\pi\sqrt{(L_1C)}$ is small compared to the operating frequency (between 5 and 10%).

Perhaps the most popular inverter configuration is the McMurray-Bedford circuit shown in *Figure 8.17*. Although *Figure 8.9* still represents the general circuit waveforms, the commutation interval has

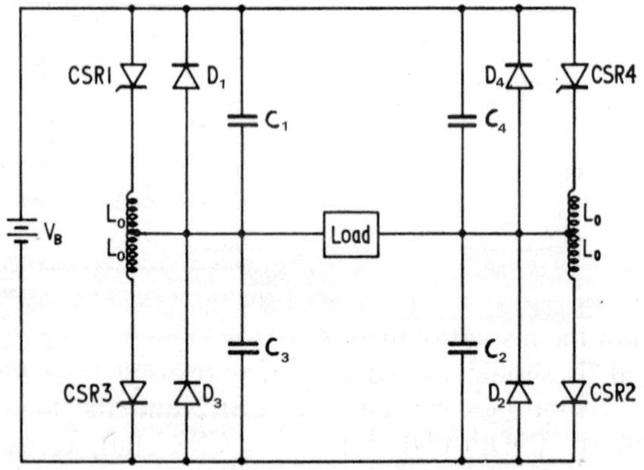

Figure 8.17. McMurray-Bedford bridge inverter circuit

been assumed too small to be distinguishable. *Figure 8.18* shows the instant of commutation for CSR3 to an enlarged time scale. Assume CSR3 and CSR4 to be conducting, the load current being at its peak value of $I_{L(max)}$. C_3 and C_4 are at zero voltage (device volt-drops and d.c. resistance of centre-tapped chokes L_0L_0 being neglected). CSR1 is now fired. If the leakage inductance of the chokes be ignored, current $I_{L(max)}$ transfers instantaneously from CSR3 to CSR1, capacitor C_1 discharging to support both this and the load current.

The current through CSR1 increases, reaching a peak of I_{pk} after time t_1 when C_1 has completely discharged. Energy stored in L_0 now free-wheels through diode D1 and is dissipated, falling to zero after a further time t_2. Load current is carried by CSR4 and D1.

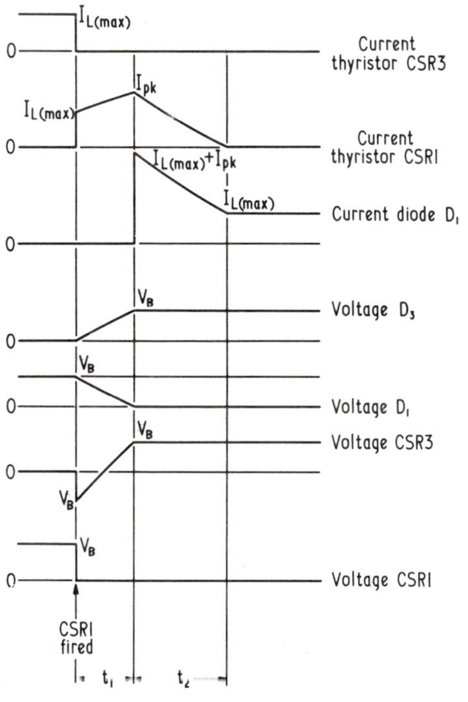

Figure 8.18. Circuit waveforms for Figure 8.17 during the commutation of thyristor CSR3

When CSR2 is fired at a later interval to commutate CSR4 the load current has decayed to $I_{L(min)}$. Therefore commutation requirements and increased device ratings are not as severe as before. In the 'worst' case, when there are no zero dwell periods in the output voltage waveform, the ratings of all devices are equal. This will be considered here.

If t_F denotes the turn-off time seen by the thyristor being commutated, and W_L is the watts loss per commutation, caused by the dissipation of the energy $\frac{1}{2}L_0(I_{pk})^2$ stored in L_0 then *Figure 8.19* shows

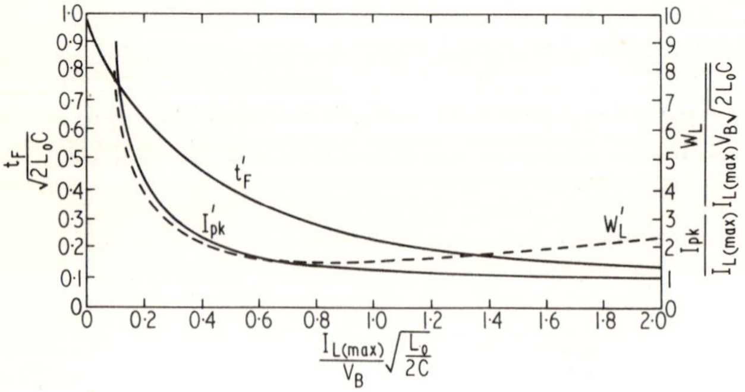

Figure 8.19. Variation of $t_F/\sqrt{(2L_0C)}$, $I_{pk}/I_{L(max)}$ and $W_L \times 2/I_{L(max)}V_B\sqrt{(2L_0C)}$ with $I_{L(max)}\sqrt{(L_0/2C)}/V_B$

the variation of turn-off, peak current and watts-loss factors with commutation factor

$$\frac{I_{L(max)}}{V_B}\left[\frac{L_0}{2C}\right]^{1/2}$$

From this graph it is seen that watts loss is minimum at a commutation factor of 0·8, the variation between 0·6 and 1·0 being slight. Below 0·6 the peak current and commutation loss increases steeply, although available turn-off time also increases. Working on the minimum loss point the value of L_0 and C can be found as:

$$L_0 = \frac{2{\cdot}76V_Bt_F}{I_{L(max)}} \tag{8.12}$$

$$C = \frac{2{\cdot}15t_FI_{L(max)}}{V_B} \tag{8.13}$$

The contribution to device ratings by the commutation interval is directly dependent on time t_2 in *Figure 8.18*. This is the period required for the current in CSR1 to decay from I_{pk} to zero due to losses across CSR1 and D1 (assumed a constant and equal to ΔV), and due to the 'effective loss' resistance of the CSR1, L_0, D1 loop (equal to R_e). To reduce this time to a minimum means introducing an external resistance in series with D1. This gives greater losses during normal operation, with a subsequent reduction in efficiency. There is no one acceptable solution for all cases. Some inverters may be using devices which are overrated, so that higher r.m.s. currents can be accepted. In others efficiency may not be important so losses across a larger R_e are tolerated. In any case it is always important to ensure that t_2 is less than half a cycle of inverter operation, to prevent commutation failures.

If I_{D_1} and I_{T_1} represent the commutation current, in r.m.s. values, through the diode and thyristor respectively, and T is the inverter periodic time, *Figures 8.20* and *8.21* show plots of $I_{D_1}\sqrt{T}/I_{L(max)}$ and

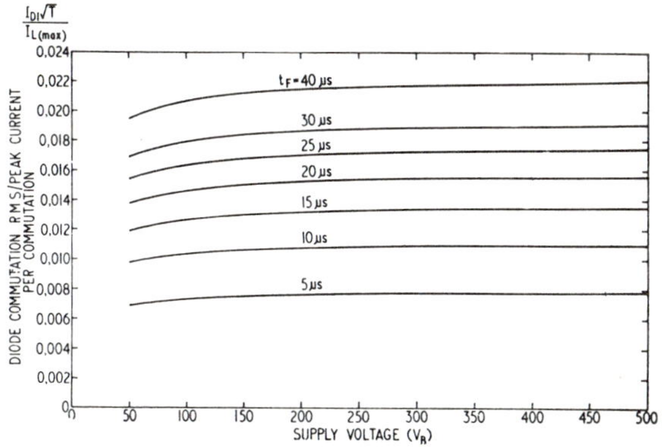

Figure 8.20. Variation of $I_{D_1}\sqrt{T}/I_{L(max)}$ with supply voltage

$I_{T_1}\sqrt{T}/I_{L(max)}$ against supply voltage. These are calculated for values of L_0 and C given by equations (8.12) and (8.13) and for $\Delta V = 2{\cdot}5$V and $R_e = 0{\cdot}2V_B/I_{L(max)}$. These graphs show the advantage of using fast turn off time devices. The total thyristor and diode ratings are

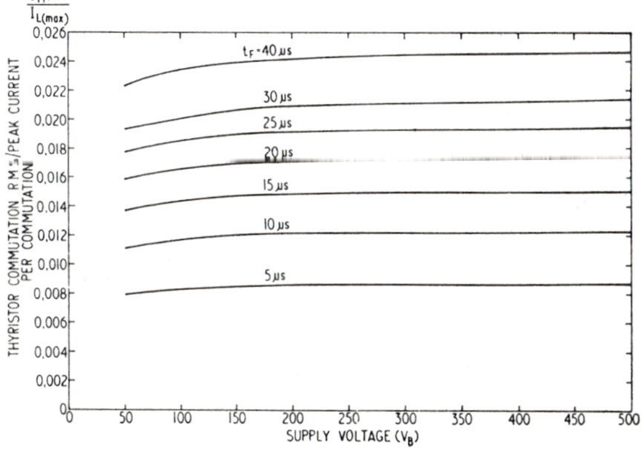

Figure 8.21. Variation of $I_{T_1}\sqrt{T}/I_{L(max)}$ with supply voltage

given by the geometric sum of values read off from *Figures 8.13, 8.14, 8.20* and *8.21*. Therefore when operating at an inverter frequency of 10Hz, $I_{T_1}/I_{L(max)}$ is nearly constant at $0{\cdot}0248/\sqrt{0{\cdot}1} = 0{\cdot}079$ whereas *Figure 8.13* gives the thyristor current without commutation as $0{\cdot}35\ I_{L(max)}$. The total current is also approximately this value. For 1kHz operation, however, $I_{T_1} = 0{\cdot}79\ I_{L\,(max)}$ and *Figure 8.14* still gives the same value as before so that total thyristor r.m.s. current with commutation is $\sqrt{(0{\cdot}79^2 + 0{\cdot}35^2)}\ I_{L(max)} = 0{\cdot}745\ I_{L\,(max)}$. Therefore this inverter is not suitable for higher-frequency operation. The same conclusions can be derived from *Figure 8.22* which gives the

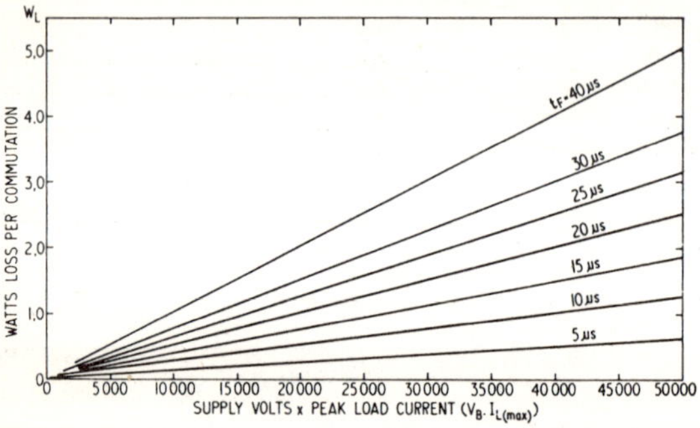

Figure 8.22. Power loss per commutation

watts loss per commutation. Now for 40μs devices and $V_B I_{L\,(max)} = 50\,000$, total energy loss per second at 10Hz is 200W whereas at 1kHz it is 20kW.

CONCLUSIONS

When designing forced-commutated thyristor circuits two factors need be considered, (i) the rating of the devices, (ii) the peak load current. The device ratings vary considerably with the operating frequency of the inverter relative to the load time constant, and with the mark-space control which determines the magnitude of the output voltage. In both choppers and inverters there are several advantages to being able to operate at a high frequency. The importance of peak load current is often overlooked. This is the current which has to be commutated in the main thyristors and it directly determines the size of the commutation components. A system which gives the smallest ratio of peak/r.m.s. load current is clearly desirable.

9

Inversion

The basic principles of an inverter have been covered in Chapter 1. It was seen that an inverter resembles a chopper in requiring forced commutation for its thyristors. It differs in the important aspect that the output from an inverter is a.c. not d.c.

This chapter looks at inverters with a view to classifying the multitude of systems that exist. A classification system based on the commutation method was selected for choppers in Chapter 5. A similar principle can be used for inverters with two additions. First, inverters fall into two major groups, *i.e.* (i) push-pull inverters, where the load must be centre-tapped, with its two halves magnetically coupled to each other, (ii) bridge inverters, in which a centre-tapped load is not essential. There was no such distinction in chopper circuits. Secondly, most inverter circuits, whether bridge or push-pull and irrespective of the commutation system used, can be operated in several voltage control modes. Apart from varying the value of the fundamental a.c. voltage this also gives a measure of control over harmonics in the waveform. Although choppers were stated as having two control modes, variable frequency and fixed frequency variable mark-space, both operate on the principle of giving unidirectional output voltage pulses, the mark to space ratio of which is controlled. They cannot be grouped according to a voltage control method, as can be done with inverters.

In the following sections inverters will be classified into two main groups, push-pull and bridge. Each group will be subdivided according to the output voltage control method used, and each sub-group will be further divided into the various thyristor commutation methods.

THE GOODNESS FACTOR

Several factors were selected to enable a comparison to be made between the various chopper systems. This section gives seven points to be considered when looking at inverters.

(1) Is it essential that the load be centre tapped, with the two halves magnetically coupled? Such a push-pull inverter severely limits the type of applications for which it may be used.

(2) What is the harmonic content of the output waveform? Sine wave inverters, with high output harmonics, will require large and bulky filters. Even if the output is not required to be sinusoidal a low harmonic content is often advantageous. For instance, in an induction motor it is the fundamental component of the waveform which produces the useful torque, whereas the harmonics result in losses.

(3) What is the complexity of the inverter and control electronics required to produce the output voltage waveform?

(4) Can the inverter operate at high frequencies with relatively high efficiency?

(5) What are the maximum and minimum values of the fundamental r.m.s. output voltage?

(6) Is the commutation voltage increased in proportion to the load being commutated? This is desirable to allow optimisation of commutation components.

(7) Is the current rating of the main thyristor increased by the commutation capacitor reset pulse?

Two other factors were considered when dealing with choppers, namely:

(i) In the advent of a commutation failure will commutation be retried, and be successful?

(ii) Does a low impedance fault current path exist across the supply?

However, for an inverter a commutation failure almost always results in a low impedance path across the source, which is protected by fast-acting fuses. Therefore commutation of the thyristors cannot be retried.

The above factors are considered again later with reference to typical systems.

PUSH-PULL AND BRIDGE INVERTERS

All inverter circuits fall into two broad groups, push-pull and bridge. A push-pull inverter is shown in *Figure 9.1*, the commutation components for the thyristors being omitted. When CSR1 is turned on, end A of the output transformer T_1 goes negative so that end D of the load is made negative. When CSR1 is turned off and CSR2 fired the load polarity reverses. Diodes D1 and D2 carry the inductive load current when the thyristors are first turned on. Clearly T_1 is not always essential. It may be possible to make the load itself centre-tapped. For instance ABC could be the stator winding of a single-phase induction motor. In either case it is important to note that when CSR1 conducts a voltage V_B is impressed across AB. If the two halves of the load are closely coupled, this raises the anode of CSR2 to a potential $2V_B$ above its cathode. Similarly CSR1 and D1 must have a voltage rating of at least $2V_B$.

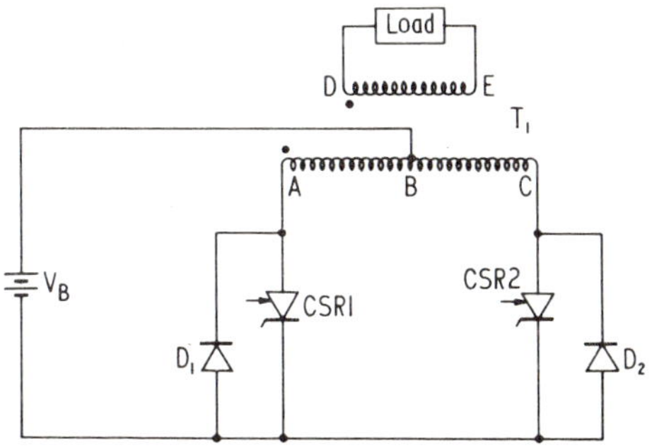

Figure 9.1. Fundamental push-pull inverter

A bridge inverter without commutation components is shown in *Figure 9.2*. It is evident that such an inverter does not need the load to be centre-tapped or connected via an intermediate centre-tapped transformer. Firing CSR1, CSR2 and CSR4, CSR3 in pairs gives an alternating load polarity. D1 to D4 carry the inductive load

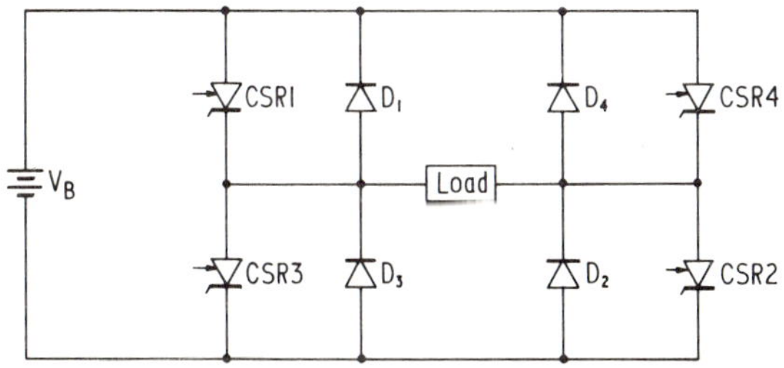

Figure 9.2. Basic bridge inverter

current. Although there are twice as many components in a bridge inverter than in a push-pull inverter, the devices now need only be rated to a voltage V_B.

Generally the choice between bridge and push-pull inverters is not difficult. If the load is not centre-tapped, or does not require a transformer, bridge inverters should be used. If the load is transformer coupled bridge inverters are used at high supply voltages, say above 200V, and push-pull inverters at lower voltages, which would not necessitate using unduly high voltage rated and expensive devices.

VOLTAGE CONTROL IN INVERTERS

There are two instances when voltage control within an inverter is required:

(*a*) When the output is to be kept at a fixed value in spite of regulation within the inverter, or fluctuations of the supply voltage.

(*b*) When the output is to be varied in a given manner, *e.g.* proportional to frequency for variable-frequency drives.

There are several ways in which this voltage control can be achieved. In all cases the a.c. output will be made up of a fundamental component and a band of harmonic frequencies. The various control methods all contrive to reduce the harmonic voltages while avoiding excessive circuit complexity.

Voltage control in inverter circuits is important. All too often the alternatives are not considered during the design stage, and traditional methods such as mark-space (or quasi-square-wave) are employed, even though another technique could be much more advantageous. Chapter 7 touched on this problem, where it was shown that by using high-frequency pulse-width modulated systems the output harmonics can be made very small.

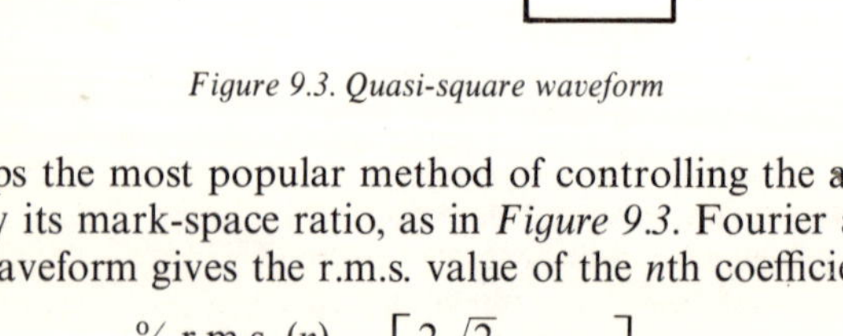

Figure 9.3. Quasi-square waveform

Perhaps the most popular method of controlling the a.c. voltage is to vary its mark-space ratio, as in *Figure 9.3*. Fourier analysis of such a waveform gives the r.m.s. value of the *n*th coefficient as

$$\frac{\%\ \text{r.m.s.}\ (n)}{V_B} = \left[\frac{2\sqrt{2}}{n\pi}\cos nD\right] \times 100 \qquad (9.1)$$

The total r.m.s. voltage of the waveform, including all harmonics, is obtained as

$$\frac{\%\ \text{r.m.s.}\ (T)}{V_B} = \left[1-\frac{2D}{\pi}\right]^{1/2} \times 100 \qquad (9.2)$$

Equations (9.1) and (9.2) are shown evaluated in Table 9.1 up to the 15th harmonic. From this table it is evident that the harmonic

Table 9.1. HARMONIC COMPOSITION OF A QUASI-SQUARE WAVE

	R.M.S. voltage as per cent of d.c. supply								
2D/T	1	3	5	7	9	11	13	15	*Total*
0·00	90·0	30·0	18·0	12·9	10·0	8·18	6·92	6·00	100
0·02	89·8	29·5	17·1	11·6	8·44	6·30	4·74	3·53	98·0
0·04	89·3	27·9	14·6	8·20	4·26	1·53	0·43	1·85	95·9
0·06	88·4	25·3	10·6	3·20	1·25	3·94	5·33	5·71	93·8
0·08	87·2	21·9	5·56	2·41	6·37	7·61	6·87	4·85	91·7
0·10	85·6	17·6	0·00	7·56	9·51	7·78	4·07	0·00	89·4
0·12	83·7	12·8	5·56	11·3	9·69	4·38	1·30	4·85	87·2
0·14	81·4	7·46	10·6	12·8	6·85	1·03	5·85	5·71	84·9
0·16	78·9	1·88	14·6	12·0	1·87	5·96	6·71	1·85	82·5
0·18	76·0	3·76	17·1	8·80	3·68	8·17	3·34	3·53	80·0
0·20	72·8	9·27	18·0	3·97	8·09	6·62	2·14	6·00	77·5
0·22	69·3	14·5	17·1	1·61	9·98	2·03	6·26	3·53	74·8
0·24	65·6	19·1	14·6	6·89	8·76	3·48	6·44	1·85	72·1
0·26	61·6	23·1	10·6	10·9	4·82	7·40	2·55	5·71	69·3
0·28	57·4	26·3	5·56	12·8	0·63	7·92	2·95	4·85	66·3
0·30	52·9	28·5	0·00	12·2	5·88	4·81	6·58	0·00	63·2
0·32	48·2	29·8	5·56	9·37	9·30	0·51	6·07	4·85	60·0
0·34	43·4	29·9	10·6	4·73	9·82	5·60	1·72	5·71	56·6
0·36	38·3	29·1	14·6	0·81	7·29	8·12	3·71	1·85	52·9
0·38	33·1	27·1	17·1	6·19	2·49	6·91	6·80	3·53	49·0
0·40	27·8	24·3	18·0	10·4	3·09	2·53	5·60	6·00	44·7
0·42	22·4	20·5	17·1	12·6	7·71	3·01	0·87	3·53	40·0
0·44	16·9	16·1	14·6	12·5	9·92	7·17	4·41	1·85	34·5
0·46	11·3	11·0	10·6	9·91	9·05	8·04	6·91	5·71	28·3
0·48	5·65	5·62	5·56	5·47	5·36	5·22	5·05	4·85	20·0
0·50	0·00	0·00	0·00	0·00	0·00	0·00	0·00	0·00	0·00

content of the output increases rapidly as the mark-space ratio of the waveform is reduced. This is illustrated more clearly by *Figure 9.4*. At low voltages various harmonics are almost equal in value to the fundamental, the total harmonic content being about ten times larger. This represents the greatest disadvantage of the quasi-square voltage control system, and normally limits the output voltage range to between 30 and 90% of the d.c. supply, *i.e.* a frequency change of 3:1. *Figure 9.3* may also be referred to as a single-pulse,

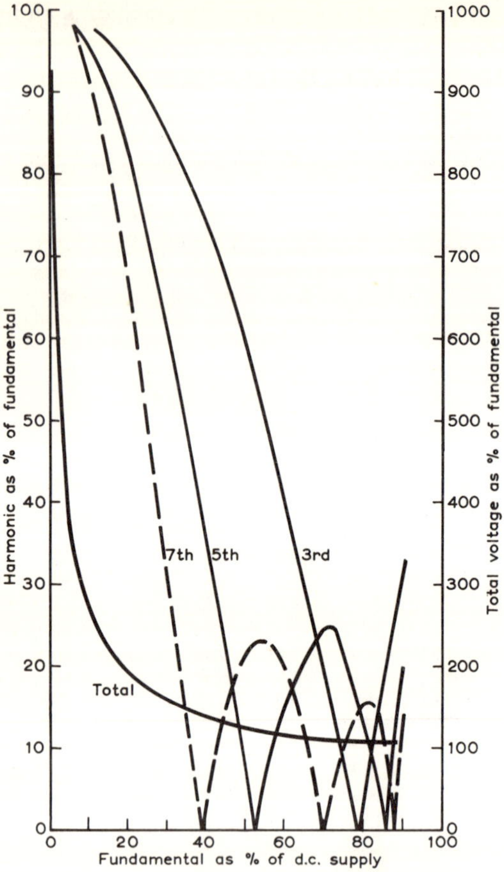

Figure 9.4. Harmonics in a quasi-square waveform

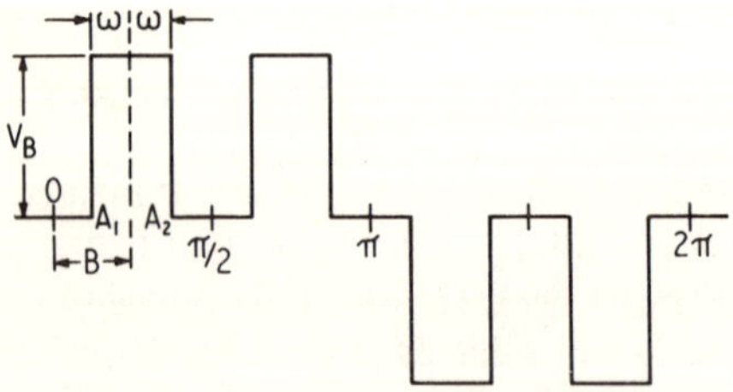

Figure 9.5. Two-pulse unidirectional wave

unidirectional wave. It is unidirectional since in any half cycle the output is either positive or negative, never both. There is also a single pulse in each half cycle. An obvious extension to this system is shown in *Figure 9.5* which illustrates a two-pulse, unidirectional waveform. Putting

$$B = \frac{A_1 + A_2}{2} \tag{9.3}$$

$$\omega = \frac{A_2 - A_1}{2} \tag{9.4}$$

the *n*th Fourier coefficient is given as

$$b_n = \frac{8V_B}{\pi n} \sin nB \sin n\omega \tag{9.5}$$

It is obviously possible to arrange the pulses symmetrically in a half cycle. This is considered in the next section which extends the technique to a large number of equally spaced pulses. In the present system the waveform will be arranged so as to eliminate, say, the *P*th harmonic from the output. Then *B* can be chosen such that

$$\sin PB = 0$$

or

$$B = \frac{\pi}{P} \tag{9.6}$$

So long as the pulses are maintained at this value of *B* their width ω can change, varying the effective voltage, but the *P*th harmonic will still be absent. From equation (9.5)

$$\frac{\%\ \text{r.m.s.}\ (n)}{V_B} = \left[\frac{4\sqrt{2}}{n\pi} \sin n\frac{\pi}{p} \sin n\omega\right] \times 100 \tag{9.7}$$

The total r.m.s. voltage is given by

$$\frac{\%\ \text{r.m.s.}\ (T)}{V_B} = 100\left[\frac{2}{\pi}(A_2 - A_1)\right]^{1/2}$$

$$= 200\left[\frac{\omega}{\pi}\right]^{1/2} \tag{9.8}$$

Equations (9.7) and (9.8) are shown evaluated in Tables 9.2, 9.3 and 9.4 for $P = 3$, 5 and 7 respectively. These show that even though lower harmonics are eliminated the total harmonic content is large. However, higher harmonics are much easier to eliminate than lower order harmonics, so that a comparison based on total content can

Table 9.2. HARMONICS IN A TWO-PULSE UNIDIRECTIONAL WAVE WITH $P = 3$

2W/T	*R.M.S. voltage as percent of d.c. supply*								
	1	3	5	7	9	11	13	15	*Total*
0·16	75·1	0·00	18·3	8·20	0·00	9·70	2·98	0·00	80·0
0·15	70·8	0·00	22·0	3·48	0·00	12·6	1·88	0·00	77·5
0·14	66·4	0·00	25·2	1·40	0·00	14·1	6·43	0·00	74·8
0·13	61·9	0·00	27·8	6·21	0·00	13·8	9·92	0·00	72·1
0·12	57·4	0·00	29·7	10·7	0·00	12·0	11·8	0·00	69·3
0·11	52·8	0·00	30·8	14·7	0·00	8·69	11·7	0·00	66·3
0·10	48·2	0·00	31·2	18·0	0·00	4·38	9·70	0·00	63·2
0·09	43·5	0·00	30·8	20·4	0·00	0·45	6·10	0·00	60·0
0·08	38·8	0·00	29·7	21·9	0·00	5·22	1·50	0·00	56·6
0·07	34·0	0·00	27·8	22·3	0·00	9·37	3·35	0·00	52·9
0·06	29·2	0·00	25·2	21·6	0·00	12·4	7·64	0·00	49·0
0·05	24·4	0·00	22·0	19·8	0·00	14·0	10·7	0·00	44·7
0·04	19·5	0·00	18·3	17·2	0·00	13·9	12·0	0·00	40·0
0·03	14·7	0·00	14·2	13·6	0·00	12·2	11·3	0·00	34·6
0·02	9·79	0·00	9·63	9·48	0·00	9·03	8·74	0·00	28·3
0·01	4·90	0·00	4·88	4·86	0·00	4·80	4·76	0·00	20·0
0·00	0·00	0·00	0·00	0·00	0·00	0·00	0·00	0·00	0·00

Table 9.3. HARMONICS IN A TWO-PULSE UNIDIRECTIONAL WAVE WITH $P = 5$

2W/T	*R.M.S. voltage as percent of d.c. supply*								
	1	3	5	7	9	11	13	15	*Total*
0·20	62·2	54·3	0·00	23·3	6·91	5·65	12·5	0·00	89·4
0·19	59·5	55·7	0·00	21·0	9·29	2·68	13·1	0·00	87·2
0·18	56·7	56·6	0·00	17·8	10·9	0·60	11·5	0·00	84·9
0·17	53·9	57·0	0·00	13·7	11·7	3·82	8·07	0·00	82·5
0·16	51·0	57·0	0·00	9·00	11·5	6·58	3·27	0·00	80·0
0·15	48·0	56·4	0·00	3·83	10·5	8·57	2·06	0·00	77·5
0·14	45·0	55·3	0·00	1·54	8·57	9·54	7·06	0·00	74·8
0·13	42·0	53·7	0·00	6·82	5·98	9·39	10·9	0·00	72·1
0·12	38·9	51·6	0·00	11·8	2·92	8·12	12·9	0·00	69·3
0·11	35·8	49·1	0·00	16·2	0·37	5·90	12·9	0·00	66·3
0·10	32·7	46·2	0·00	19·8	3·63	2·97	10·7	0·00	63·2
0·09	29·5	42·8	0·00	22·4	6·61	0·30	6·70	0·00	60·0
0·08	26·3	39·1	0·00	24·0	9·06	3·54	1·65	0·00	56·6
0·07	23·1	35·0	0·00	24·4	10·8	6·36	3·67	0·00	52·9
0·06	19·8	30·6	0·00	23·7	11·7	8·43	8·39	0·00	49·0
0·05	16·6	25·9	0·00	21·8	11·6	9·50	11·7	0·00	44·7
0·04	13·3	21·0	0·00	18·8	10·6	9·45	13·1	0·00	40·0
0·03	9·96	15·9	0·00	15·0	8·82	8·28	12·4	0·00	34·6
0·02	6·64	10·7	0·00	10·4	6·30	6·13	9·60	0·00	28·3
0·01	3·32	5·37	0·00	5·33	3·28	3·26	5·23	0·00	20·0
0·00	0·00	0·00	0·00	0·00	0·00	0·00	0·00	0·00	0·00

Table 9.4. HARMONICS IN A TWO-PULSE UNIDIRECTIONAL WAVE WITH $P = 7$

2W/T	R.M.S. voltage as percent of d.c. supply								
	1	3	5	7	9	11	13	15	*Total*
0·14	33·3	56·7	22·8	0·00	11·4	15·8	3·22	1·61	74·8
0·13	31·0	55·0	25·1	0·00	7·96	15·6	4·97	0·81	72·1
0·12	28·8	52·9	26·8	0·00	3·89	13·5	5·90	3·06	69·3
0·11	26·5	50·3	27·8	0·00	0·49	9·78	5·86	4·64	66·3
0·10	24·1	47·3	28·1	0·00	4·83	4·93	4·86	5·21	63·2
0·09	21·8	43·9	27·8	0·00	8·79	0·50	3·06	4·64	60·0
0·08	19·4	40·0	26·8	0·00	12·0	5·87	0·75	3·06	56·6
0·07	17·0	35·9	25·1	0·00	14·4	10·6	1·68	0·81	52·9
0·06	14·6	31·3	22·8	0·00	15·5	14·0	3·83	1·61	49·0
0·05	12·2	26·6	19·9	0·00	15·4	15·8	5·35	3·68	44·7
0·04	9·79	21·5	16·5	0·00	14·1	15·7	6·00	4·95	40·0
0·03	7·35	16·3	12·8	0·00	11·7	13·7	5·65	5·14	34·6
0·02	4·90	11·0	8·70	0·00	8·38	10·2	4·38	4·21	28·3
0·01	2·45	5·50	4·40	0·00	4·36	5·40	2·39	2·36	20·0
0·00	0·00	0·00	0·00	0·00	0·00	0·00	0·00	0·00	0·00

often be misleading. Perhaps the most serious disadvantage of this control method is that the maximum output voltage is limited by the necessity of not allowing adjacent pulses to merge. With quasi-square waves the maximum voltage is 90% of the d.c. supply, when the output waveshape has no zero voltage periods. In two-pulse systems the maximum fundamental obtainable is 75·1%, 62·2% or 33·3% of the d.c. according to whether the third, fifth or seventh harmonic is being eliminated. Therefore in a variable frequency drive, a 240V, 50Hz motor could run at a maximum frequency of $340 \times 0{\cdot}333 \times 50/240 = 24$Hz and still remain fully fluxed ($340 = 240 \times \sqrt{2}$, *i.e.* the peak rectified mains voltage) when the seventh harmonic is eliminated.

A second extension to quasi-square control is to have several equally spaced unidirectional pulses in a half cycle. This is conveniently obtained by feeding a high-frequency carrier triangular wave and a low-frequency reference square wave into a comparator as in *Figure 9.6(a)*. The output voltage swings between V_B and zero volts according to the relative magnitudes of the two waveforms, as in *Figure 9.6(b)*. The width of the output pulses is determined by the ratio of A_S/A_T and there are as many pulses per cycle as the ratio f_T/f_S. The Fourier coefficient is given by

$$\frac{\%\ \text{r.m.s.}\ (n)}{V_B} = \sum_{M=0,2,4\ldots} \frac{2\sqrt{2}}{n\pi}[\cos nx(M) - \cos nx(M+1)] \quad (9.9)$$

and

$$\frac{\%\ \text{r.m.s.}\ (T)}{V_B} = 100\left[\frac{f_T}{2\pi f_S}\{x(0)-x(1)\}\right]^{1/2} \tag{9.10}$$

Equations (9.9) and (9.10) have been evaluated in Tables 9.5, 9.6, 9.7 and 9.8 for $f_T/f_S = 4$, 6, 10 and 20 respectively. These tables indicate an increase in total harmonic content with pulse numbers where

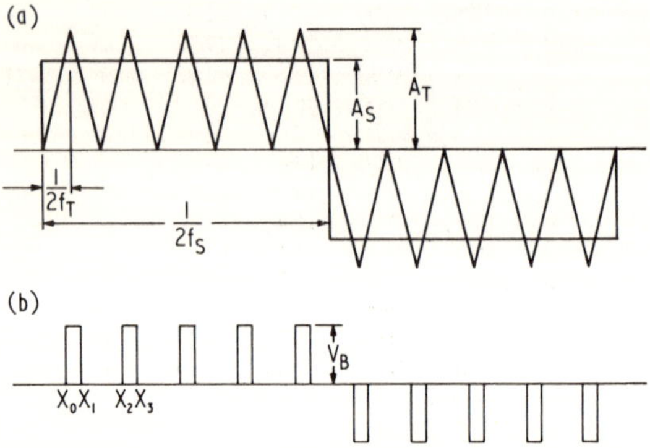

Figure 9.6. Pulse-width modulation with a square wave using unidirectional switching

$f_T/f_S = 2$ corresponds to a quasi-square wave. However, the lower harmonics are now considerably attenuated. This is illustrated in *Figure 9.7* where the third harmonic at high pulse numbers varies almost linearly with the fundamental, being 30% of its value. This is

Table 9.5. HARMONIC CONTENT OF A SQUARE MODULATED UNIDIRECTIONAL WAVE WITH $f_T/f_S = 4$

A_S/A_T	*R.M.S. voltage as percent of d.c. supply*								
	1	3	5	7	9	11	13	15	*Total*
0	90·0	30·0	18·0	12·9	10·0	8·18	6·92	6·00	100
0·1	82·7	36·2	9·74	17·7	1·11	11·5	2·29	7·84	94·9
0·2	74·8	40·3	0·00	17·3	8·31	6·80	9·31	0·00	89·4
0·3	66·5	42·3	9·74	11·8	13·8	2·70	7·44	7·84	83·7
0·4	57·8	41·9	18·0	2·84	12·6	10·3	1·53	6·00	77·5
0·5	48·7	39·2	23·5	6·96	5·41	10·7	9·05	3·25	70·7
0·6	39·3	34·3	25·5	14·7	4·37	3·58	7·92	8·49	63·2
0·7	29·7	27·6	23·5	18·1	12·1	6·05	0·77	3·25	54·8
0·8	19·9	19·3	18·0	16·2	14·0	11·4	8·72	6·00	44·7
0·9	9·99	9·90	9·74	9·50	9·18	8·80	8·35	7·84	31·6
0·98	2·00	2·00	2·00	2·00	1·99	1·99	1·99	1·98	14·1

clearly a highly desirable state of affairs since filter designs can be simplified. There is also a vast reduction from the 100% third-harmonic content obtained at low voltages with quasi-square wave

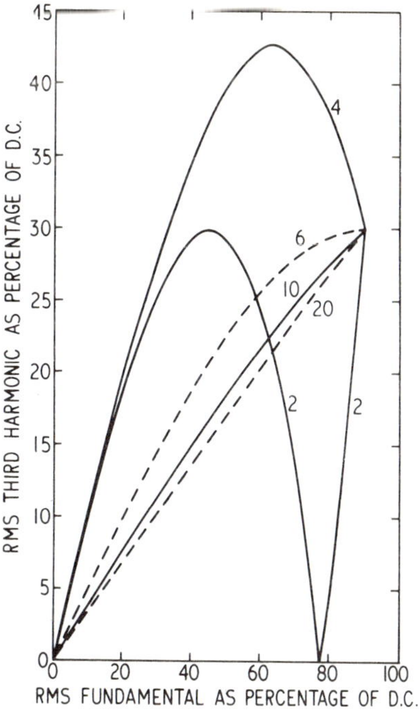

Figure 9.7. Third-harmonic content of a square-modulated unidirectional wave

Table 9.6. HARMONIC CONTENT OF A SQUARE MODULATED UNIDIRECTIONAL WAVE WITH $f_T/f_S = 6$

A_S/A_T	*R.M.S. voltage as percent of d.c. supply*								
	1	3	5	7	9	11	13	15	*Total*
0	90·0	80·0	18·0	12·9	10·0	8·18	6·92	6·0	100
0·1	81·7	29·6	25·5	4·00	8·93	14·6	2·15	4·22	94·9
0·2	73·2	28·5	31·2	5·36	5·90	16·3	10·3	0·02	89·4
0·3	64·5	26·7	34·8	14·0	1·58	12·7	13·8	4·26	83·7
0·4	55·6	24·3	36·0	20·8	3·07	5·04	11·2	6·02	77·5
0·5	46·6	21·2	34·8	24·9	7·05	4·25	3·57	4·26	70·7
0·6	37·4	17·6	31·2	25·6	9·49	12·2	5·65	0·02	63·2
0·7	28·2	13·6	25·5	22·9	9·86	16·2	12·4	4·22	54·8
0·8	18·8	9·25	18·0	17·2	8·07	15·0	13·6	5·98	44·7
0·9	9·44	4·67	9·34	9·23	4·52	8·93	8·73	4·22	31·6
0·98	1·90	0·92	1·90	1·90	0·92	1·90	1·90	0·92	14·1

Table 9.7. HARMONIC CONTENT OF A SQUARE MODULATED UNIDIRECTIONAL WAVE WITH $f_T/f_S = 10$

A_S/A_T	*R.M.S. voltage as percent of d.c. supply*								
	1	3	5	7	9	11	13	15	*Total*
0	90·0	30·0	18·0	12·9	10·0	8·16	6·94	6·00	100
0·1	81·3	27·8	17·8	14·6	18·2	0·85	4·37	5·33	94·9
0·2	72·4	25·4	17·1	15·6	25·0	9·77	1·09	3·51	89·4
0·3	63·6	22·7	16·1	15·9	29·7	17·5	2·37	0·92	83·7
0·4	54·6	19·9	14·6	15·4	32·1	23·2	5·44	1·87	77·5
0·5	45·6	16·8	12·7	14·1	32·0	26·2	7·61	4·26	70·7
0·6	36·5	13·6	10·6	12·2	29·3	26·0	8·52	5·72	63·2
0·7	27·4	10·3	8·19	9·72	24·3	22·8	8·03	5·94	54·8
0·8	18·3	6·93	5·58	6·75	17·4	16·9	6·22	4·87	44·7
0·9	9·17	3·47	2·83	3·45	9·05	8·99	3·38	2·74	31·6
0·98	1·85	0·68	0·58	0·68	1·85	1·85	0·68	0·58	14·1

control methods. For higher harmonics it is necessary to go to a larger pulse number to obtain appreciable harmonic reduction. For instance the curves for the seventh harmonics are shown in *Figure 9.8*. From these it is seen that with ten pulses per cycle there is still a large and variable harmonic content. With twenty pulses the variation is from 14% to 17% of the fundamental for the seventh harmonic at 90% and 10% of the fundamental voltage points. This can be compared with 14% and 90% for the same two points when

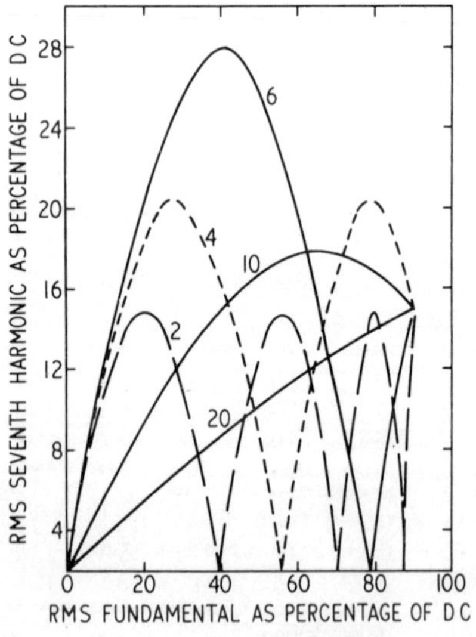

Figure 9.8. Seventh-harmonic content of a square-modulated unidirectional wave

Table 9.8. HARMONIC CONTENT OF A SQUARE MODULATED UNIDIRECTIONAL WAVE WITH $f_T/f_S = 20$

A_S/A_T	*R.M.S. voltage as percent of d.c. supply*								
	1	3	5	7	9	11	13	15	*Total*
0·1	81·1	27·2	16·5	12·1	9·68	8·28	7·49	7·23	94·9
0·2	72·1	24·3	15·0	11·1	9·16	8·14	7·75	8·07	89·4
0·3	63·1	21·4	13·3	10·0	8·46	7·75	7·69	8·46	83·7
0·4	54·1	18·4	11·6	8·84	7·59	7·13	7·31	8·38	77·5
0·5	45·1	15·4	9·74	7·54	6·58	6·30	6·62	7·84	70·7
0·6	36·1	12·4	7·87	6·14	5·42	5·28	5·66	6·86	63·2
0·7	27·1	9·31	5·94	4·67	4·17	4·10	4·47	5·51	54·8
0·8	18·1	6·22	3·98	3·15	2·82	2·81	3·09	3·85	44·7
0·9	9·04	3·11	2·00	1·58	1·43	1·42	1·58	1·98	31·6
0·98	1·81	0·62	0·40	0·32	0·29	0·29	0·32	0·40	14·1

using quasi-square wave control. Therefore overall this control system is superior to quasi-square wave control provided a sufficiently high ratio of f_T/f_S, say 20, is chosen. The maximum d.c. voltage is still 90% of the d.c. supply. The only disadvantage of this system is that the inverter must now run at twenty times the output frequency required, which can result in extra losses.

In the above system a square wave was used as the reference source. A greater reduction in harmonic content would be obtained if a sine wave was used as the reference as in *Figure 9.9*. The output

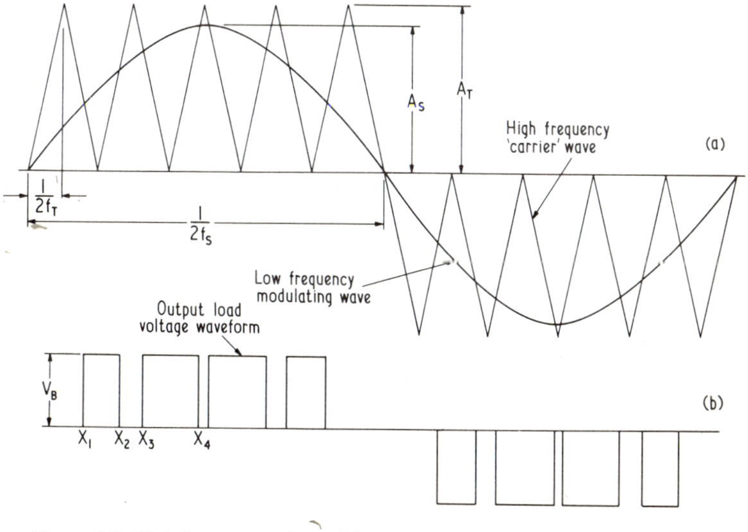

Figure 9.9. High frequency pulse-width modulation with unidirectional switching

pulses are not of constant width, but provided the values of X_1, X_2, X_3 etc., are known the magnitude of the harmonic coefficient can again be found from equation (9.9). However, equation (9.10) is now modified to take account of the unequal pulse widths and is given by

$$\frac{\% \text{ r.m.s. } (T)}{V_B} = \left[\frac{2}{\pi}\sum \{x(M+1)-x(M)\}\right]^{1/2} \quad (9.11)$$

Equations (9.9) and (9.11) are shown evaluated in Tables 9.9 to 9.12. Voltage control is again effected by changing the ratio of A_S/A_T.

Table 9.9. HARMONIC CONTENT OF A SINE MODULATED UNIDIRECTIONAL WAVE WITH $f_T/f_S = 4$

A_S/A_T	*R.M.S. voltage as percent of d.c. supply*								
	1	3	5	7	9	11	13	15	*Total*
0	0	0	0	0	0	0	0	0	0
0·1	9·93	9·77	9·46	8·99	8·40	7·69	6·88	6·00	26·5
0·2	16·6	15·9	14·4	12·4	9·97	7·32	4·64	2·14	34·4
0·3	23·2	21·1	17·4	12·5	7·18	2·26	1·65	4·13	40·7
0·4	29·6	25·4	17·9	9·19	1·25	4·24	6·48	5·68	46·2
0·5	33·9	27·5	16·9	5·54	3·19	7·27	6·60	2·87	49·5
0·6	40·0	29·5	13·4	1·02	8·41	7·66	2·01	3·52	54·2
0·7	44·0	30·0	10·0	5·36	9·91	5·05	2·41	5·90	57·0
0·8	47·9	29·8	6·01	9·03	9·44	0·95	5·85	5·13	59·7
0·9	51·6	29·0	1·63	11·6	7·08	3·44	6·92	1·61	62·3
0·98	53·4	28·3	0·62	12·4	5·33	5·33	6·38	0·62	63·6

Table 9.10. HARMONIC CONTENT OF A SINE MODULATED UNIDIRECTIONAL WAVE WITH $f_T/f_S = 6$

A_S/A_T	*R.M.S. voltage as percent of d.c. supply*								
	1	3	5	7	9	11	13	15	*Total*
0	0	0	0	0	0	0	0	0	0
0·1	6·98	0·29	7·20	6·57	0·84	7·20	5·98	1·32	24·0
0·2	14·7	0·60	14·7	13·1	1·60	13·1	10·2	2·09	34·8
0·3	21·2	0·25	19·9	18·0	0·59	14·4	11·4	0·55	41·7
0·4	27·8	0·45	23·9	21·5	0·88	12·7	9·08	0·39	47·7
0·5	34·3	1·46	26·7	23·4	2·17	8·47	3·59	0·38	52·9
0·6	40·8	2·74	28·1	23·1	2·72	3·07	3·48	2·69	57·7
0·7	48·8	5·52	27·2	20·4	2·22	2·73	10·9	6·50	63·0
0·8	55·4	7·28	25·7	15·6	0·32	5·67	13·7	6·41	67·1
0·9	60·8	10·4	22·5	10·8	3·59	5·06	12·6	4·51	70·2
0·98	66·3	13·6	18·7	5·12	7·86	2·74	8·82	0·21	73·1

Table 9.11. HARMONIC CONTENT OF A SINE MODULATED UNIDIRECTIONAL WAVE WITH $f_T/f_S = 10$

A_S/A_T	*R.M.S. voltage as percent of d.c. supply*								
	1	3	5	7	9	11	13	15	*Total*
0	0	0	0	0	0	0	0	0	0
0·1	7·26	0·28	0·24	0·88	7·53	6·70	0·43	0·68	25·0
0·2	13·6	0·54	0·12	0·11	13·7	12·3	1·48	0·29	34·6
0·3	20·8	0·95	0·18	0·12	19·9	17·4	2·72	0·27	42·8
0·4	27·5	0·49	0·57	0·02	24·1	21·2	3·63	0·92	49·1
0·5	34·7	0·98	0·69	0·80	27·7	23·0	5·70	0·66	55·2
0·6	42·1	0·60	0·35	3·67	27·8	24·7	6·21	1·17	60·7
0·7	48·9	0·17	0·93	4·90	27·0	23·2	8·51	2·16	65·2
0·8	55·9	1·31	0·91	8·17	24·9	20·8	9·71	2·89	69·9
0·9	62·5	1·43	1·57	11·0	20·7	17·5	10·5	4·11	73·9
0·98	68·5	0·93	1·57	13·4	16·6	12·7	11·9	4·70	77·3

Table 9.12. HARMONIC CONTENT OF A SINE MODULATED UNIDIRECTIONAL WAVE WITH $f_T/f_S = 20$

A_S/A_T	*R.M.S. voltage as percent of d.c. supply*								
	1	3	5	7	9	11	13	15	*Total*
0	0	0	0	0	0	0	0	0	0
0·1	7·58	0·68	1·04	0·22	0·33	0·26	0·38	1·26	25·9
0·2	14·9	0·60	0·09	0·50	0·35	0·30	0·62	0·48	36·2
0·3	21·5	0·15	0·15	0·77	0·53	0·52	0·88	0·24	43·7
0·4	28·2	0·03	0·56	0·89	0·00	0·01	1·02	0·55	50·2
0·5	35·3	0·05	0·00	0·41	0·53	0·15	0·47	0·50	56·2
0·6	42·4	0·16	0·05	0·21	0·22	0·13	0·82	0·53	61·6
0·7	49·2	0·10	0·22	0·39	0·29	0·15	0·85	0·97	66·3
0·8	56·8	0·53	0·59	0·26	0·06	0·36	0·28	0·09	71·2
0·9	64·0	0·55	0·02	0·15	0·49	0·21	0·13	0·76	75·5
0·98	69·0	0·26	0·42	0·19	0·10	0·16	0·32	1·25	78·6

The maximum value of this is limited to 0·98 rather than unity to prevent adjacent pulses from merging into each other. The following factors can be noticed:

(1) Unlike all the previous systems this method of voltage control results in severe attenuation of frequencies below a certain value. The two-pulse asymmetrical wave technique could be referred to as 'selected harmonic reduction', since in this system the pulse spacing is so chosen as to eliminate any required harmonic. The present system can be called 'lower harmonic reduction' since it works to reduce harmonics below a certain frequency. On examining the tables it will be seen that the harmonic numbers with the largest amplitude occur at $f_T/f_S \pm 1$. Therefore, for example, with $f_T/f_S = 10$

the harmonics are largest at the ninth and 11th. This is logical since the tenth harmonic is the 'carrier' wave itself, and no attempt is made to eliminate it. Quite clearly the higher the ratio of f_T/f_S the more effective this system becomes. The same statement was made when considering modulation with a square wave, but there the effect of higher carrier frequencies was to keep the proportion of the harmonics constant (and equal to the value for a square wave) as the fundamental was varied, and not to eliminate it.

(2) This system has two disadvantages. Firstly the high inverter frequency required to give effective lower-harmonic reduction leads to smaller efficiencies. Secondly the maximum output is well below 90% of d.c. supply, as obtained with a square wave. This would limit the maximum operating frequency when running with some types of loads which need to be fully fluxed, as described before.

(3) There is a characteristic increase in total harmonic content, with higher operating frequencies. As explained before, this is not normally serious since higher harmonics can be more easily filtered out than lower-order harmonics.

In the above discussions all the waveforms looked at have been unidirectional, that is the instantaneous voltage in any half cycle has been either positive or negative, never both. For bidirectional switching the output swings positive and negative. The simplest unidirectional system was the quasi-square waveform shown in *Figure 9.3*. It is interesting to note that this can be obtained by combining two square waves phase shifted from each other, as in *Figure 9.10*, the output being zero when the waveforms are in anti-

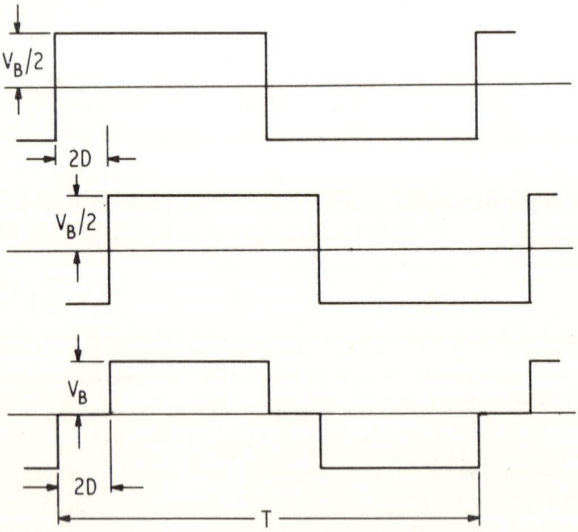

Figure 9.10. Production of a quasi-square wave by combining two phase-shifted square waves

phase. This single-pulse system was then extended to two-pulse waveforms, as in *Figure 9.5*. A two-pulse bidirectional wave is shown in *Figure 9.11*. The Fourier coefficient for this is given by equation

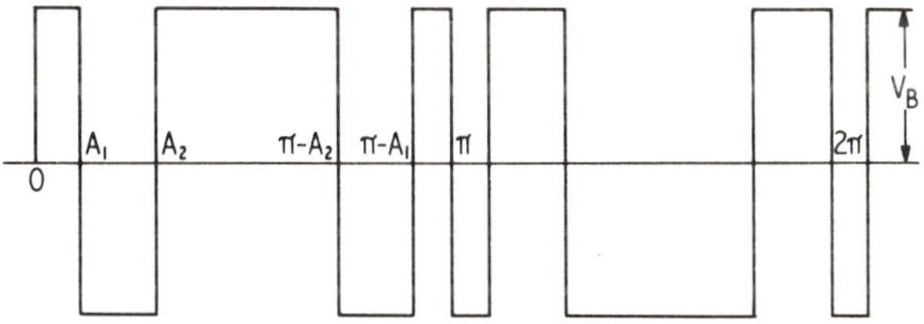

Figure 9.11. Two-pulse bidirectional wave

(9.12). Quite clearly the value of B can be chosen to eliminate any harmonic while ω is varied to control the fundamental voltage.

If
$$B = \frac{A_1 + A_2}{2}$$
$$\omega = \frac{A_2 - A_1}{2}$$

Then
$$\frac{\%\ \text{r.m.s.}\ (n)}{V_B} = \frac{2\sqrt{2}}{n\pi}[1 - 4 \sin nB \sin n\omega] \tag{9.12}$$

It is of course possible to combine two phase-shifted waveforms of *Figure 9.11* together, as was done in *Figure 9.10*. This would give a modified result

$$\frac{\%\ \text{r.m.s.}\ (n)}{V_B} = \frac{2\sqrt{2}}{n\pi}[1 - 4 \sin nB \sin n\omega] \cos nD \tag{9.13}$$

where $2D$ is the phase shift between the two. Since the square of the voltage ignores signs, the r.m.s. of the total waveform is given by

$$\frac{\%\ \text{r.m.s.}\ (T)}{V_B} = 100\left[1 - \frac{2D}{\pi}\right]^{1/2} \tag{9.14}$$

In equation (9.13) there are now three variables. Two of these, say ω and B, can be used to eliminate any two harmonics, while the third controls the magnitude of the fundamental voltage. Therefore if B and ω are chosen to satisfy equations (9.15) and (9.16) then harmonics P_1 and P_2 will be absent from the output over the whole range of variation of D.

$$1 - 4\sin(P_1 B)\sin(P_1 \omega) = 0 \tag{9.15}$$
$$1 - 4\sin(P_2 B)\sin(P_2 \omega) = 0 \tag{9.16}$$

Equations (9.13) and (9.14) have been used to obtain the harmonic analysis of waveforms where the third and fifth, the third and seventh and the fifth and seventh harmonics have been eliminated. The results are given in Tables 9.13–9.15.

This is a method of 'selected harmonic reduction' so comparison should be made with the quasi-square control and two-pulse unidirectional waveforms, *i.e.* Tables 9.1–9.4. From these it is clear that the maximum fundamental output voltage is less than that obtainable from the quasi-square wave. The lower the harmonics eliminated the smaller this maximum voltage. This is not the case for unidirectional switching, where the reverse is true; but it must be remembered that the limitation was then fixed by the necessity of not allowing adjacent pulses to merge. This is not necessary for bidirectional switching since in effect positive and negative pulses are already merged in the primary wave, and control is achieved by phase shifting two such identical waveforms up to the full half cycle. For example the quasi-square wave gives a peak fundamental which

Table 9.13. HARMONIC CONTENT OF A TWO-PULSE BIDIRECTIONAL WAVE WITH THIRD AND FIFTH HARMONIC ELIMINATION

$2D/T$	*R.M.S. voltage as percent of d.c. supply*								
	1	3	5	7	9	11	13	15	*Total*
0	75·5	0	0	22·4	36·8	27·2	2·40	15·2	100
0·02	75·4	0	0	20·3	31·1	21·0	1·65	8·92	98·0
0·04	74·9	0	0	14·3	15·7	5·10	0·15	4·69	95·9
0·06	74·2	0	0	5·58	4·61	13·1	1·85	14·4	93·8
0·08	73·1	0	0	4·21	23·5	25·3	2·39	12·3	91·7
0·10	71·8	0	0	13·2	35·0	25·9	0·41	0·00	89·4
0·12	70·2	0	0	19·7	35·7	14·6	0·45	12·3	87·2
0·14	68·3	0	0	22·4	25·2	3·41	2·03	14·4	84·9
0·16	66·2	0	0	20·9	6·90	19·8	2·33	4·69	82·5
0·18	63·8	0	0	15·4	13·6	27·1	1·16	8·92	80·0
0·20	61·1	0	0	6·91	29·8	22·0	0·74	15·2	77·5
0·22	58·2	0	0	2·81	36·7	6·76	2·18	8·92	74·8
0·24	55·0	0	0	12·0	32·3	11·6	2·24	4·69	72·1
0·26	51·7	0	0	19·0	17·7	24·6	0·89	14·4	69·3
0·28	48·1	0	0	22·3	2·31	26·3	1·02	12·3	66·3
0·30	44·4	0	0	21·3	21·6	16·0	2·29	0·00	63·2
0·32	40·5	0	0	16·4	34·2	1·71	2·11	12·3	60·0
0·34	36·4	0	0	8·26	36·2	18·6	0·60	14·4	56·6
0·36	32·2	0	0	1·41	26·8	27·0	1·29	4·69	52·9
0·38	27·8	0	0	10·8	9·16	23·0	2·36	8·92	49·0
0·40	23·3	0	0	18·2	11·4	8·40	1·95	15·2	44·7
0·42	18·8	0	0	22·1	28·4	10·0	0·30	8·92	40·0
0·44	14·2	0	0	21·7	36·5	23·8	1·53	4·69	34·6
0·50	0	0	0	0	0	0	0	0	0

is 90% of the d.c. value. For unidirectional switching this reduces to 75·1%, 62·2% and 33·3% when eliminating the third, fifth and seventh harmonics respectively. For bidirectional switching comparable figures are 75·5%, 77·5% and 84·0% when the third and fifth, the third and seventh or the fifth and seventh harmonics are eliminated respectively.

A comparison of the harmonic tables shows that in general unidirectional switching has a higher total harmonic content than bidirectional switching, the quasi-square wave having the least.

An interesting aspect of harmonic elimination is that the amplitude of the harmonics increases with the harmonic number. This is contrary to that found with quasi-square waves. For bidirectional switching the ninth harmonic is the largest when the third and fifth harmonics are eliminated, whereas when the third and seventh are eliminated the eleventh predominates. The same is also true for

Table 9.14. HARMONIC CONTENT OF A TWO-PULSE BIDIRECTIONAL WAVE WITH THIRD AND SEVENTH HARMONIC ELIMINATION

2D/T	*R.M.S. voltage as percent of d.c. supply*								
	1	3	5	7	9	11	13	15	*Total*
0	77·5	0	11·8	0	20·0	33·2	31·3	16·7	100
0·02	77·4	0	11·3	0	16·9	25·6	21·5	9·81	98·0
0·04	76·9	0	9·57	0	8·51	6·22	1·97	5·16	95·9
0·06	76·2	0	6·95	0	2·51	16·0	24·2	15·9	93·8
0·08	75·1	0	3·66	0	12·7	30·8	31·1	13·5	91·7
0·10	73·7	0	0·00	0	19·0	31·5	18·4	0·00	89·4
0·12	72·1	0	3·66	0	19·4	17·8	5·87	13·5	87·2
0·14	70·1	0	6·95	0	13·7	4·16	26·5	15·9	81·9
0·16	67·9	0	9·57	0	3·75	24·2	30·4	5·16	82·5
0·18	65·5	0	11·3	0	7·36	33·1	15·1	9·81	80·0
0·20	62·7	0	11·8	0	16·2	26·8	9·69	16·7	77·5
0·22	59·7	0	11·3	0	20·0	8·25	28·4	9·81	74·8
0·24	56·5	0	9·57	0	17·5	14·1	29·1	5·16	72·1
0·26	53·1	0	6·95	0	9·63	20·0	11·5	15·9	69·3
0·28	49·4	0	3·66	0	1·25	32·1	13·3	13·5	66·3
0·30	45·6	0	0·00	0	11·7	19·5	29·8	0·00	63·2
0·32	41·5	0	3·66	0	18·6	2·08	27·5	13·5	60·0
0·34	37·3	0	6·95	0	19·6	22·7	7·80	15·9	56·6
0·36	33·0	0	9·57	0	14·6	32·9	16·8	5·16	52·9
0·38	28·5	0	11·3	0	4·97	28·0	30·8	9·81	49·0
0·40	24·0	0	11·8	0	6·18	10·3	25·4	16·7	44·7
0·42	19·3	0	11·3	0	15·4	12·2	3·93	9·81	40·0
0·44	14·5	0	9·57	0	19·8	29·1	20·0	5·16	34·6
0·46	9·72	0	6·95	0	18·1	32·6	31·3	15·9	28·3
0·48	4·87	0	3·66	0	10·7	21·1	22·8	13·5	20·0
0·50	0	0	0	0	0	0	0	0	0

Table 9.15. HARMONIC CONTENT OF A TWO-PULSE BIDIRECTIONAL WAVE WITH FIFTH AND SEVENTH HARMONIC ELIMINATION

2D/T	*R.M.S. voltage as percent of d.c. supply*								
	1	3	5	7	9	11	13	15	*Total*
0	84·0	14·5	0	0	8·16	17·6	23·1	21·6	100
0·02	83·8	14·3	0	0	6·89	13·5	15·8	12·7	98·0
0·04	83·3	13·5	0	0	3·48	3·29	1·45	6·67	95·9
0·06	82·5	12·3	0	0	1·02	8·47	17·8	20·5	93·8
0·08	81·3	10·6	0	0	5·20	16·3	22·9	17·5	91·7
0·10	79·8	8·54	0	0	7·76	16·7	13·5	0·00	89·4
0·12	78·1	6·18	0	0	7·91	9·42	4·32	17·5	87·2
0·14	76·0	3·61	0	0	5·59	2·20	19·5	20·5	84·9
0·16	73·6	0·91	0	0	1·53	12·8	22·3	6·67	82·5
0·18	70·9	1·82	0	0	3·01	17·5	11·1	12·7	80·0
0·20	67·9	4·49	0	0	6·60	14·2	7·12	21·6	77·5
0·22	64·7	7·00	0	0	8·15	4·37	20·9	12·7	74·8
0·24	61·2	9·26	0	0	7·15	7·48	21·4	6·67	72·1
0·26	57·5	11·2	0	0	3·93	15·9	8·49	20·5	69·3
0·28	53·5	12·7	0	0	0·51	17·0	9·81	17·5	66·3
0·30	49·3	13·8	0	0	4·80	10·3	21·9	0·00	63·2
0·32	45·0	14·4	0	0	7·59	1·10	20·2	17·5	60·0
0·34	40·4	14·5	0	0	8·02	12·0	5·73	20·5	56·6
0·36	35·7	14·1	0	0	5·95	17·4	12·4	6·67	52·9
0·38	30·9	13·1	0	0	2·03	14·8	22·6	12·7	49·0
0·40	25·9	11·8	0	0	2·52	5·43	18·6	21·6	44·7
0·42	20·9	9·94	0	0	6·29	6·47	2·89	12·7	40·0
0·44	15·7	7·78	0	0	8·10	15·4	14·7	6·67	34·6
0·46	10·5	5·35	0	0	7·39	17·3	23·0	20·5	28·3
0·48	5·27	2·72	0	0	4·37	11·2	16·8	17·5	20·0
0·50	0·00	0·00	0	0	0·00	0·00	0·00	0·00	0·00

unidirectional switching, but to a lesser extent and at higher harmonic numbers. It should also be noted that for unidirectional switching every harmonic which is an odd multiple of the harmonic being eliminated is also absent from the output waveform. Therefore the third, ninth, 15th, etc., are absent; or the fifth, 15th, etc., are absent; or the seventh, 21st, etc., are absent, and so on. For bidirectional switching only the two harmonics chosen are eliminated.

Generally harmonic elimination causes the remaining harmonics to be of much larger magnitude. Perhaps, however, it is worth noting the phenomenal increase in the third harmonic when unidirectional switching is used to eliminate either the fifth or the seventh harmonic. The results given in Table 9.15 also deserve special mention. Although the fifth and seventh harmonics have been completely eliminated from the output, the third and ninth harmonics are also less than those obtained when using the quasi-square method of control. Higher harmonics are considerably greater. However, this waveform has special merit when used in three-phase inverters, as interconnections of the output transformer can then be used to eliminate triplen harmonics, making the 11th harmonic the lowest present.

The two-pulse bidirectional wave can be extended to several equi-spaced bidirectional pulses by mixing a triangular and square wave, as in *Figure 9.12.* This waveform will contain odd and even

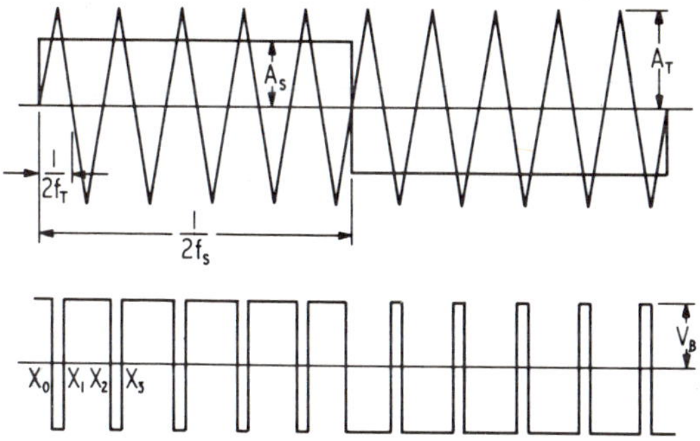

Figure 9.12. Pulse-width modulation with a square wave using bidirectional switching

harmonics. Knowing the value of intersection points X_0, X_1, X_2 etc. the r.m.s. of the nth harmonic can readily be obtained by the arithmetic sum of the individual pulses over the 0 to π interval. Therefore

$$\frac{\%\ \text{r.m.s.}\ (n)}{V_B} = \frac{\sqrt{2}}{n\pi}100\Big\{1-\cos nx(0) + \sum[\cos nx(2M+1)-\cos nx(2M+2)] - \sum[\cos nx(2M)-\cos nx(2M+1)]\Big\} \quad (9.17)$$

The r.m.s. value of the total harmonic is constant, irrespective of the depth of modulation and the operating frequency, and is equal to that of a square wave since there are no zero periods in the output.

The harmonics, as calculated from equation (9.17) are given in Tables 9.16–9.19 for values of f_T/f_S = 4, 6, 10 and 20 respectively.

It should be noted that for bidirectional switching, even harmonics are absent from the output, apart from the inverter operating frequency and its triplen harmonics. For unidirectional switching all even harmonics are missing.

The lower harmonic content with bidirectional switching, when the inverter operating frequency is higher than the output frequency, is less than for a quasi-square wave. It is also less than that obtained with unidirectional switching and is due to the absence of any 'zero' periods in the waveform. For maximum modulation the output

Table 9.16. HARMONIC CONTENT OF A SQUARE MODULATED BIDIRECTIONAL WAVE WITH $f_T/f_S = 4$

Harmonic number	*R.M.S. voltage as percent of d.c.*									
	0·1	0·2	0·3	0·4	0·5	0·6	0·7	0·8	0·9	1·0
1	8·6	17·3	26·2	35·1	44·1	53·2	62·4	71·6	80·8	90·0
2	0	0	0	0	0	0	0	0	0	0
3	1·67	3·73	6·16	8·91	12·0	15·3	18·8	22·4	26·2	30·0
4	88·9	85·6	80·2	72·8	63·7	52·9	40·9	27·8	14·1	0
5	1·11	1·48	1·11	0	1·80	4·22	7·18	10·5	14·2	18·0
6	0	0	0	0	0	0	0	0	0	0
7	7·94	14·3	18·7	20·6	20·1	17·1	11·8	4·70	3·74	12·9
8	0	0	0	0	0	0	0	0	0	0
9	8·98	18·1	26·2	32·3	35·6	35·8	32·8	27·0	19·0	10·0
10	0	0	0	0	0	0	0	0	0	0
11	2·17	5·45	9·22	12·8	15·5	16·9	16·7	14·9	11·9	8·18
12	26·7	17·6	4·69	9·27	21·2	28·5	29·6	24·3	13·6	0
13	0·52	0·86	3·79	7·51	11·1	13·6	14·4	13·3	10·6	6·92
14	0	0	0	0	0	0	0	0	0	0
15	7·04	9·68	7·04	0	9·06	17·1	21·4	20·5	14·7	6·00

Table 9.17. HARMONIC CONTENT OF A SQUARE MODULATED BIDIRECTIONAL WAVE WITH $f_T/f_S = 6$

Harmonic number	*R.M.S. voltage as percent of d.c.*									
	0·1	0·2	0·3	0·4	0·5	0·6	0·7	0·8	0·9	1·0
1	8·82	17·7	26·6	35·6	44·6	53·7	62·7	71·8	80·9	90·0
2	0	0	0	0	0	0	0	0	0	0
3	2·45	5·06	7·83	10·7	13·8	16·9	20·1	23·4	26·7	30·0
4	0	0	0	0	0	0	0	0	0	0
5	0·78	1·86	3·22	4·82	6·66	8·68	10·9	13·2	15·6	18·0
6	88·9	85·6	80·2	72·8	63·7	52·9	40·9	27·8	14·1	0
7	0·41	0·38	0·10	1·00	2·30	3·95	5·90	8·09	10·4	12·9
8	0	0	0	0	0	0	0	0	0	0
9	2·06	3·45	4·10	3·97	3·07	1·44	0·82	3·58	6·70	10·0
10	0	0	0	0	0	0	0	0	0	0
11	8·34	15·3	20·4	23·0	23·2	20·7	15·9	9·04	0·80	8·18
12	0	0	0	0	0	0	0	0	0	0
13	9·02	17·8	25·3	30·8	33·4	33·1	29·7	23·8	15·9	6·92
14	0	0	0	0	0	0	0	0	0	0
15	2·75	6·00	9·25	12·0	13·8	14·5	13·8	12·0	9·25	6·00

reduces to a square wave and the harmonic content is identical to comparable quasi-square and unidirectional systems.

It should also be noted that with bidirectional switching the odd-harmonic content generally increases as the operating frequency is raised. The reverse was true for unidirectional switching. However, the presence of the large-valued even harmonics still makes it desirable to operate this system at an inverter frequency of the order of twenty times the desired output frequency.

The square reference wave may of course be replaced by a sine

Table 9.18. HARMONIC CONTENT OF A SQUARE MODULATED BIDIRECTIONAL WAVE WITH $f_T/f_S = 10$

Harmonic number	*R.M.S. voltage as percent of d.c.*									
	0·1	0·2	0·3	0·4	0·5	0·6	0·7	0·8	0·9	1·0
1	8·94	17·9	26·9	35·9	44·9	53·9	62·9	71·9	81·0	90·0
2	0	0	0	0	0	0	0	0	0	0
3	2·81	5·67	8·60	11·6	14·6	17·6	20·7	23·8	26·9	30·0
4	0	0	0	0	0	0	0	0	0	0
5	1·47	3·04	4·70	6·44	8·26	10·1	12·1	14·0	16·0	18·0
6	0	0	0	0	0	0	0	0	0	0
7	0·80	1·74	2·82	4·01	5·32	6·71	8·18	9·71	11·3	12·9
8	0	0	0	0	0	0	0	0	0	0
9	0·32	0·84	1·54	2·40	3·42	4·57	5·83	7·18	8·57	10·0
10	88·9	85·6	80·2	72·8	63·7	52·9	40·9	27·8	14·1	0
11	0·10	0·05	0·44	1·05	1·88	2·90	4·08	5·38	6·76	8·18
12	0	0	0	0	0	0	0	0	0	0
13	0·57	0·83	0·77	0·39	0·30	1·26	2·46	3·84	5·35	6·92
14	0	0	0	0	0	0	0	0	0	0
15	1·23	2·07	2·46	2·38	1·84	0·86	0·49	2·15	4·02	6·00

Table 9.19. HARMONIC CONTENT OF A SQUARE MODULATED BIDIRECTIONAL WAVE WITH $f_T/f_S = 20$

Harmonic number	*R.M.S. voltage as percent of d.c.*									
	0·1	0·2	0·3	0·4	0·5	0·6	0·7	0·8	0·9	1·0
1	8·99	18·0	27·0	36·0	45·0	54·0	63·0	72·0	81·0	90·0
2	0	0	0	0	0	0	0	0	0	0
3	2·95	5·92	8·90	11·9	14·9	17·9	20·9	24·0	27·0	30·0
4	0	0	0	0	0	0	0	0	0	0
5	1·72	3·47	5·23	7·02	8·83	10·6	12·5	14·3	16·2	18·0
6	0	0	0	0	0	0	0	0	0	0
7	1·17	2·38	3·62	4·89	6·18	7·49	8·82	10·2	11·5	12·9
8	0	0	0	0	0	0	0	0	0	0
9	0·85	1·75	2·69	3·66	4·67	5·71	6·76	7·83	8·91	10·0
10	0	0	0	0	0	0	0	0	0	0
11	0·63	1·32	2·06	2·85	3·68	4·54	5·43	6·33	7·25	8·18
12	0	0	0	0	0	0	0	0	0	0
13	0·47	1·00	1·60	2·25	2·96	3·70	4·48	5·28	6·10	6·92
14	0	0	0	0	0	0	0	0	0	0
15	0·33	0·75	1·23	1·78	2·39	3·05	3·75	4·48	5·24	6·00

wave. The output is very similar to *Figure 9.12* except that the modulation depth varies linearly along the cycle. This waveform will also contain odd and even sine terms, and since $X_0 = 0$ the r.m.s. voltage of the nth harmonic can be derived from equation (9.18).

$$\frac{\%\ \text{r.m.s.}\ (n)}{V_B} = \frac{\sqrt{2}}{n\pi}100\left\{\sum[\cos nx(2M+1)-\cos nx(2M+2)] - \sum[\cos nx(2M)-\cos nx(2M+1)]\right\} \qquad (9.18)$$

The solutions of equation (9.18) for four values of f_T/f_S are given in Tables 9.20–9.23. It is seen from these that the harmonic content of the waveform is very similar to unidirectional switching as in Tables 9.9 to 9.12. The harmonic with the largest amplitude is that which occurs close to the chopping frequency f_T, both odd and even har-

Table 9.20. HARMONIC CONTENT OF A SINE MODULATED BIDIRECTIONAL WAVE WITH $f_T/f_S = 4$

Harmonic number	*R.M.S. voltage as percent of d.c.*									
	0·1	0·2	0·3	0·4	0·5	0·6	0·7	0·8	0·9	0·98
1	8·34	16·0	21·4	29·4	34·9	44·4	49·8	57·2	63·4	70·9
2	0·09	0·61	1·03	6·58	7·28	11·1	12·0	14·9	19·4	22·2
3	2·52	1·23	2·16	0·11	0·64	0·01	0·67	1·22	1·79	1·80
4	89·2	87·2	84·9	80·6	77·0	69·4	64·4	57·0	50·7	42·4
5	2·50	1·25	2·18	3·04	4·38	6·81	8·17	11·2	12·7	15·2
6	0·28	1·79	3·00	2·79	5·18	9·21	12·4	15·8	20·7	25·5
7	8·24	15·3	19·8	22·3	24·7	25·1	25·2	22·6	17·9	13·6
8	0·93	0·87	2·01	1·73	3·34	5·29	7·50	11·6	10·2	13·4
9	8·17	14·8	18·7	24·7	26·2	26·7	24·7	21·1	16·1	8·92
10	0·47	2·88	4·70	10·7	11·9	13·6	13·5	11·0	9·10	5·19
11	2·39	1·41	2·29	0·39	0·57	0·94	2·11	1·44	4·55	5·86
12	27·5	22·0	16·2	7·43	1·14	7·07	10·5	11·8	8·42	4·28
13	2·34	1·47	2·33	6·24	7·08	8·65	7·44	3·31	3·97	1·47
14	0·65	3·82	5·99	10·6	13·6	18·1	18·4	18·9	15·8	11·0
15	7·87	12·9	14·3	8·93	6·41	0·89	3·78	5·53	4·63	1·00

Table 9.21. HARMONIC CONTENT OF A SINE MODULATED BIDIRECTIONAL WAVE WITH $f_T/f_S = 6$

Harmonic number	*R.M.S. voltage as percent of d.c.*									
	0·1	0·2	0·3	0·4	0·5	0·6	0·7	0·8	0·9	0·98
1	6·98	14·1	21·0	28·7	35·6	42·7	49·1	56·1	63·5	70·5
2	0·02	0·14	1·29	1·13	0·15	0·11	0·24	0·10	1·86	1·46
3	0·44	0·41	0·82	0·29	0·75	0·88	0·22	0·44	0·95	1·28
4	0·20	0·80	2·72	4·34	6·71	8·52	12·6	14·8	19·6	22·1
5	0·26	2·11	1·46	0·65	0·21	1·66	0·36	1·47	0·04	0·62
6	89·5	87·8	85·1	81·1	76·4	70·8	65·2	58·2	50·4	42·5
7	0·26	2·05	0·87	1·34	0·36	1·58	0·82	1·62	2·97	3·33
8	0·41	1·58	1·71	4·68	6·11	9·46	11·4	15·3	17·1	21·2
9	0·41	0·21	2·22	0·78	3·89	5·49	7·89	10·3	13·0	15·6
10	0·09	0·65	1·08	1·62	0·44	0·12	0·30	0·21	1·14	2·17
11	6·91	13·4	18·7	23·2	25·5	26·8	24·6	22·9	16·8	12·8
12	0·09	1·29	1·07	1·19	0·48	2·87	1·26	3·46	3·85	5·39
13	6·88	13·2	19·0	23·0	26·0	25·9	25·8	22·3	20·0	13·9
14	0·13	0·86	0·70	0·62	2·26	2·42	5·38	6·60	9·29	10·9
15	0·37	0·17	0·76	2·59	3·23	6·16	6·65	10·2	10·7	13·1

monics being considered. As an example when operating at $f_T/f_S = 6$ the sixth harmonic is very large. For zero modulation depth the fundamental is zero and the sixth harmonic has a value 90% of the d.c. supply, since the output is a square wave at this frequency. As the modulation depth increases the fundamental also increases in

Table 9.22. HARMONIC CONTENT OF A SINE MODULATED BIDIRECTIONAL WAVE WITH $f_T/f_S = 10$

Harmonic number	*R.M.S. voltage as percent of d.c.*									
	0·1	0·2	0·3	0·4	0·5	0·6	0·7	0·8	0·9	0·98
1	6·94	13·3	21·1	27·4	35·2	43·2	48·9	57·5	63·8	71·6
2	0·01	0·46	0·84	1·25	0·01	0·75	0·04	0·11	0·14	0·72
3	0·42	0·26	0·12	0·05	0·44	0·05	0·16	0·12	0·38	0·26
4	0·04	0·90	0·66	0·41	0·21	0·95	1·45	0·60	0·81	1·08
5	0·26	0·63	0·25	0·62	0·22	0·17	0·43	0·29	0·41	0·23
6	0·02	0·76	0·45	0·26	0·17	0·17	0·20	0·81	1·45	1·01
7	1·31	0·11	1·44	0·32	1·06	0·66	2·09	0·84	0·06	1·38
8	0·24	1·35	2·77	4·55	5·23	9·38	11·4	16·4	19·3	22·4
9	1·89	2·50	1·26	1·64	1·03	1·00	1·07	0·83	0·55	0·11
10	89·4	88·0	85·1	81·8	76·7	70·4	65·3	56·9	50·1	41·4
11	1·88	2·67	1·66	2·41	0·96	1·69	0·99	0·68	0·84	0·63
12	0·35	0·70	2·15	3·55	7·53	9·74	12·8	15·4	18·3	23·2
13	1·31	0·28	0·71	0·14	0·94	0·83	0·68	0·37	0·79	0·59
14	0·04	1·05	0·59	0·54	0·51	0·48	0·60	0·70	0·23	1·99
15	0·25	0·50	0·13	0·53	0·10	0·65	1·27	1·58	1·72	2·99

Table 9.23. HARMONIC CONTENT OF A SINE MODULATED BIDIRECTIONAL WAVE WITH $f_T/f_S = 20$

Harmonic number	*R.M.S. voltage as percent of d.c.*									
	0·1	0·2	0·3	0·4	0·5	0·6	0·7	0·8	0·9	0·98
1	7·23	13·9	21·1	28·1	35·0	42·1	49·9	56·5	63·9	70·7
2	0·01	0·02	0·02	0·10	0·34	0·07	0·79	0·11	0·03	0·27
3	0·05	0·56	0·24	0·19	0·35	0·16	0·14	0·10	0·49	0·22
4	0·01	0	0·06	0·71	0·43	0·03	0·02	0·18	0·86	0·03
5	0·50	0·13	0·05	0·42	0·12	0·13	0·43	0·61	0·75	0·14
6	0	0·04	0·01	0·37	0·11	0·01	0·35	0·28	0·76	0·62
7	0·63	1·02	0·32	0·29	0·80	0·40	0·17	0·17	0·78	0·62
8	0·02	0	0·02	0·76	0·28	0·22	0·70	0·54	0·25	0·01
9	1·03	0·98	0·63	0·86	0·19	0·03	0·19	0·03	0·32	0·22
10	0·02	0·09	0·04	0·31	0·42	0·10	0·12	0·06	0·37	0·22
11	0·35	0·18	0·25	0·13	0·48	0·36	0·12	0·12	0·03	0·14
12	0·03	0·18	0·14	0·07	0·35	0·11	0·11	0·45	0·10	0·10
13	0·22	0·03	0·15	1·05	0·13	0·78	0·14	0·65	0·73	0·33
14	0·06	0·14	0·06	0·05	0·14	0·17	0·34	0·66	0·50	0·30
15	0·13	0·76	0·29	1·14	0·28	0·75	0·07	0·48	0·70	0·56

value and the sixth harmonic reduces (whereas adjacent even harmonics, *i.e.* the fourth and the eighth increases in value). When operating with $f_T/f_S = 20$ the harmonic content up to the 15th is very similar to unidirectional methods except that there are now odd and even terms. However, the total harmonic content, which for bidirectional switching is 100% of the d.c. supply irrespective of the modulation depth, is much higher. As in previous methods using 'lower-harmonic reduction' it is clear that the inverter frequency should be several orders larger than the output frequency for effective harmonic reduction. This is therefore a disadvantage of the system as it can lead to lower efficiencies.

Before leaving this section it would be useful to look at two other techniques for voltage control which have been specifically used as methods for harmonic reduction. The first of these is called staggered phase carrier cancellation. It essentially consists in combining several high-frequency modulated waves in which the carriers are out of phase, whereas the modulating (low-frequency) wave is in phase. This results in a strengthening of the low-frequency and a weakening of the high-frequency carrier. This is evident when the waveforms of *Figure 9.13* are considered. Here the carriers in (*a*) and (*b*) are 180°

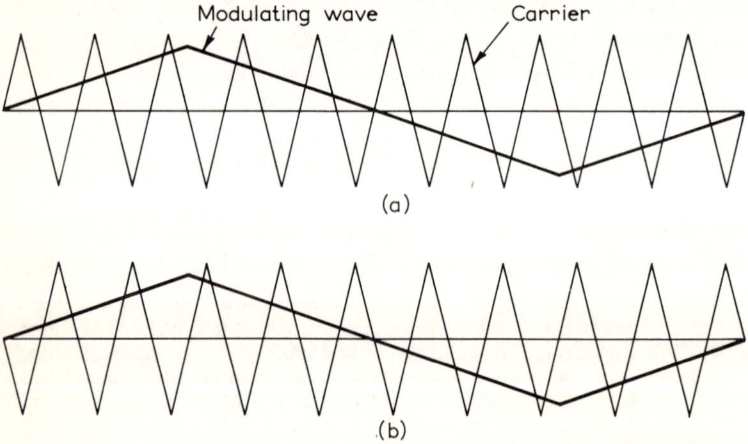

Figure 9.13. Staggered phase carrier cancellation

out of phase, whereas the low frequency signal is in phase. Combining the two waveforms will give the lowest carrier harmonic at twice the carrier frequency.

This system can be extended. For instance combining four waveforms with their low frequencies in phase but carriers phase shifted by 90°, 180° and 270° would result in the carrier and its first, second and third harmonics being eliminated from the output.

The disadvantage of staggered phase-carrier cancellation is the extra hardware needed. When combining the two waves shown in *Figure 9.13* two inverters, as in *Figure 9.2*, are required, doubling the power and control circuits. Its prime use is in fixed-frequency sine wave inverters where the extra cost can be justified on account of the reduction in size of the output filter.

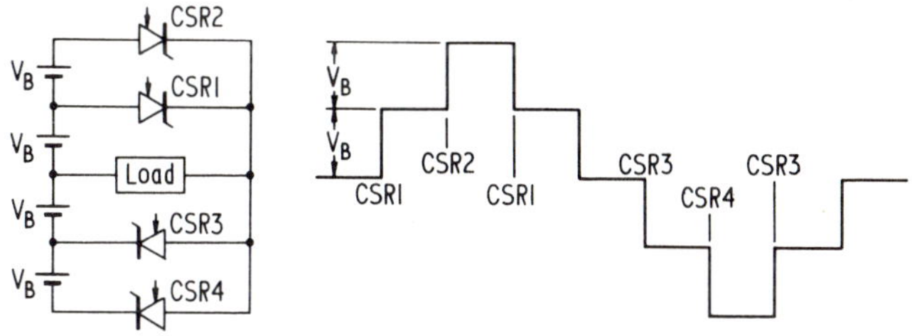

Figure 9.14. Waveform synthesis using a tapped supply

The second technique worth mentioning is that of waveform synthesis. This is frequently used, especially in larger installations, where several inverters are run in parallel but phase shifted, their outputs being summed by a transformer to produce a stepped waveform with reduced harmonic content. The same effect can be obtained by using a tapped supply, as in *Figure 9.14*, the thyristor firing sequence being indicated alongside the output waveform. Alternatively, instead of

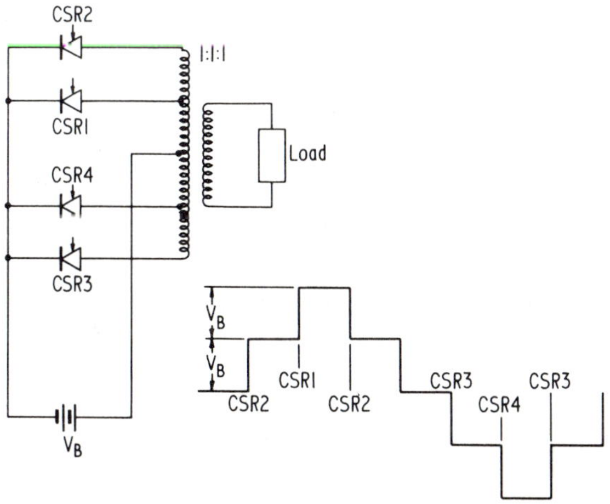

Figure 9.15. Waveform synthesis with a tapped primary transformer

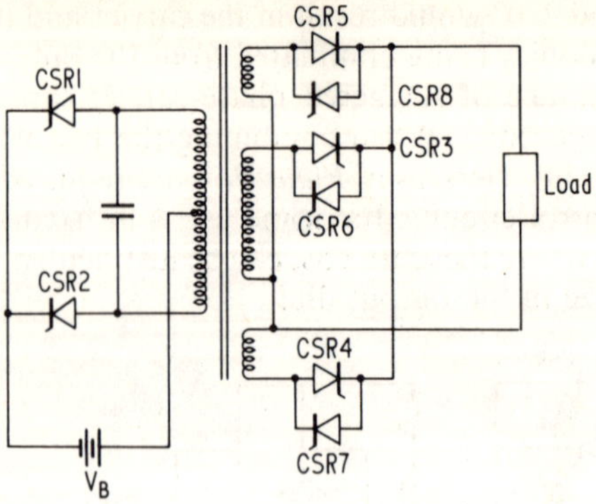

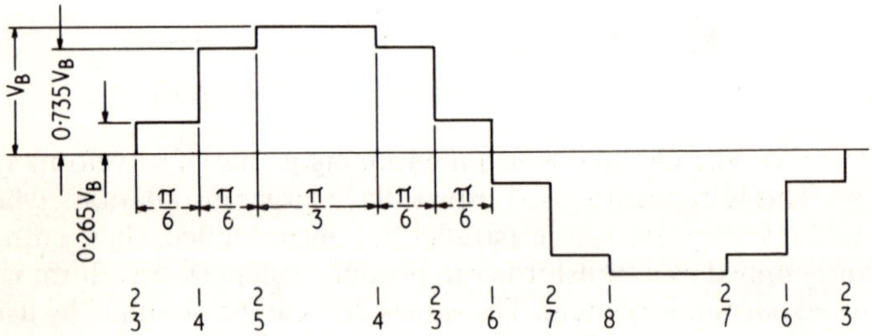

Figure 9.16. Waveform synthesis with a tapped secondary transformer

tapping the supply, the primary of the transformer can be tapped as shown in *Figure 9.15*. *Figure 9.16* shows an inverter arrangement with a tapped secondary, the primary being the normal form of a push-pull inverter. Firing sequence of the thyristors is also shown. Provided the tappings on the secondary are such as to give the waveform indicated, it can be calculated that the output contains no harmonics below the 11th.

INVERTER COMMUTATION CIRCUITS

Irrespective of whether the inverter is push-pull or bridge, and the voltage control mode adopted, the commutation systems used can be grouped into four classes, as was done for chopper circuits in Chapter 5.

(1) *Parallel-capacitor commutation*, the commutation capacitor discharging directly into the thyristors being turned off.

(2) *Parallel capacitor-inductor commutation*, where a series capacitor and inductor are connected across the thyristors being commutated.

(3) *Series-capacitor commutation*, the commutation capacitor being in series with the load conduction path.

(4) *Coupled-pulse*, the commutation capacitor pulse being coupled to the thyristors via a transformer or auto-transformer.

These four methods are examined below. With reference to typical bridge and parallel inverter circuits.

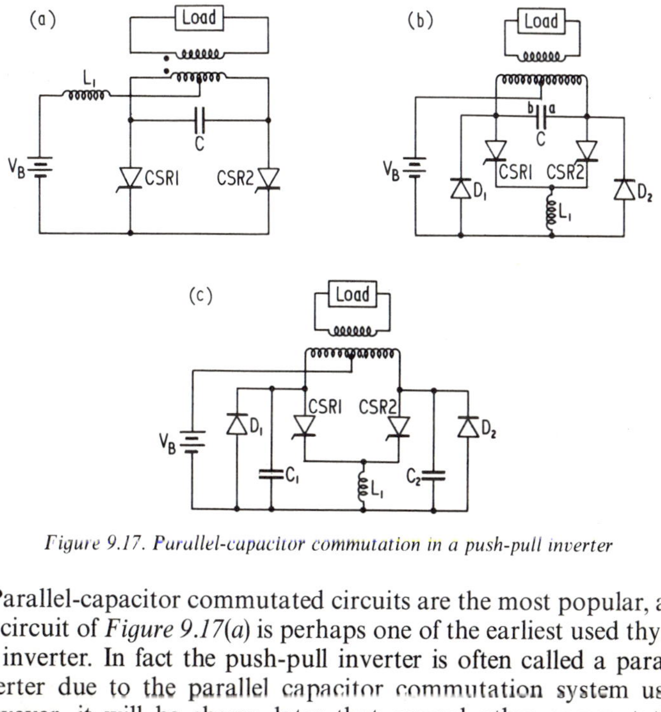

Figure 9.17. Parallel-capacitor commutation in a push-pull inverter

Parallel-capacitor commutated circuits are the most popular, and the circuit of *Figure 9.17(a)* is perhaps one of the earliest used thyristor inverter. In fact the push-pull inverter is often called a parallel inverter due to the parallel capacitor commutation system used. However, it will be shown later that several other commutation methods can also be used with push-pull inverters.

The circuit of *Figure 9.17(a)* is similar to *Figure 9.1*, choke L_1 being added and diodes D1 and D2 being omitted. *Figure 9.17(b)* shows a more efficient system. Firing CSR1 charges capacitor C with plate a positive to a voltage $2V_B$. When CSR2 is fired capacitor C is connected across CSR1 turning it off. The capacitor now discharges to zero voltage, its stored energy then being dissipated in the L_1-D1-C-CSR2 conduction path. After this C charges to $2V_B$ with plate b positive, ready to turn CSR2 off when CSR1 is fired.

Figure 9.17(c) shows an alternative arrangement of the commutation capacitors. Although two are now required they need only have a d.c. rating.

The simple push-pull capacitor circuits have a disadvantage in that one or another of the main thyristors must be triggered to commutate the conducting device. This means that mark-space voltage control schemes are not possible. Even varying the d.c. supply voltage would reduce the commutation energy, which is undesirable. Although h.f. pulse-width modulation systems can be used, either with a rectangular or sine reference, the inverter losses, caused by the circulation of the commutation current via L_1-D1-C-CSR2, make it inefficient at high frequencies. The most usual voltage control system consists of two parallel inverters, with a common secondary, whose square wave outputs are phase shifted to give a mark-space controlled output.

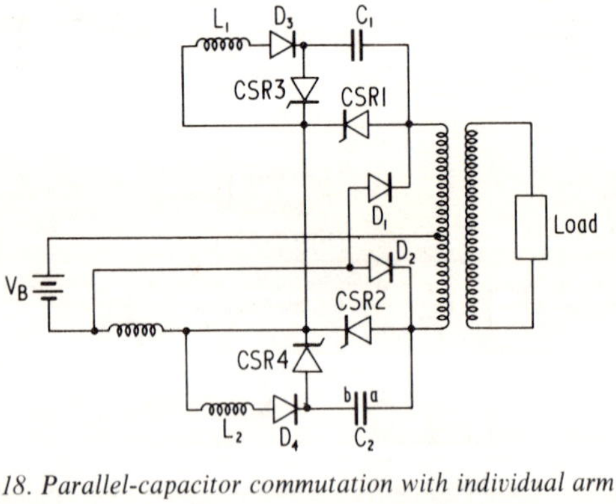

Figure 9.18. Parallel-capacitor commutation with individual arm control

Figure 9.18 shows a commutation system which permits individual thyristor turn-off. While CSR1 is conducting thyristor CSR4 is fired. This charges C_2 to $2V_B$ with plate *a* positive. Now when CSR2 turns on, thyristor CSR1 being previously turned off, C_2 discharges through it and L_2D4, recharging with plate *b* positive. To turn CSR2 off thyristor CSR4 is again fired to connect C_2 across it. Thyristor CSR1 can be fired later in the cycle as required, the greater the delay between the turning off of one thyristor and the firing of the next, the lower the output voltage. As mentioned before, it is important to bear in mind that this is not true mark-space control since the inductive load current would normally be decaying through feedback diodes D1 and D2. Therefore the load voltage at any inverter setting is determined by the magnitude of the load current and its power factor.

Figure 9.19(a) shows a single-phase bridge inverter circuit. With CSR1 and CSR4 conducting, capacitor C is charged to V_B with plate a positive. When CSR2 and CSR3 are fired to commence the next step of the output, capacitor C is connected across CSR1 and CSR4 and turns them off. The circuit is therefore an example of parallel-capacitor commutation. Inductor L_1 prevents the supply from being instantaneously short-circuited during commutation. *Figure 9.19(b)* shows an extension to a three-phase inverter. Firing any thyristor with positive anode voltage will commutate all other conducting thyristors in its row.

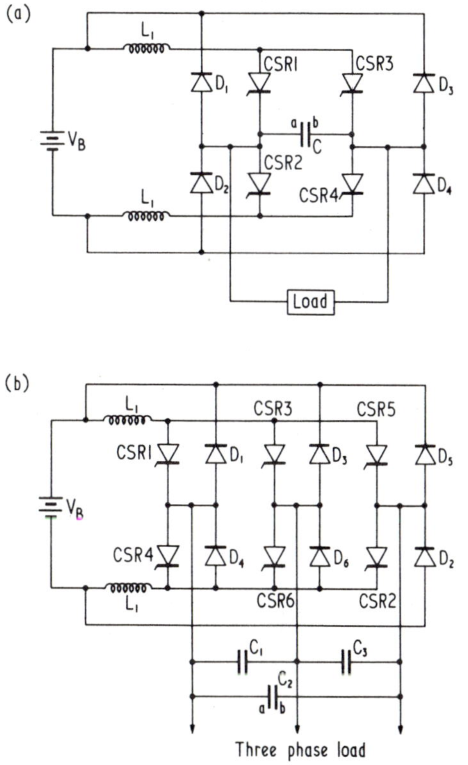

Figure 9.19. Parallel-capacitor commutation in a bridge inverter

An alternative parallel-capacitor commutated bridge is illustrated in *Figure 9.20*. An auxiliary d.c. supply is shown connected to the commutation circuit, to enable the main supply to be varied for voltage control. Where this is not required a separate supply is not essential and can be shorted out. When CSR7 is fired C is charged to

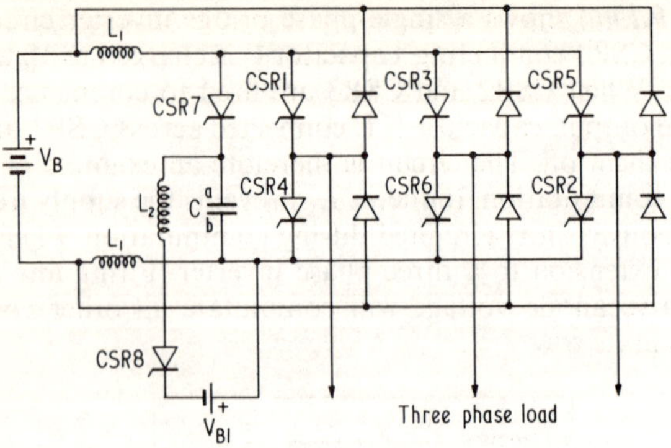

Figure 9.20. Alternative form of bridge parallel-capacitor commutation with provision for an external commutation voltage source

V_B with plate *a* positive. Firing CSR8 will cause the capacitor to resonate through L_2 and recharge with *b* positive (the voltage being increased if V_{B_1} is included). When CSR7 is next fired capacitor *C* short circuits the input to the bridge inverter and causes all the conducting thyristors in both arms of the bridge to turn off. It is important once again to bear in mind that turning off all devices will not make the load voltage zero. This can only be obtained by refiring appropriate devices to allow the current to free-wheel through them.

An example of parallel capacitor-inductor commutation is given in *Figure 9.21*. This differs from *Figure 9.18* in two aspects. First,

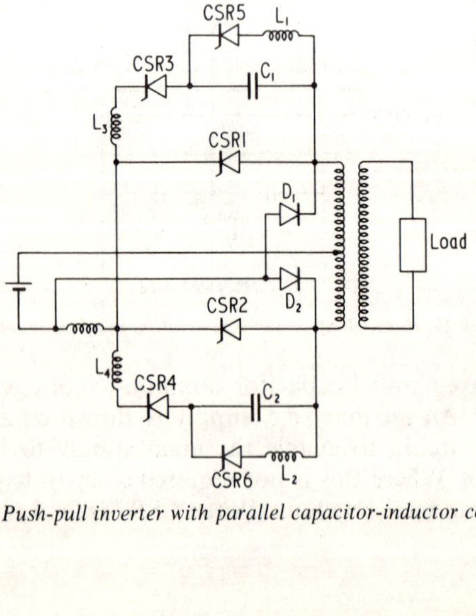

Figure 9.21. Push-pull inverter with parallel capacitor-inductor commutation

inductors L_3 and L_4 are in series with C_1 and C_2 reducing the rating of the commutation devices. Secondly, the resonant pulse of the capacitors does not now pass through the main thyristors, so reducing their current rating. Apart from this the two circuits have similar performance, CSR5 and CSR6 being normally fired simultaneously as CSR1 and CSR2 respectively.

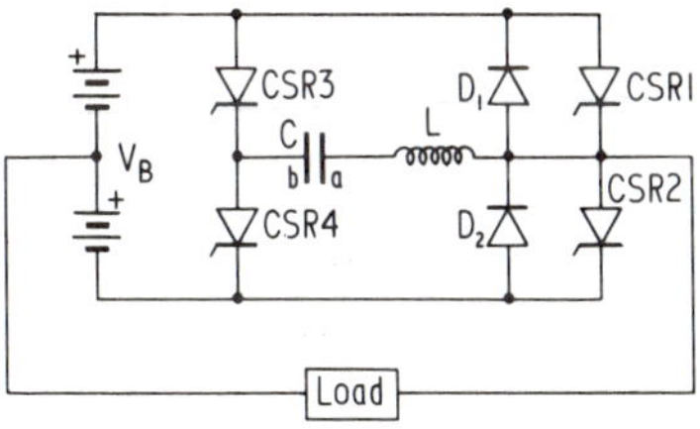

Figure 9.22. Parallel capacitor-inductor commutation in a bridge with tapped supply

Figure 9.22 shows parallel capacitor-inductor commutation used in a bridge circuit. A centre-tapped supply is illustrated, which reduces the number of components needed. When this is not available a conventional inverter is adopted, all components now being duplicated. Referring to *Figure 9.22* with CSR1 conducting CSR4 is fired to charge C with plate a positive. To turn CSR1 off CSR3 is turned on. C now resonates through D1 and CSR3 commutating CSR1. This circuit is extremely versatile since it allows individual commutation of all devices. It has also theoretically zero commutation loss, so can be operated at high frequencies. Its only disadvantage is the relatively large number of commutation devices required.

The inverter of *Figure 9.23*, although similar in principle to *Figure 9.20*, has been designed for aero-space applications. Therefore chokes L_1, which can be bulky, have been replaced by CSR7 and CSR8. With these devices, and appropriate bridge thyristors conducting, CSR9 is fired to prime C ready for commutation. When CSR10 is now turned on capacitor C discharges via its inductor L, turning CSR7 and CSR8 off and so eventually commutating the inverter thyristors. This circuit also has high operating efficiencies.

In series-capacitor commutated inverters the commutation devices are normally in the main thyristor current path. *Figure 9.24* shows a typical circuit with its waveforms. The inverter is operated at a frequency determined by the oscillatory frequency of the load and C, so that these inverters find most frequent application for producing fixed frequency sine wave output.

Figure 9.25 is another example of a simple series capacitor commutated inverter. Once again the trigger frequency of CSR1 is

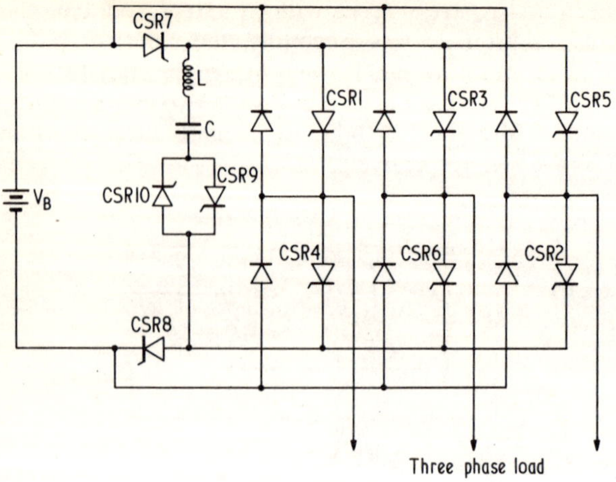

Figure 9.23. Lightweight parallel capacitor-inductor commutated bridge inverter

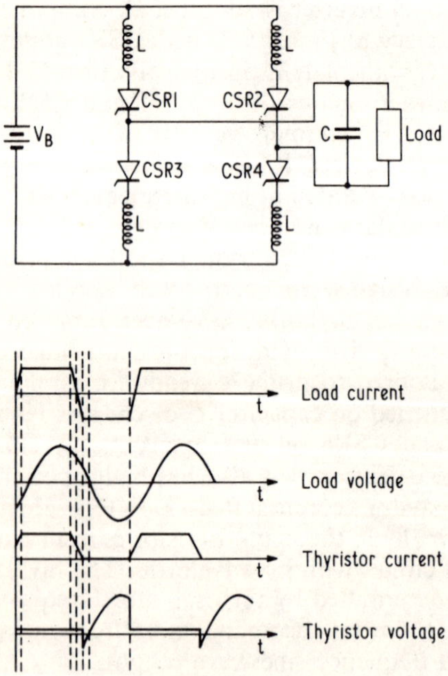

Figure 9.24. Series-capacitor commutation bridge inverter

chosen to coincide with the oscillatory frequency of C_2 and L_1. Capacitor C_1 is large valued and acts as a pivot for the see-saw action of the output. This circuit is very useful as a simple sine wave fixed-frequency generator.

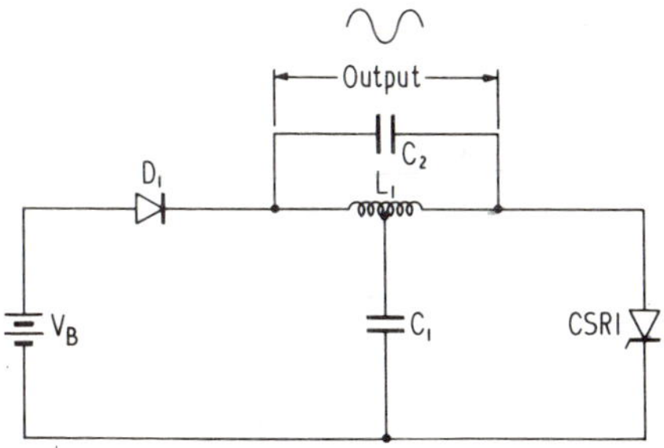

Figure 9.25. Inexpensive series-capacitor-commutated inverter

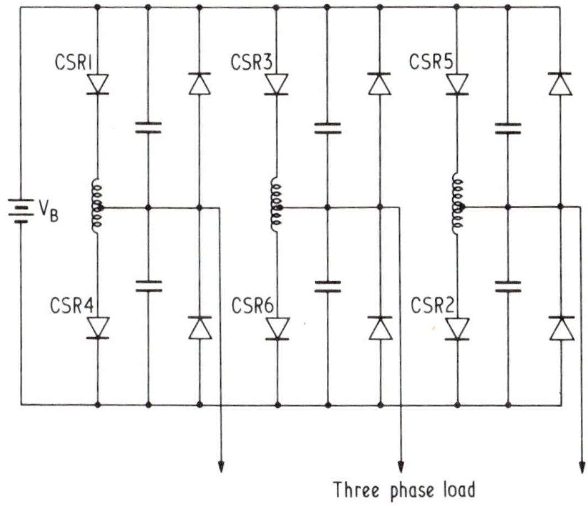

Figure 9.26. Coupled-pulse commutated inverter

The coupled pulse inverter shown in *Figure 9.26*, which was analysed in Chapter 8, is perhaps the most frequently used and reliable circuit. With any device conducting the commutation capacitor in the corresponding half of the bridge is charged to V_B. When the thyristor in that leg is fired the capacitor discharges through half of the auto-transformer, coupling a turn-off pulse to the conducting thyristor via the other half of the transformer. The circuit is versatile in that firing one device automatically commutates the other half of the bridge. It can be used to produce outputs with zeros in the waveform with no additional circuitry. It does, however, suffer from low efficiency, due to commutation losses at higher frequencies, and is therefore not suited for pulse-width modulated voltage control methods. It cannot be used in circuits which have d.c. supply variations to change the output, since the commutation voltage is now affected. *Figure 9.27* overcomes this by using an auxiliary fixed voltage low power commutation source V_{B_1}. The diodes in series with the main thyristors enable the commutation capacitors to charge to a voltage which can be greater than V_B.

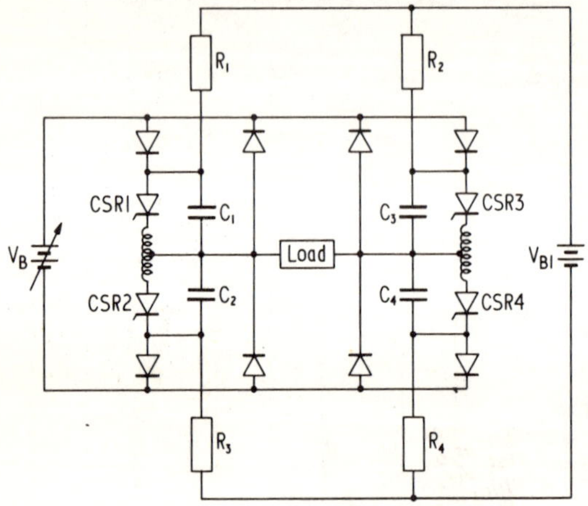

Figure 9.27. Coupled-pulse inverter with an auxiliary commutation voltage source

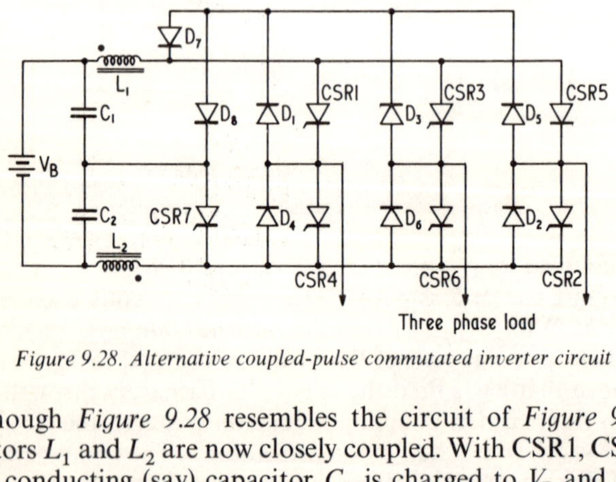

Figure 9.28. Alternative coupled-pulse commutated inverter circuit

Although *Figure 9.28* resembles the circuit of *Figure 9.20* the inductors L_1 and L_2 are now closely coupled. With CSR1, CSR5 and CSR6 conducting (say) capacitor C_2 is charged to V_B and C_1 is at zero voltage. Firing CSR7 couples a pulse in L_1 via L_2 which blocks the conducting thyristors. Load current can continue to free-wheel via feedback diodes and D8, CSR7, therefore true mark-space control is obtained without additional circuitry. Capacitor C_1 is now charged so that during the next part of the cycle, when CSR1 is fired (say), it discharges into L_1 and couples a pulse to L_2 turning CSR7 off.

CONCLUSIONS

Bridge inverters are used more frequently than push-pull inverters since they do not require a transformer-coupled load. However, where load isolation or a step-up voltage between d.c. supply and load is essential, a push-pull inverter should be considered for low supply voltages. When the supply voltage is high push-pull inverters may require devices with large voltage ratings. Bridge inverters are then preferable.

Although mark-space (quasi-square) control is most commonly used to change the output from an inverter, it produces a very large harmonic content at low voltages. It is also not always possible to obtain zero periods in the voltage waveform with some types of inverters. Square-wave output, with control over the d.c. input to the inverter, overcomes both these disadvantages but requires an auxiliary supply for the commutation circuit.

High-frequency pulse-width modulation techniques produce low harmonics, and for bidirectional switching there are no zero dwell periods in the output. However, several of the most popular inverter commutation circuits suffer from relatively high commutation losses, and give poor efficiency with this method of control.

When choosing commutation circuitry the same considerations apply as with choppers. Parallel-capacitor commutation has universal application, although when sufficient commutation voltage is available series capacitor-inductor or coupled-pulse often gives a simpler circuit configuration with higher efficiency. When the output is a sine wave at a fixed frequency, series-capacitor commutation should be investigated.

From what has been said it is clear that it would be foolish to try to determine the ideal circuit for all applications. As a guide, when the output harmonics are not critically important, d.c. supply control with square wave output, and the use of a relatively simple inverter, perhaps with inherent large commutation loss, may prove most economical. For better performance high-frequency pulse-width modulation and a low-loss inverter would normally be required.

10

Special modifications to choppers and inverters

Thyristor chopper and inverter circuits have been discussed and classified in previous chapters. Although these classification systems represent rather rigid boundaries between different groups of circuits, it is sometimes required to modify the circuit within a group, to obtain an enhancement of one of its parameters. Such modifications are discussed in this chapter. It is important to appreciate that none of these alterations is sufficient in itself to change the classification group of a circuit.

Some modifications to the basic circuits have already been discussed. *Figure 9.27* showed an alteration to the standard coupled-pulse inverter, which enabled the d.c. supply voltage to be lowered, without reducing the commutation voltage below a certain minimum value. Similarly, *Figure 10.1* gives an addition to *Figure 9.22* to

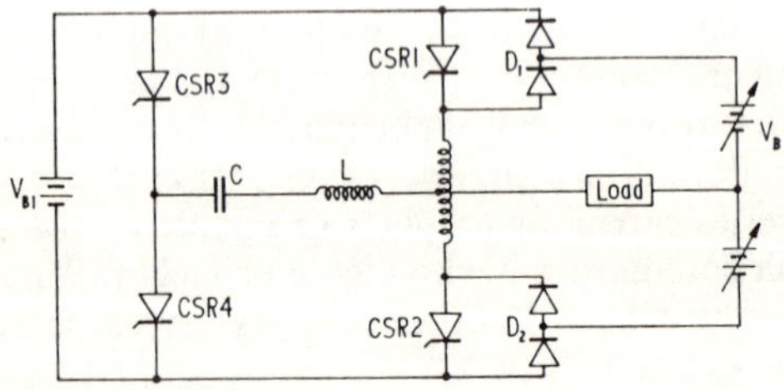

Figure 10.1. Parallel-capacitor-inductor commutated inverter with an auxiliary commutation voltage source

enable it to be used in a d.c. supply voltage control mode. When CSR2 and CSR3 or CSR1 and CSR4 conduct, the commutation capacitor is charged to at least V_{B_1} irrespective of the value of V_B. The reverse-biased action of one of the pair of diodes D1 and D2 prevents the commutation voltage source from supplying any load current.

This chapter is primarily concerned with examining the changes required to accomplish three functions:

(i) *Higher-frequency operation.* The maximum and minimum output voltage from a chopper or inverter, operating in a mark-space control mode, is limited by the set and reset times of the commutation capacitor. This limits the maximum operating frequency which can be used and still give a useful output voltage range.

(ii) *Reduction of commutation losses.* This again limits the operating frequency. A chopper or inverter may be designed to give small minimum on and off times, and adequate output voltage control range at high frequencies. But excessive commutation losses reduce the efficiency obtained and may make it unsuitable for use at these frequencies.

(iii) *Commutation voltage boosting.* The higher the commutation capacitor voltage is made, the smaller its size, for the same commutation energy output. This may lead to cheaper components. If the voltage boost is proportional to load current, optimisation of components at all outputs can be obtained.

HIGH FREQUENCY

Two problems can arise when a chopper or inverter is operated at high frequencies. First, the finite time required to set and reset the commutation capacitor limits the maximum and minimum voltage at any frequency. Secondly, the necessity of ensuring that a thyristor is reverse-biased for the duration of its turn-off time (typically 20μs) limits the operating frequency unless some form of sequencing is used, as explained below.

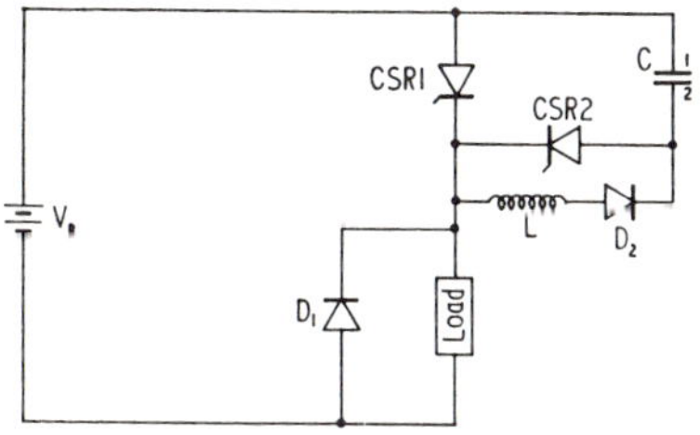

Figure 10.2. Parallel-capacitor commutated chopper

Figure 10.2 shows a chopper circuit, which has been described in Chapter 5. Suppose *C* is charged to V_B with plate 1 positive, due to a preceding cycle. Thyristor CSR1 is now fired to supply load current. This causes *C* to resonate via *L* and D2 and recharge with plate 2 positive. The resonant time is given by $\pi\sqrt{(LC)}$ and represents the 'minimum on' period for CSR1, since CSR2 cannot be fired to commutate it until the capacitor has been set. When CSR2 is fired, CSR1 turns off and C discharges at constant (assumed) load current, I_L recharging to V_B in time $2CV_B/I_L$ if the resonance loss in L is ignored. CSR1 cannot be refired until this is completed, therefore it represents its 'minimum off' period. Capacitor C is chosen to ensure that CSR1 is reverse-biased for longer than its turn-off time, under maximum load conditions. If the load current now falls, the reset time can be very large, severely limiting the peak output voltage.

Several techniques exist for resetting the capacitor rapidly, even under light load conditions. *Figure 10.3* shows a modification to

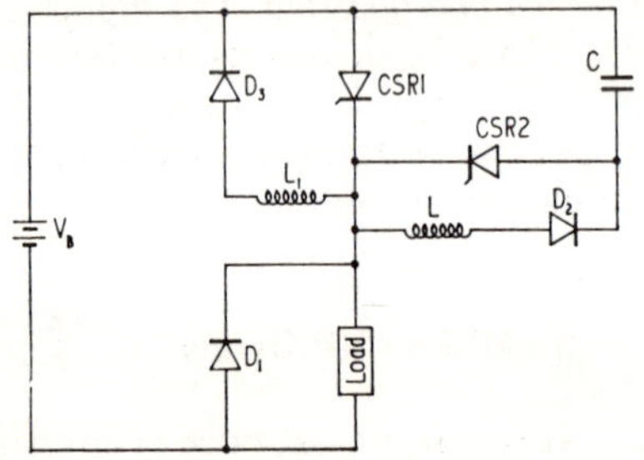

Figure 10.3. Addition of an auxiliary capacitor reset path to Figure 10.2 for higher-frequency operation

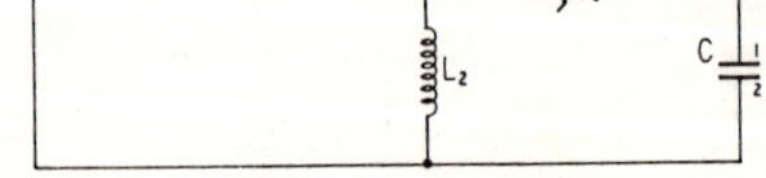

Figure 10.4. Chopper circuit with inherent high-frequency properties

Figure 10.2, with the addition of reset choke L_1 and diode D3. Now even under open-circuit load conditions the maximum reset time (neglecting resonance losses) is $\pi\sqrt{(L_1C)}$ which can be made relatively short.

Not all chopper circuits need additions to reduce the commutation capacitor reset time. *Figure 10.4* shows a parallel-capacitor commutated circuit with inherent high-frequency capabilities. CSR3 is fired simultaneously with CSR1 and sets C within time $\pi\sqrt{(L_1C)}$. When CSR2 is fired, to turn CSR1 off, the capacitor resets within $\pi\sqrt{(L_2C)}$. The circuit is an example of parallel-capacitor commutation.

The majority of inverter circuits are not affected by limitation on set and reset times of the commutation capacitor. Of the circuits described in Chapter 9, only a few need modifications. The push-pull inverter reproduced in *Figure 10.5(a)* is one of them. When CSR1 or CSR2 is fired, the capacitor has to discharge and reset through the

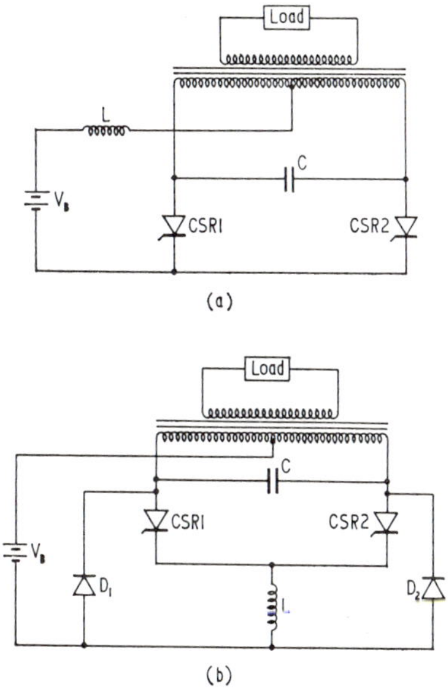

Figure 10.5. Push-pull parallel-capacitor commutated inverters suitable for lower-frequency operation

load. This time can be very large on light loads. Modifying the circuit as in *Figure 10.5(b)* allows C to reset partially by resonance through L and the feedback diodes. Resonance losses are made up by charging the capacitor through the load. For low-loss circuits reset time is considerably reduced.

Both choppers and inverters suffer from the maximum-frequency limitation imposed by the turn-off time requirements of the thyristors. Returning to *Figure 10.4* the values of L_2 and C must be large enough to ensure that CSR1 is reverse-biased for longer than its turn-off time. This puts a limit on the minimum time between the firing of CSR2 and the refiring of CSR1. For a device with 20μs turn-off, the minimum inverter period would typically be about 200μs giving a maximum operating frequency of 5kHz. For higher frequencies sequencing must be used.

Figure 10.6 shows a three-stage sequential chopper based on the basic circuit shown in *Figure 10.4*. The suffixes *a*, *b* and *c* refer to the separate inverter components. Suppose CSR1*a* was conducting and CSR2*a* has been fired to turn it off. Then before the turn-off pulse has been completed CSR1*b* can be fired to recommence the load cycle, which ends when CSR2*b* is fired. To recommence the load current, assuming CSR1*a* and CSR1*b* are still reverse biased, CSR1*c* is fired. And so on. Clearly any number of stages can be connected in sequence to obtain any required operating frequency.

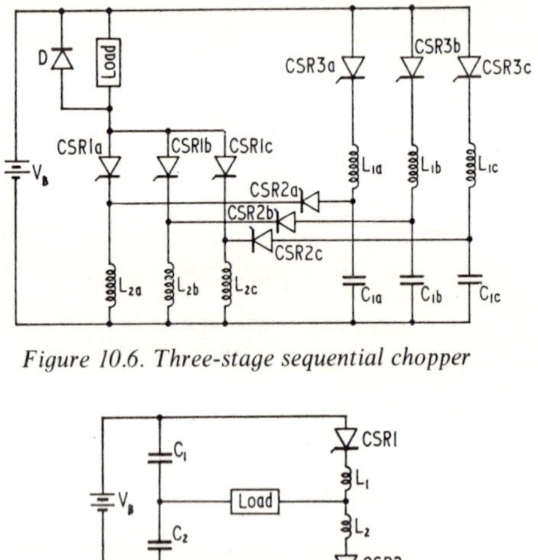

Figure 10.6. Three-stage sequential chopper

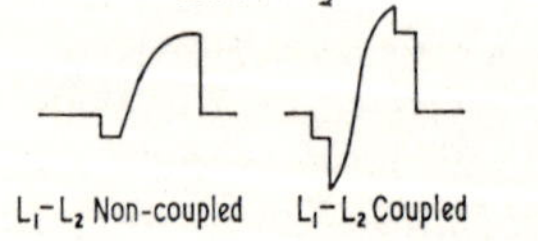

Figure 10.7. Series-capacitor commutated inverter with device waveforms

Sequencing can also be used with inverters. *Figure 10.7* shows a basic two-thyristor inverter which does not use sequencing. CSR1 is fired which charges C_2 to voltage V_x. The current through the device attempts to reverse and it turns off. CSR2 is then fired which discharges C_2 and charges C_1 to V_x. The thyristor turns off and CSR1 is refired. For a low-impedance load the output is a sine wave provided the device trigger rate corresponds to the resonance time of the inverter. The thyristor waveforms shown for L_1 and L_2 coupled and non-coupled clearly illustrate the larger turn-off pulse obtained if the chokes are coupled. This is normally used.

The operating frequency of this inverter is limited by the requirement of ensuring that devices are reverse biased for longer than their turn-off time. *Figure 10.8* shows a three-section sequential inverter which overcomes this limitation. Its operation is essentially as for the sequential chopper, device conduction periods being indicated in the figure.

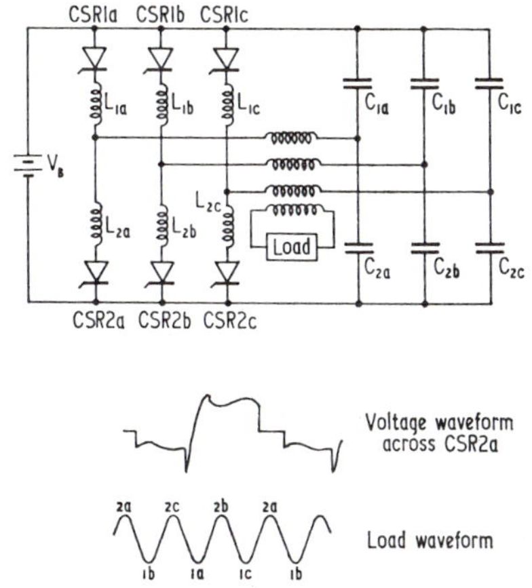

Figure 10.8. Three-stage sequential inverter with thyristor and load waveforms

It has been seen that series-capacitor commutated inverters are most efficient when used with sine-wave converters in which the inverter frequency is related to the resonant load frequency. In the discussions so far the two frequencies have been maintained equal. However, this does not necessarily have to be so. In *Figure 10.9* the output load is a tuned circuit L_2C_2 with relatively high Q. Once it is set into oscillation it will therefore produce a sine wave output whose amplitude decreases slowly due to losses. Periodically CSR1 and CSR2 are fired to compensate for this. *Figure 10.9* also shows various circuit waveforms. Firing CSR1 will charge C_1 with plate *a* positive. CSR1 now goes off and CSR2 is fired. Capacitor C_1 resonates with the load and recharges with plate *b* positive. CSR2 will turn off and remain reverse-biased so long as the peak amplitude of the load oscillations is arranged to be lower than the capacitor voltage. This is shown clearly in the thyristor waveforms. Such an inverter is very useful since the output frequency can be several times

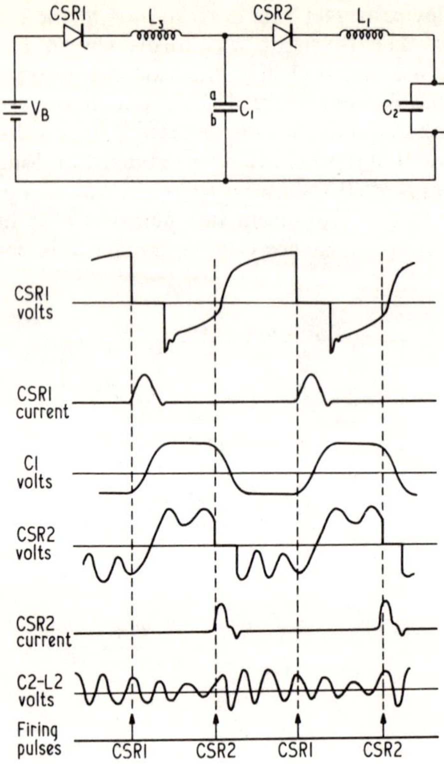

Figure 10.9. Series-capacitor-commutated inverter in which the load frequency is several times that at which the devices operate

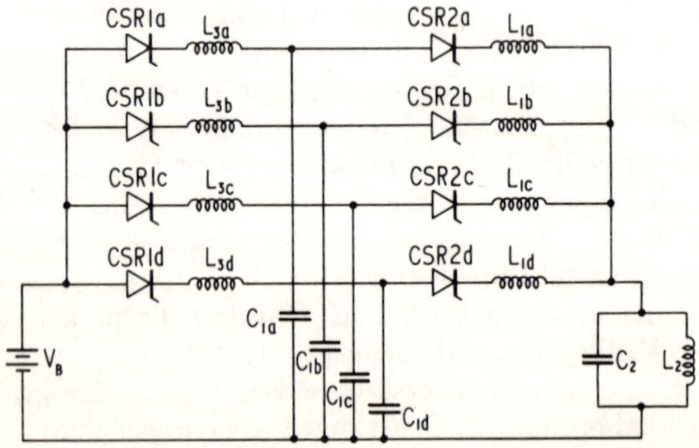

Figure 10.10. Four-stage sequential inverter using the basic arrangement of Figure 10.9

the inverter frequency, allowing slower turn-off devices to be used. However, if the Q of the load circuit is low, the inverter frequency approaches the output frequency and in the limit must be equal to it.

Figure 10.10 shows a method of sequencing *Figure 10.9.* The operation of the circuit is illustrated by the waveforms of *Figure 10.11.* In the example shown the thyristors are fired at one eighth the tuned output frequency and the bursts of energy are fed to the output at half the output frequency.

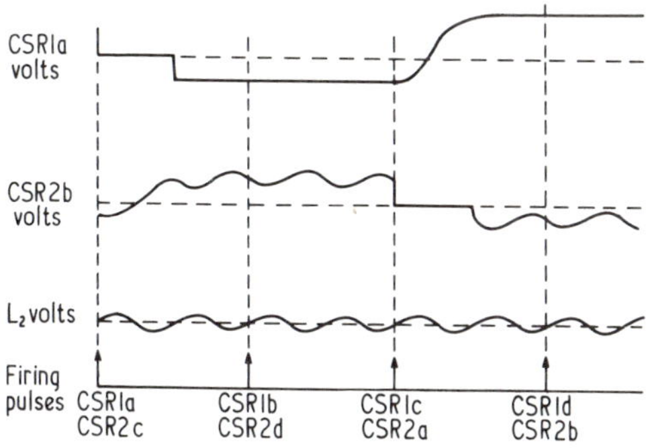

Figure 10.11. Circuit waveforms of the inverter of Figure 10.10

Although theoretically there is no limitation on the operating frequency of sequential choppers and inverters, in practice this is not so. A thyristor shows appreciable switching losses which occur principally during device turn on, when there is a large voltage across it while the current is increasing. This can cause considerable heating, which is serious at high frequencies. Therefore depending on the thyristor switching rate and the inverter efficiency required, the maximum operating frequency of an inverter is fixed, although this is well above that obtained when sequencing is not used.

REDUCTION OF COMMUTATION LOSSES

Losses normally occur in commutation circuitry due to the transfer of energy from the commutation capacitor to a commutation choke, during the turn-off process, and the subsequent dissipation of the energy stored in the choke. The commutation loss is proportional to the load current, which determines the capacitor size for any voltage, and the frequency of operation. There is a fixed watts loss per commutation, so that the higher the operating frequency the larger the energy loss, and the lower the system efficiency.

Choppers do not normally have inherent commutation loss problems. For instance referring to *Figures 10.2* and *10.4* the commutation energy is interchanged between choke and capacitor during the turn-off period, so that there should theoretically be no efficiency loss. Referring to *Figure 10.4* and assuming C to be charged to V_{CM} with plate 1 positive, then when CSR2 is fired, CSR1 turns off. C now resonates with L_2 and discharges to zero voltage. The commutation energy has transferred from C to L_2. However, the commutation interval proceeds beyond this point, with the transfer of energy back from L_2 to C, so that the capacitor recharges to V_{CM} with plate 2 positive. When CSR3 is next fired C will resonate through the supply and L_1 and be reset. There is therefore zero commutation loss. However, it can be seen that the capacitor voltage builds up over several cycles to the value V_{CM} which may be several times the supply V_B, depending on the losses in the resonant paths. For zero loss the voltage would reach an infinite value. To limit this voltage boost free-wheeling diodes are sometimes connected across L_2. This means that when C has discharged to zero volts, the energy stored in L_2 is dissipated in the free-wheel path and there is no build-up of capacitor voltage. The price paid is loss of efficiency.

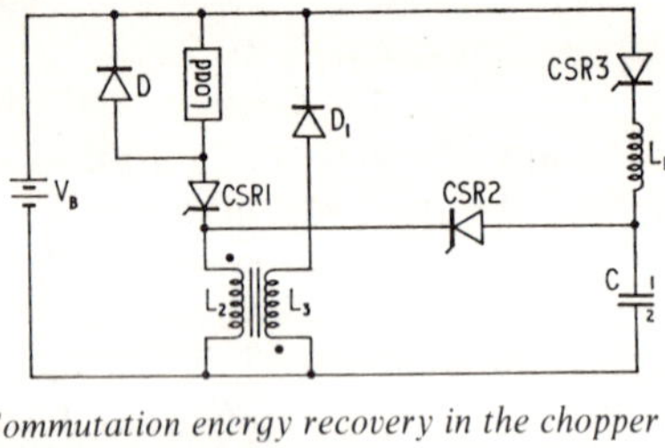

Figure 10.12. Commutation energy recovery in the chopper of Figure 10.4

Figure 10.12 shows a modification to the circuit of *Figure 10.4*, which prevents excessive commutation capacitor voltage build-up (which increases device voltage ratings) while still maintaining high efficiency. Winding L_3 is closely coupled to L_2 and normally has a voltage induced in it such that D1 is reverse-biased. During the commutation interval, however, when the voltage on L_2 reverses, D1 will conduct, clamping L_3 to V_B and giving an overshoot across L_2 of $V_B\sqrt{(L_2/L_3)}$, which can be made small. There is theoretically no commutation loss since the commutation energy stored in L_2 is fed back to the supply by L_3.

Inverters are normally more prone to commutation losses than choppers, probably due to their greater complexity. There are many circuits which are usually loss free, but which exhibit losses when free-wheeling diodes are added across chokes, to prevent excessive voltage build up in the system. As with choppers feedback windings can now be used to overcome this.

Figure 10.5(b) illustrates a popular push-pull inverter configuration. It is unsuitable for high-frequency operation owing to its large commutation loss. For instance when CSR1 fires, capacitor *C* discharges through *L* and recharges to twice the supply voltage. Then diode D1 conducts, feeding back inductive load current to the supply; and the energy stored in *L* during the commutation interval is dissipated in the *L*-D1-CSR1 path. *Figure 10.13* shows a modified

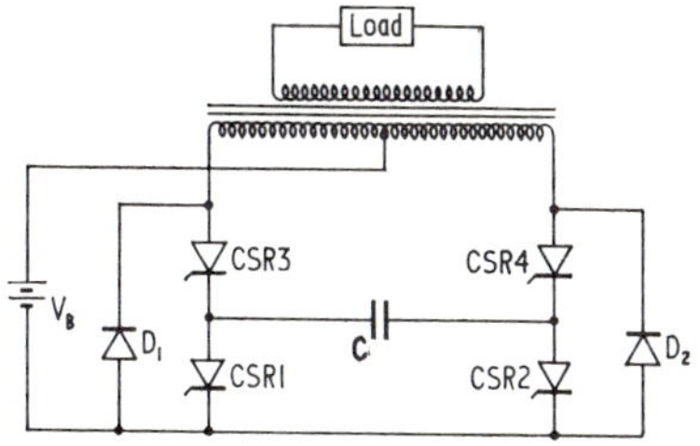

Figure 10.13. 'Lossless' push-pull inverter

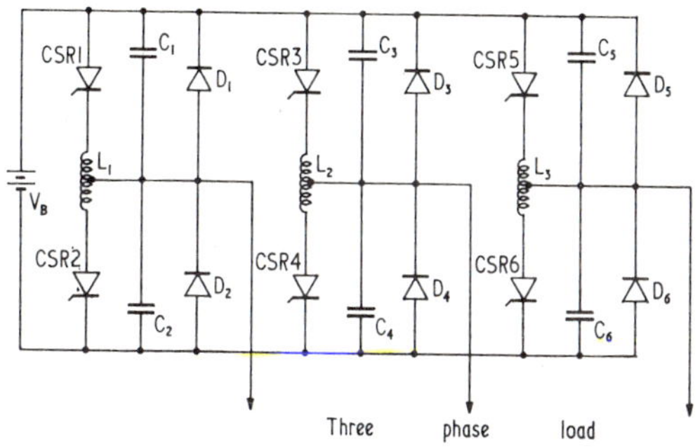

Figure 10.14. Coupled-pulse three-phase bridge inverter with high commutation losses

push-pull circuit which has zero commutation loss. Four thyristors are used so that there is no possibility of a short circuit of the commutation capacitor during the turn-off process, and the series choke of *Figure 10.5* can be eliminated. Suppose CSR1 and CSR3 are conducting, CSR4 is fired to prime the capacitor, after which it turns off. To reverse the load current CSR2 is fired. The capacitor discharges and recharges with reverse polarity through CSR3, after which this thyristor also goes off. There is clearly no commutation loss due to the absence of stored energy in a commutation choke. Bridge inverter coupled-pulse circuits of the McMurray-Bedford type, as shown in *Figure 10.14*, are probably the most frequently used. They suffer from high commutation losses. For instance, if CSR1 is conducting C_1 is discharged and C_2 charged to V_B. To turn CSR1 off CSR2 is fired. This couples a pulse to the top device turning it off. Capacitor C_2 then discharges to zero voltage, after which time the energy stored in L_1 is dissipated in the free-wheeling path CSR2-D2-L_1.

Several modifications have been used to reduce this circulating current, for instance by tapping the load. *Figure 10.15* shows the

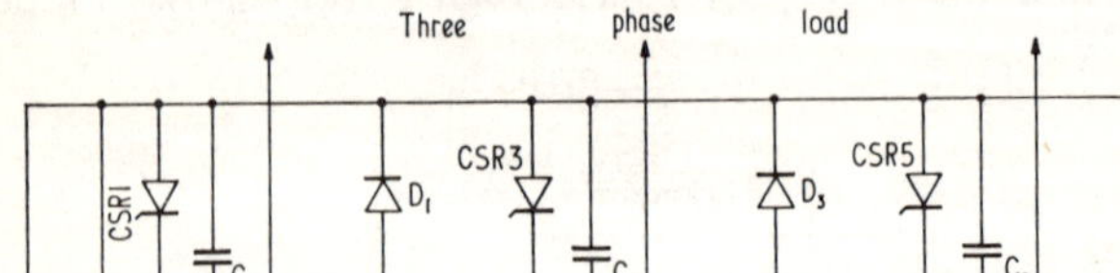

Figure 10.15. Recovery of commutation energy in the inverter of Figure 10.14

addition of an auxiliary transformer with feedback winding which gives much better performance. During the commutation interval, for instance when CSR2 is fired as above, the energy stored in L_1, in circulating through D2, induces a voltage in winding 1 which feeds back the power from L_1 to the supply via step up winding 4 and bridge rectifier D7. There is an overshoot voltage on C_2 due to this feedback action, as seen with chopper circuits. However, it is small. Therefore for a 1 :20 turns ratio between windings 1 and 4 it will be 1/20th of the supply voltage.

VOLTAGE BOOSTING

If the commutation voltage is high, the commutation capacitor can have a low value and still provide the same turn-off energy required for the thyristors. For instance in the chopper circuit of *Figure 10.2* if t_F is the turn-off time of CSR1 then to commutate I_L the capacitor size required is:

$$C = \frac{I_L t_F}{V_C} \tag{10.1}$$

V_C is the commutation capacitor voltage prior to commutation, and here it can be a maximum of V_B. This equation shows that to commutate 500A using a capacitor voltage of 20V and 20μs turn-off time thyristors would require a capacitor of 500μF, 20V rating. If the value of V_C is increased by a factor of 10, by some method of voltage boosting, a commutation capacitor of 50μF, 200V will be needed, which is smaller and cheaper. Another disadvantage of using high-valued capacitors is that the resonant time of C through L is increased. Suppose this is to be limited to 0·1ms to give sufficient output voltage control at high operating frequencies. Then L must be 2μH, which is impracticable since normally the inductance of connecting leads would be greater than this. Clearly a voltage system which increased V_C and reduced C would be advantageous.

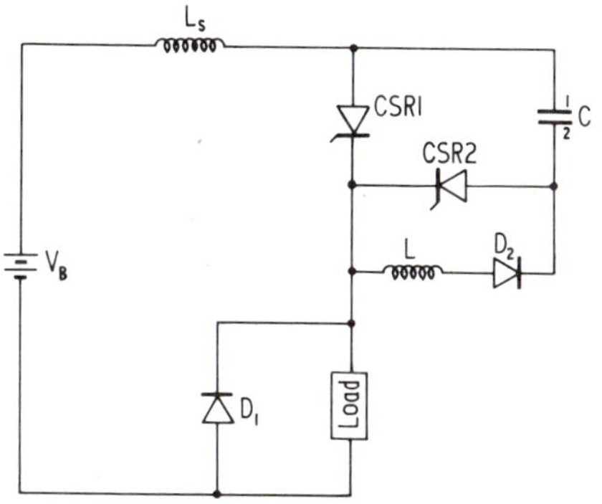

Figure 10.16. Chopper with series-inductor voltage boosting

Figure 10.16 shows a modification to *Figure 10.2* where a series inductor L_S is used to increase the commutation voltage. The operation of this is as follows. When CSR1 is conducting suppose that the load current reaches a value I_L. CSR2 is fired to turn it off. Capacitor C discharges at constant current I_L, assuming the load inductance is sufficiently large to maintain the current at this value during the short capacitor discharge period. C recharges to V_B with plate 1 positive. Normally if L_S were not present CSR2 would turn off at this point and the load current would free-wheel in D1. However, owing to the presence of L_S there is energy $\frac{1}{2}L_S I_L^2$ stored in it which causes CSR2 to remain on, the voltage on C increasing as the current through it decreases; the difference between this and the load, free-wheeling in D1. When C has charged to $V_B + I_L\sqrt{(L_S/C)}$ the current through it has fallen to zero and D1 carries the total load current. CSR2 now goes off. When CSR1 is next fired C resonates through D2 and recharges to an increased value of commutation voltage.

The value of L_S need not be large. For instance, in the previous example considered, for $I_L = 500$A, $C = 50\mu$F a value of L_S equal to only 8μH gives a voltage boost of 200V. The principal disadvantage of this system is that voltage boost is not effective during the first cycle, when CSR2 is first fired without any load current flowing. Thereafter the voltage boosting increases proportionally to the load current, a valuable asset.

Figure 10.17 shows a further modification to the basic parallel-capacitor commutated chopper which overcomes the disadvantage

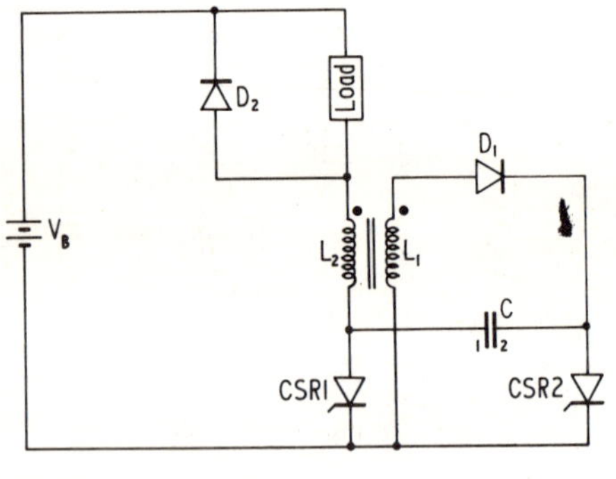

Figure 10.17. Transformer voltage boost in a chopper circuit

of not providing a voltage boost during the first commutation, as experienced in *Figure 10.16*. Initially CSR2 is fired to charge C to V_B with plate 1 positive. When CSR1 is fired C resonates through it and $L_1 - $D1, recharging with plate 2 positive. In addition to this discharge, the build up of load current in L_2 induces a voltage in winding L_1 which further increases the commutation capacitor voltage. This voltage boost is clearly proportional to the load current. It is caused by the change of load current during the cycle in which commutation is to occur, and not in a previous cycle; therefore it is effective from the very first commutation.

In inverters, as in choppers, there are several circuits where voltage boosting occurs as part of the circuit function. In these it is necessary to reduce the boost, if it is undesirable, by feed-back windings to the supply, as discussed in the previous section. There are many inverters where the capacitor voltage cannot exceed that of the supply unless auxiliary circuits are used.

Figure 10.15 showed a coupled-pulse inverter which used a feed-back transformer to increase its high-frequency performance, by reducing commutation losses. The commutation capacitor voltage in this circuit is clamped, by the action of the free-wheeling diodes D1–D6, to V_B. To increase the capacitor voltage inductors can be placed in series with the supply or in the free-wheeling diode paths. Alternatively, boost transformers can be used as shown in *Figure 10.18*. The primary of transformer T_1 is connected in the main load current path as shown. Four secondaries feed the commutation

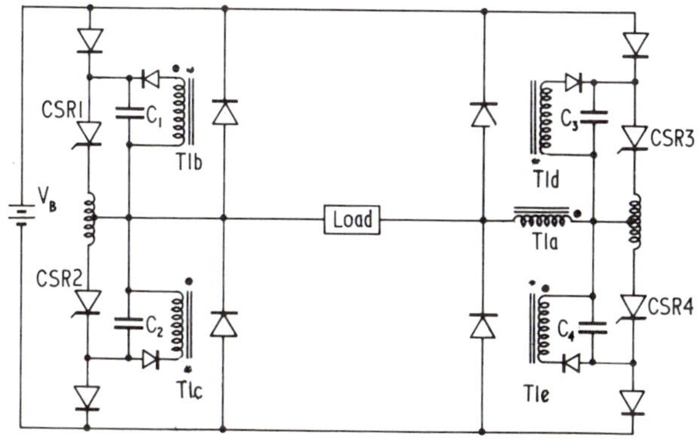

Figure 10.18. Transformer voltage boost in an inverter

capacitors, their polarities being as indicated. Since the voltage on the capacitors is to exceed the supply voltage, diodes in series with the main thyristors are required. Suppose CSR1 and CSR4 are conducting. The load current build up in T_{1_a} induces a voltage in T_{1_c} and T_{1_d} which charges C_2 and C_3 to boost voltages proportional to the magnitude of the load current change. This voltage is held on the capacitors until CSR2 and CSR3 are fired, and they form part of the inverter commutation voltage.

Figure 10.19 shows a modification to a standard inverter which has been discussed in *Figure 9.19*. Inductors L_d are required as part of the commutation process. Inductors L_x in the free-wheel diode path are added to give voltage boost proportional to the load current. Since the commutation capacitors are to charge to a voltage greater than that of the supply, auxiliary thyristors are required in series with them, to prevent a loss of charge when boosting stops. CSR1*a* is fired with CSR1, CSR2*a* with CSR2 and so on, so that apart from boosting the circuit operation is as before.

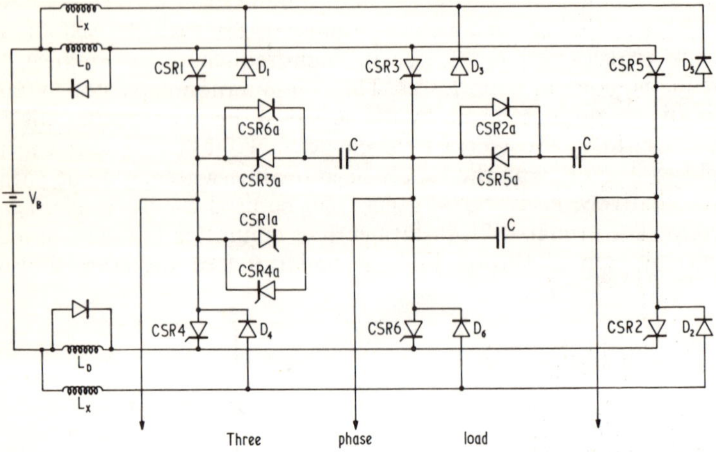

Figure 10.19. Series-inductor voltage boost in a bridge inverter

Figure 10.20 shows circuit additions to *Figure 10.5(b)* for voltage boosting. Once again a series inductor L_2 is used. After C has charged to $2V_B$ at the end of a commutation the energy stored in L_2 during a previous load cycle transfers to C boosting its voltage. Diodes D1 and D2 can conduct, carrying the inductive load current. However,

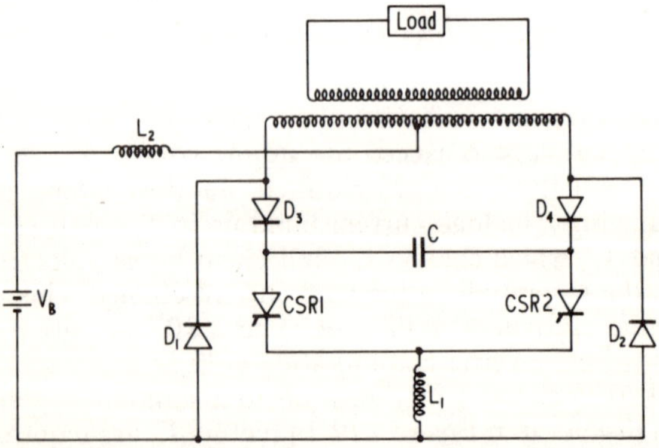

Figure 10.20. Push-pull inverter with series-inductor voltage boost

owing to the reverse-biased action of series diodes D3 and D4, the boost voltage on C is not lost, being available for the next commutation.

CONCLUSIONS

Modifications to standard choppers and inverters are often required. For instance, if the circuit is to operate at a high frequency, under variable load conditions, the value of the commutation capacitor is chosen to handle the peak expected current. But on light loads the capacitor reset times can be excessive unless an auxiliary circuit, or sequencing, is used. High frequency also means greater losses, especially during commutation. It is sometimes desirable to recover this loss by feeding it back into the supply.

For circuits operating from low-voltage supplies, voltage boosting gives reduced commutation components, but increases device voltage ratings. In some circuits where the current is large and the supply very low, voltage boosting may be essential to obtain commutation device values which can be practically obtained.

As has been shown, none of the above modifications changes the commutation group of the chopper or inverter, since they do not affect the basic mode in which the commutation capacitor is connected to the thyristor being turned off. The additional circuits also increase the cost and complexity of the chopper or inverter, and therefore should not be used unless it is required for the application under consideration.

11

Machine control

In previous chapters the properties of thyristors have been explained and typical power applications given. Since its first fabrication, in the latter part of 1957, the thyristor has grown in popularity and now finds use in widely differing fields. However, this device was intended originally as a power semiconductor and nowhere has its effect been felt more than in machine control. Owing to its small size, ruggedness and relatively low cost, the thyristor has given practical reality to systems which were before only an experimenter's dream.

Control electronics bridges the gap between two widely differing sciences. On the one hand the thyristor, with its associated control circuitry, may be regarded as a tool of the electronic engineer, whereas the machine, which differs only slightly from conventional designs, is the instrument of the electrical engineer. But it is not possible to treat thyristors and machines separately if they are to work together as a system. The characteristics of the machine affect performance of the thyristor circuit and vice-versa. However, the rapid growth of thyristor control systems has resulted in electronic engineers developing systems, while having no knowledge of machine fundamentals, or electrical engineers blundering into a thyristor control scheme, even when unfamiliar with transistor circuitry.

A control engineer is a unique animal who must have a good working knowledge of both electronics and electrical machines. The previous chapters have dealt with thyristors at some length. This chapter considers the principles of machines and how they can be controlled by thyristors.

Machines fall into two groups, motors and generators. To explain the difference between them it is necessary to discuss the operating principles of an elementary machine, as in *Figure 11.1.* The essential requirement for production of magnetic force is interaction of two magnetic fields. The force of attraction and repulsion between two bar magnets is well known. A similar force would result if one or both magnets were replaced by a current carrying conductor.

In *Figure 11.1* magnet N-S is fixed (stator) whereas coil *a-b* is mounted on a drum free to rotate (rotor). If current flows into the rotor at *a* and out at *b*, as indicated, this will produce a magnetic field so that side Y is a north pole and X a south pole. The drum will rotate, the north pole at Y aligning itself with the stator south pole. The machine is an elementary motor, the electrical energy in the coil being converted to mechanical work on the drum.

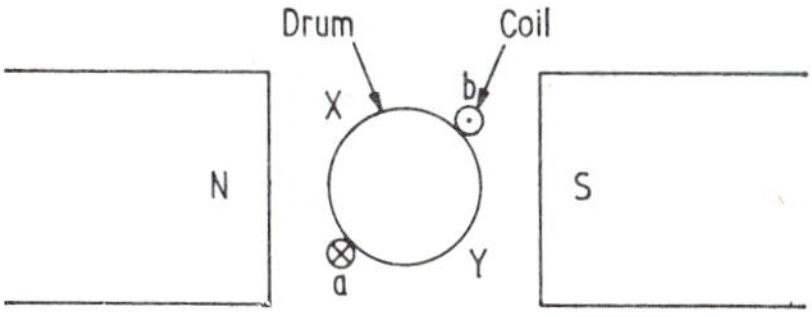

Figure 11.1. Elementary electrical machine

Now assume that no external voltage is applied to coil *a-b*, but the drum is rotated clockwise. Work done in rotating the drum must be opposed by current flowing in the coil. (If it were to aid the external force this would create the well-known paradox of a perpetual motion machine.) The direction of coil current will be as indicated in *Figure 11.1*. Mechanical work is done in overcoming the repulsion between similar poles on stator and rotor; this is converted to electrical energy in the coil. The machine is a generator.

It is clear that the construction of a motor and generator for a given type are very similar; it is only the terms of reference which differ. For a motor, energy input is electrical and output is mechanical. For a generator, mechanical input energy is converted to an electrical output.

Motors and generators can be further divided into various groups as shown in *Figure 11.2*. These will be discussed in the following sections.

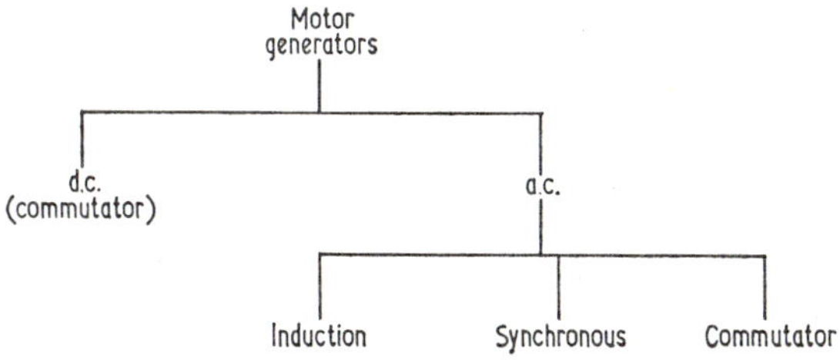

Figure 11.2. Types of electrical machines

Figure 11.3 shows the construction of an elementary d.c. machine where the rotor is a single coil and the commutator consists of two half cylinders. During a complete coil rotation the commutators are in contact with brushes A and B for half the period. During motor operation a d.c. voltage is applied to brushes A and B, with the polarity shown. Current will flow up coil side *a* and down side *b*. This

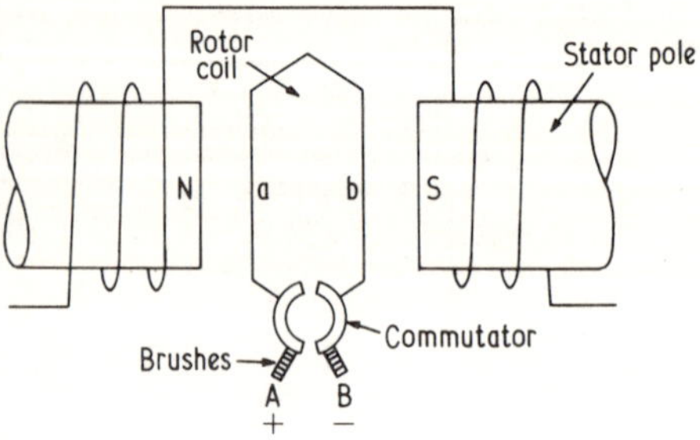

Figure 11.3. Basic d.c. machine

current produces rotor poles, which interact with the stator flux to produce a torque on the coil in an anticlockwise direction. The function of the commutator is to switch the rotor poles, by reversing coil current, as shown in *Figure 11.4*, and so maintain unidirectional motion.

In *Figure 11.3* if brushes A and B were connected to a load and the armature rotated in a clockwise direction, by an external mechanical force, an e.m.f. would be induced in coil sides *a* and *b*, forcing current down side *b* and up side *a*. The commutator ensures that this current is unidirectional in the load. The machine is now a d.c. generator.

During motor action, when the rotor revolves, the changing flux induces in it a voltage which will produce poles to oppose the stator flux. This is referred to as the motor back e.m.f. It is a function of the

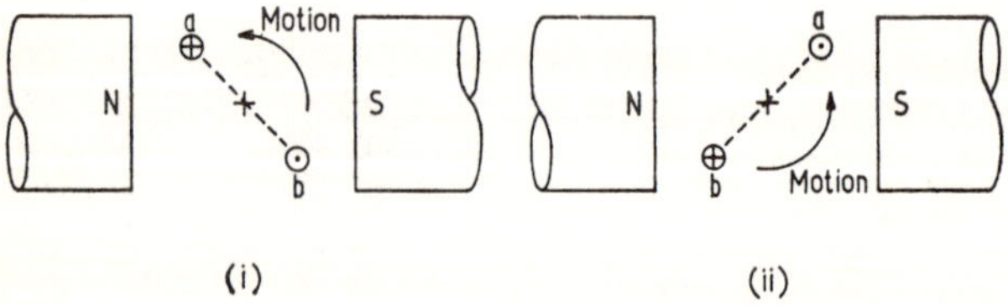

Figure 11.4. The switching action of a commutator

motor speed and the strength of the field flux, and is secondary generator action in a motor.

The function of the commutator must be clearly noted. It senses the rotor position (by virtue of its construction) and switches the rotor current, at the appropriate instant, to ensure that torque is unidirectional in a motor and that the generator output is d.c.

Figure 11.5 illustrates an elementary a.c. motor, which differs from *Figure 11.3* in that the commutator and brushes are removed, and the ends of the rotor short-circuited. Assume magnetising current to flow and produce field poles as illustrated. Now if the poles were caused to rotate they would induce a current in the rotor coil. This current, by Lenz's law, would oppose the rotating stator field. *Figure 11.5* shows two positions of the stator field, the rotor being

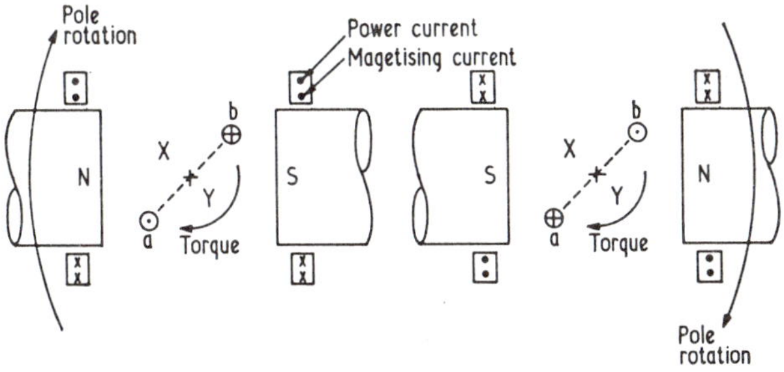

Figure 11.5. Induced rotor polarity during rotation of the stator field

assumed at a standstill, and illustrates that the rotor current is such that the torque is always unidirectional. A.C. is induced in the rotor by the stator flux, the action being identical to that of a transformer with short-circuited secondary. An equal and opposite current flows in the stator to balance the rotor ampere turns. This is the power current.

This machine is called an induction motor. In an actual machine the stator poles are not physically rotated. Instead they are supplied with magnetising currents displaced in amplitude and time, to produce a rotating stator flux. The stator will carry this input a.c. magnetising current, with an input a.c. power current, which produces mechanical motor output.

The rotor of an induction machine may be driven by an external force at a speed greater than that of the stator field. This is shown in *Figure 11.6* where the stator poles are assumed almost stationary. Applying the principle that the poles induced in the rotor by the stator field will oppose this stator field, and that the rotor ampere turns are balanced by flow of stator power current, current directions will be as shown. It is seen that the power current is reversed. The machine converts mechanical input into electrical output and is now a generator.

Since current is induced in the rotor due to a changing flux linkage with the stator field, it is evident that an induction machine must always run below (motor) or above (generator) the speed of the rotating stator field, never at the same speed. It was also seen that in a d.c. machine power and magnetising currents are carried by two separate windings, whereas for an induction machine they both flow in the same winding. This is the fundamental difference between the two. It is essential in both systems that stator poles be maintained during both motor and generator action.

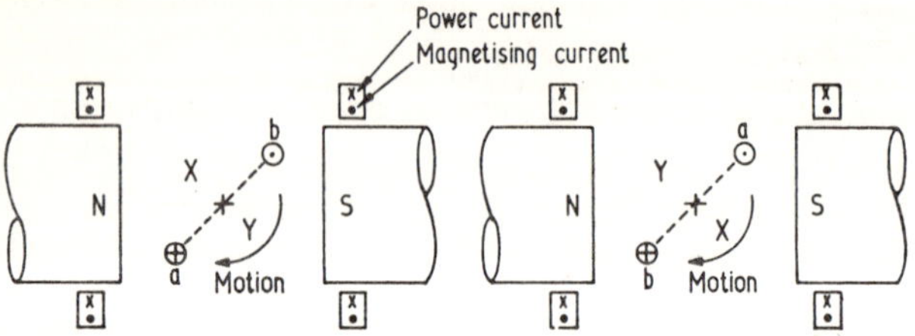

Figure 11.6. Generator action in an induction machine

If the rotor of *Figure 11.3* is replaced by a magnet, the result is a synchronous machine. Like the induction machine the stator field is rotating which enables the rotor to 'lock on' and follow at the same speed. Any other speed would cause alternate attraction and repulsion, making the machine unstable and causing it to stop. For motor action the input at the stator provides excitation for the revolving field and supplies the power component necessary to overcome the load torque and cause the rotor to follow the stator field. In this respect the machine is similar to an induction motor. However, when generating, no input is required to the stator, since excitation is provided independently to the rotor. The action is now more like that in a d.c. generator. In fact the only difference between the synchronous and the d.c. generator is that the commutator (rectifier) is not present in a synchronous machine, resulting in an a.c. output. In many synchronous machines the rotor poles are produced by an electromagnet. The commutator in *Figure 11.3* is replaced by slip rings so that the current in coil sides *a* and *b* is unidirectional and independent of rotation.

The a.c. commutator machine is very similar to an induction machine. The stator field is revolving, the rotor being connected to a commutator as in a d.c. machine. This is illustrated in *Figure 11.7*. The rotating field induces alternating current in the rotor. The frequency of these currents depends on the frequency with which the stator field cuts the rotor conductors and on the speed difference between stator field and rotor. This is as for normal induction machines. However, the rotor coil currents flow through the brushes, and since the brushes are stationary the stator field always cuts them at a fixed speed (that of the rotating field), so that the frequency at the brushes is constant. The commutator acts as a frequency converter, between the speed-dependent frequency in the rotor conductors and the fixed frequency at the brushes. This introduces several advantages during machine control, as seen later.

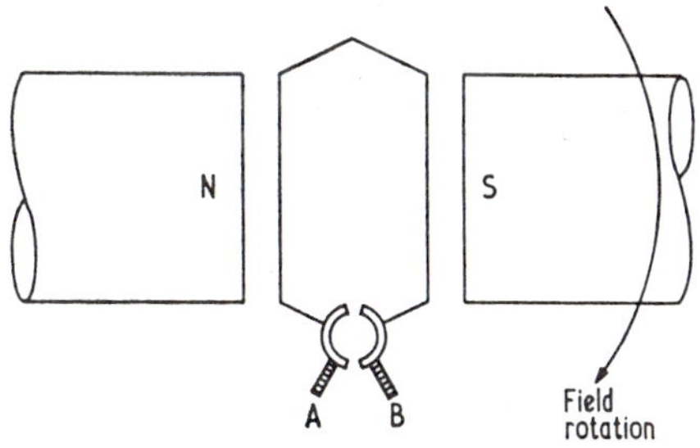

Figure 11.7. Elementary a.c. commutator machine

Having looked at basic types of machines, the methods by which they can be controlled are now examined. In general, there are three considerations for motors, *i.e.* starting, speed control, and braking.

D.C. MOTORS

The armature and field coils of a d.c. motor can be arranged in several ways as shown in *Figure 11.8*. The arrangement used determines the overall performance of the motor. In all cases the following equations hold:

$$E = V - I_a R \tag{11.1}$$

$$N = \frac{E}{K_n \phi} \tag{11.2}$$

$$T = K_t \phi I_a \tag{11.3}$$

$$\phi = K_f I_f \tag{11.4}$$

where V is the applied d.c. voltage, E is motor back e.m.f. at speed N and field flux ϕ, I_a is the value of the armature current which gives motor torque T, I_f is the field current and K_f, K_n and K_t are called the flux, speed and torque constants respectively. R is the series resistance of the motor and is equal to either armature resistance alone or sum of armature (R_a) and field (R_f) resistances, depending on the type of motor connection used.

The performance of the motor can be readily derived by the above equations. For example, equations (11.1) and (11.2) show that the motor speed is roughly proportional to supply voltage V (ignoring secondary effects due to volt drop caused by armature current), provided the motor flux does not also vary in the same manner with voltage. This can be satisfied with series and separately excited machines but not with shunt field windings in which the current is determined by the supply voltage and field resistance. Therefore shunt motors cannot be used for speed control using supply voltage variation.

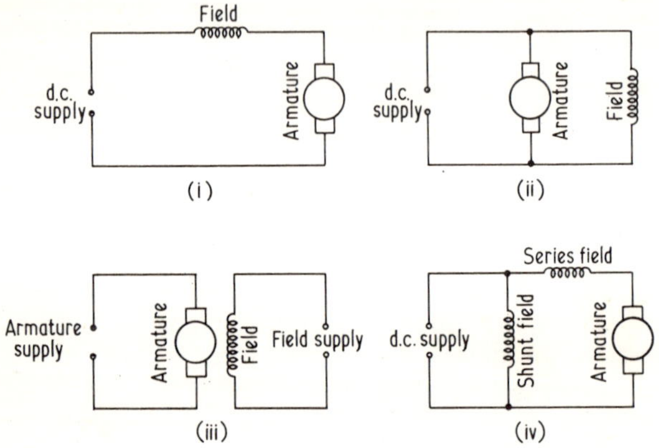

Figure 11.8. Armature and field arrangements in a d.c. machine: (i) series excitation, (ii) shunt excitation, (iii) separate excitation, (iv) compound excitation

The methods by which a d.c. motor can be started, controlled and stopped will now be examined with reference to some typical thyristor control circuits. These circuits are used for illustration purposes only, and any of the circuits described in this book may be used, provided their output corresponds to that required by the motor.

Starting

Equation (11.2) shows that when a motor is starting, its back e.m.f. is very small at low speeds. Therefore, from equation (11.1) the armature current will be large since the armature resistance is small. The high starting current would give a good motor starting torque, and this is one of the advantages of a d.c. motor. However, excessive currents can cause machine damage over a period of time, and should be avoided. For large motors the peak current drawn at starting is also limited by the supply authorities, in order to prevent reductions in the voltage of the supply to other consumers. Since the armature current is proportional to the difference between supply voltage and motor back e.m.f., and inversely proportional to the armature resistance, there are two methods which may be used to limit motor starting current. In one technique an external resistance can be connected in series with the armature to increase the effective armature resistance. As the motor speeds up the resistance is progressively reduced, either manually or by automatic methods such as servo-driven potentiometers. This is the system most commonly used with conventional d.c. motor starters. With thyristor drives it is much easier to arrange for the supply voltage to the motor to be progressively increased as the machine speeds up, so that the peak armature current is held at a predetermined value which would give good starting torque and long motor life. Depending on whether the motor is supplied from an a.c. or d.c. source it is now possible to use a variety of thyristor circuits to give controlled rectification (Chapter 6), or d.c. line control (Chapter 5) respectively. The problems of static contactors have been discussed in Chapter 3.

Control

The principal parameter of interest in d.c. motor control systems is the speed of the machine. Generally this parameter is linked in a secondary loop to the motor current. Such an arrangement is shown in *Figure 11.9*. The d.c. motor is mechanically connected to a tachogenerator which provides a feedback signal proportional to the motor speed. This feedback signal is compared with the reference (or demand) speed signal and the error, after amplification if required, can be used to adjust the thyristor controller so that the motor speed equals that of the reference. *Figure 11.9* shows a system in which the speed error signal is first passed into another servo point where it is controlled by a current feedback signal. This inner loop has overriding command such that, even though the reference speed may be abruptly changed, the error signal driving the thyristor controller is modified to ensure that motor current is always below a predetermined maximum value. Although a tachogenerator has been shown in *Figure 11.9* as providing the feedback speed signal, it is not the only system that can be used. For instance equations (11.1) and (11.2) indicate that the motor speed is proportional to the supply voltage provided allowance is made for the drop in the motor armature resistance. Since this resistance can be measured, all that is required is a knowledge of the armature current at any supply voltage. Therefore the feedback speed signal can be made proportional to the supply voltage on which is superimposed an $I_a R$ drop compensation signal. Since this signal is proportional to armature current it may also be used in the current limit loop if required.

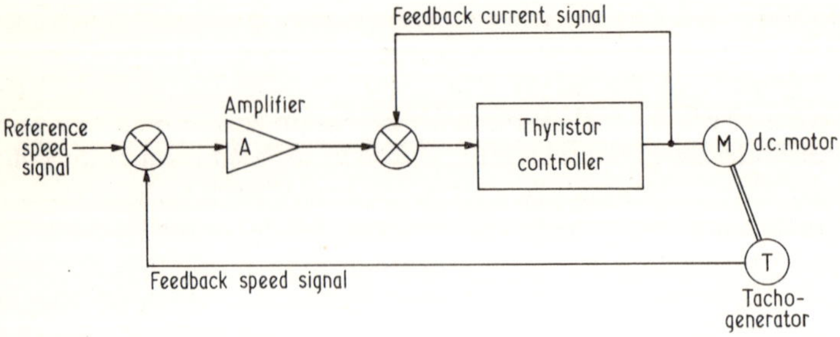

Figure 11.9. Closed-loop d.c. motor speed-control system

For a certain load torque, motor speed will be such that the resultant current interacting with the field will give a machine torque just sufficient to overcome the load, plus internal losses. Speed control is therefore a form of torque control. In a d.c. motor, if the armature voltage is reduced, but the load torque kept constant, speed will fall so that the back e.m.f. reduces sufficiently to increase the current to almost its original value (flux being kept constant) and give an unchanged machine torque. Similarly a reduction in field flux would cause a reduction in speed so that the increased current can give the same torque. The flux-current product would tend to be maintained constant. If the load torque was allowed to fall as flux decreased, motor speed would increase to compensate for lost torque and so keep the power constant.

Depending on the type of supply, either a form of controlled rectification or d.c. line control may be used. *Figure 11.10* shows a system which uses a unidirectional thyristor converter to control the voltage

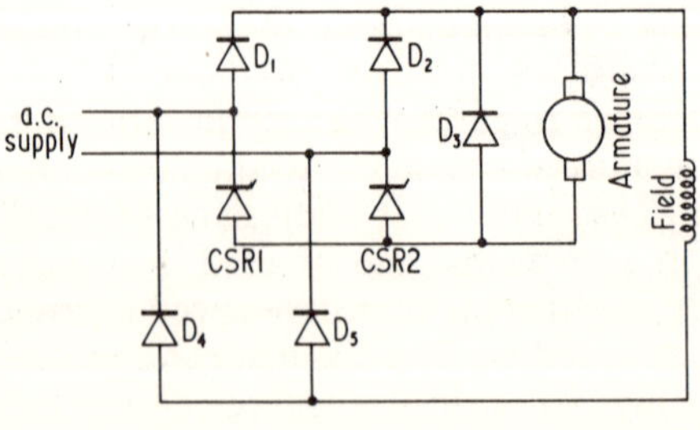

Figure 11.10. Separately excited motor speed-control system

applied to the motor armature. The field may be supplied either from a separate bridge rectifier, or as in *Figure 11.10* where diodes D1 and D2 are used to pass current to both the armature and the field. The field is supplied by the diode bridge D1-D2-D4-D5. Although this circuit economises on the number of devices used, the ratings of diodes D1 and D2 have been increased. For large machines a separate field rectifier system is to be preferred.

Figure 11.11 shows a simple half-wave series motor drive which has found use in small domestic control systems such as food mixers. The circuit also has an element of feedback speed control, as will be

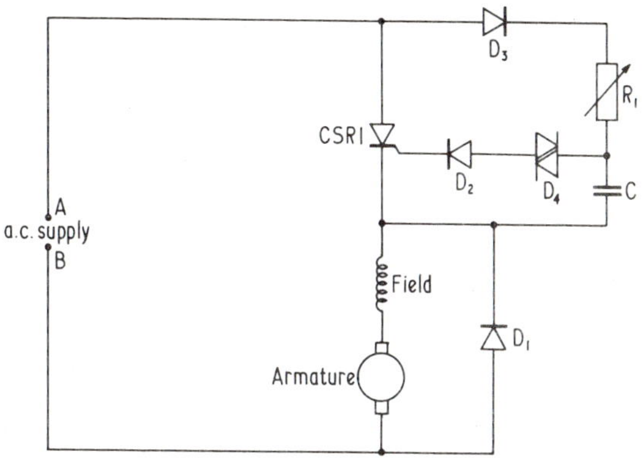

Figure 11.11. Simple half-wave series motor drive

evident from the description of its operation. When line A of the input goes positive to B, capacitor C begins to change via R_1 and the motor. As soon as the capacitor voltage reaches the breakdown value of trigger diode D4 it breaks over and fires CSR1. Load current now flows from the a.c. supply to the motor for the remaining half cycle. When the supply polarity reverses CSR1 turns off and motor load current free-wheels in D1. If the motor is heavily loaded then diode D1 is forward-biased for the whole time that CSR1 is off. This means that when C commences to charge, its time constant is determined primarily by the value of R_1 (motor resistance being small). Since the motor speed of a series motor varies substantially with load this would mean that when the load decreased the motor would speed up even though the value of R_1 remained unchanged. However, the load current would now tend to go discontinuous during a half cycle, so that D1 no longer clamped the motor voltage. This indicates that the charging time of C will be affected by the motor back e.m.f. (which is developed on residual motor flux). Since this e.m.f. increases with speed the capacitor would take longer to reach the trigger voltage of D4. This effectively increases the firing delay of the half-wave cycle and reduces the motor voltage. The motor speed will therefore fall and remain relatively stable about the value determined by the setting of R_1. Therefore R_1 represents the demand speed and motor back e.m.f. is the feedback speed signal.

Although it is usual to control the armature voltage on a d.c. motor, equation (11.2) shows that speed variation can also be obtained by keeping the armature voltage fixed but changing the field voltage, and hence flux. *Figure 11.12* shows such a system where

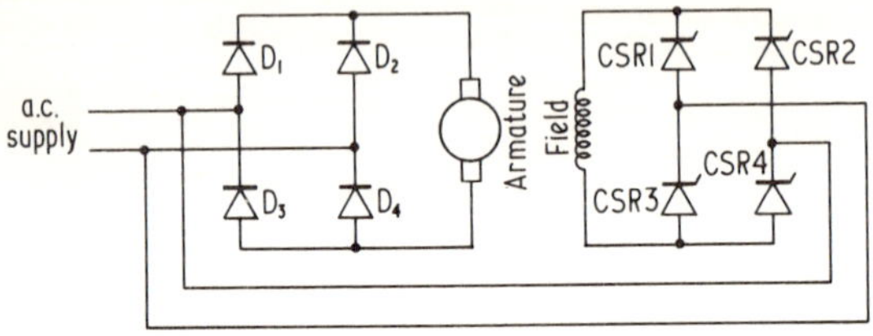

Figure 11.12. Field control for limited motor speed variation

a bidirectional converter is used to supply the field. This has the advantage of being capable of regenerating the highly inductive current stored in the field back to the supply. This facility is often required in fast response drives, such as in Ward Leonard schemes with thyristor control of the generator field current. Generally a field control is used only where a narrow speed range, say 4:1, is required. This is because at high flux densities the motor poles would saturate, whereas at low densities trouble would be experienced with motor commutation owing to the predominant effect of the armature reactance. No such restriction is placed on armature voltage control, and with this method drives having speed ranges of 100:1 are quite common.

In the discussion so far it has been assumed that the motor is being supplied from an a.c. source. This is the most common industrial drive since an a.c. supply is almost universally available. However, for many applications, especially in battery vehicles, only a d.c. supply is present, and the motor must now be driven by some form of chopper arrangement. Such a system was described in the appendix to Chapter 5. *Figure 11.13* shows an alternative arrangement of a chopper which, although very popular for general-purpose use, should not be employed for motor control. Referring to the load voltage waveform suppose that at t_0 thyristor CSR1 is fired. The load voltage rises to V_B and power is supplied to the motor. From a previous cycle C was charged to $V_B - E$ with plate a positive, where E is the motor back e.m.f. at this speed. When CSR1 turns on, the capacitor discharges through D1 and resonates with L, recharging with plate b positive to $V_B - E$, if resonance losses be neglected. When CSR2 is now fired, CSR1 is reverse-biased by $V_B - E$ and turns off.

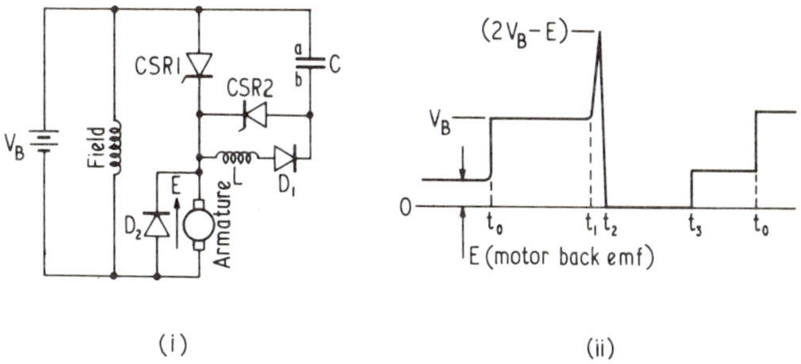

Figure 11.13. Chopper control of a shunt motor on light load: (i) circuit diagram, (ii) load-voltage waveform

Motor current flows in D2, and C charges to V_B with a positive (drop in D2 being ignored). The load voltage is zero as at t_2. On light loads, however, the inductive load current will decay to zero (it is flowing against the motor back e.m.f.) before CSR1 is refired. When this occurs, say at t_3, the capacitor will discharge through D1 and its voltage will fall to $V_B - E$. The motor back e.m.f. has therefore had two effects. Firstly it has distorted the load-voltage waveform between heavy loads and light loads. Since the magnitude of the load voltage is determined by its waveform this means a change in the motor voltage, and hence its speed, even though the firing period of the thyristors has been unaltered. This, however, is not serious and can be compensated for by using closed-loop speed control, as in *Figure 11.9*. A much more serious effect of the back e.m.f. has been a reduction in the available thyristor commutation voltage. Even though this voltage is reduced most on light loads, when commutation is least demanding, it is a very serious disadvantage of this circuit. A much better solution is to replace D1 by a thyristor, which turns off as soon as capacitor resonance has been completed, and so prevents the motor back e.m.f. from affecting the capacitor voltage. Back e.m.f. effects in phase-control circuits have been discussed in Chapter 6.

Apart from controlling the speed, it is often necessary to be able to reverse the direction of rotation of a d.c. motor. This can, of course, be done by electromechanical contactors if required. *Figure 11.14* shows a reversible converter arrangement in which the motor direction can be readily changed by operating one or other of the converters. It is now necessary to have all elements of the converters controllable (*i.e.* bidirectional converters only) to avoid short circuits between supply lines. This converter arrangement was introduced in Chapter 7 in relation to cycloconverters, and it was shown

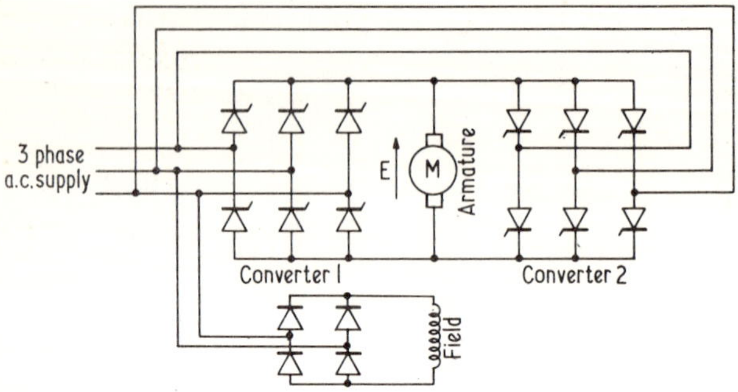

Figure 11.14. Reversible three-phase thyristor d.c. motor controller

there that there are two ways in which it may be operated. In the first method both converters are fired simultaneously but with delay angles $\alpha_1 + \alpha_2 = 180°$. This gives equal mean output voltages from both converters and they can readily be changed from one direction of operation to another by adjusting the delay angles. However, owing to instantaneous voltage differences between the converters it is necessary to introduce a reactor between them in order to limit the circulating-current flow. A better control method, in the present application, is to sense the converter current direction and to ensure that only one converter is fired at any time. *Figure 11.15* shows the

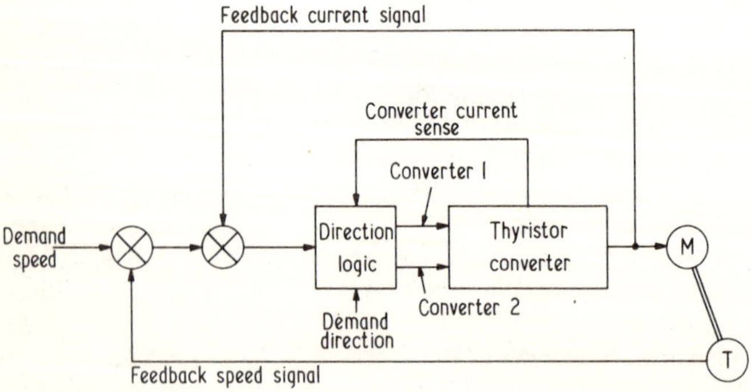

Figure 11.15. Closed-loop control for the converter of Figure 11.14

closed-loop speed-control arrangement for such a system. In addition to demand speed there is also now a demand direction. This is used to operate one of the two converters to change the polarity of the armature voltage and hence the motor direction. Feeding into the same direction-logic circuitry is a signal which senses the direction of current flow in the converters, and ensures that the changeover from one converter to another does not occur unless the current in the previously conducting converter has fallen to zero. In addition the magnitude of the motor current can be sensed to provide current limiting as before.

Braking

In a purely mechanical method of braking the kinetic energy of a revolving motor may be dissipated as heat by friction brakes. However, electrical methods often provide considerable advantages. There are three techniques for braking which are then commonly used.

(i) Plugging

The direction of rotation of the motor is rapidly changed by reversing the supply voltage polarity. The kinetic energy of the motor is now dissipated as heat in the motor itself. *Figure 11.16* illustrates the principles of plugging using field reversal. When the motor field current is suddenly reversed while the machine is still revolving in the same direction, the d.c. source will oppose the motor rotation and try to turn it in the reverse direction. Since the motor back e.m.f.

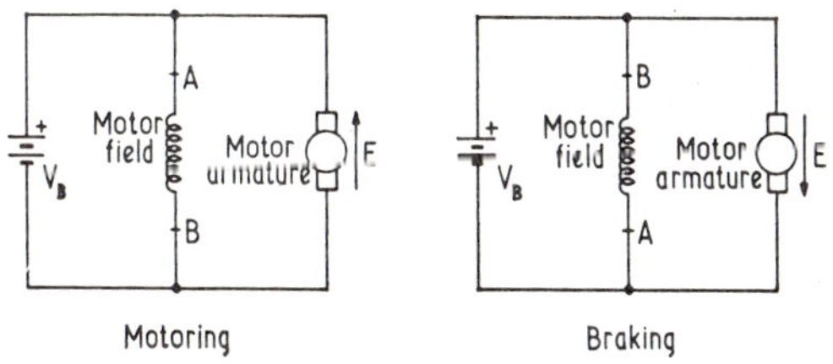

Figure 11.16. Plugging a d.c. motor

has reversed with field current it will help the d.c. source in circulating a large current through the motor. This current exceeds the stalled motor current and gives rise to severe heat losses in the motor, bringing it to a rapid halt. If the supply is still maintained the motor will then commence to rotate in the reverse direction.

The dual converter system shown in *Figure 11.14* can readily be extended to plugging. Suppose converter 1 was on and the motor was rotating in the forward direction; the motor back e.m.f. would be as shown. To plug the machine converter 1 is switched off and converter 2 operated such that the converter voltage aids the motor back e.m.f. in circulating a large current through it. The converter firing-angle delay should not exceed 90° as regeneration would then occur, as explained below. Controlled plugging is clearly possible by varying the delay angle, it being greatest at $\alpha = 0°$.

(ii) Dynamic braking

The kinetic energy of the motor is dissipated as heat in the motor itself, or in external resistors. Dynamic braking differs from plugging since the motor now acts as a generator and feeds current into an external resistor or circulates it round its own windings. Such a system is shown in *Figure 11.17*. During motoring CSR4 is off. The

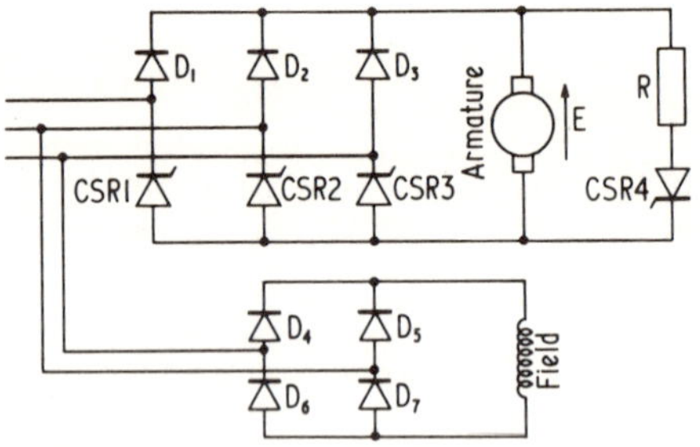

Figure 11.17. D.C. motor speed control, with provision for dynamic braking

three-phase bridge is used to control the motor speed. To brake the motor the bridge converter is switched off but the motor field current is maintained. Since the motor is still revolving in the same direction its back e.m.f. is as shown. CSR4 is now fired. The motor acts as a generator and feeds a current through *R*, the motor kinetic energy being dissipated as heat in this current path. A dynamic braking system for a chopper controlled drive was described in the appendix to Chapter 5.

(iii) Regeneration

The kinetic energy of the motor is recovered and may be used again during motoring. During regeneration the motor again acts as a generator. Referring to *Figure 11.14* suppose the motor is running in the forward direction, converter 1 being on. To generate, converter 1 is turned off and 2 turned on. If the delay angle of this converter exceeds 90° there is net regeneration, this being maximum for $\alpha = 180°$.

Figure 11.18 shows a regenerative system for a chopper controller. During motoring, CSR2 is off and CSR1 operates in the usual on-off mode. Arrows on both thyristors indicate that they are

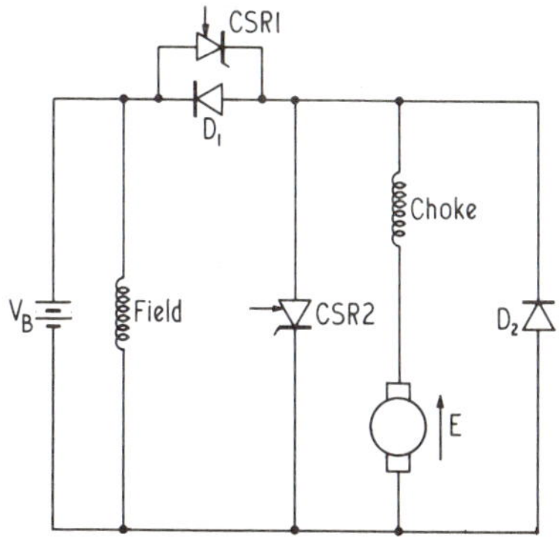

Figure 11.18. Motoring and regenerative braking of a d.c. motor

forced-commutated by circuitry which is not shown. Diode D1 is also reverse-biased throughout the motoring period, back e.m.f. E being less than the supply voltage V_B. Although regeneration could be obtained by switching off CSR1 and strengthening the motor field to increase the value of E, such a system has a limited control range since soon the field would run into saturation. A better system is to maintain the field constant but to operate CSR2 as a chopper during regeneration. CSR1 and D2 now no longer play a part in the system, which is essentially a step-up chopper, as described in Chapter 1.

Although regeneration is the best system of braking, owing to its high efficiency, it is much more complex and expensive to obtain. Dynamic braking is inefficient, but by the use of external resistors the motor losses are minimised. However, it is important to appreciate that since both these systems rely on a generated output from the motor they are not effective at low speeds, where the motor back e.m.f. is small. Field strengthening can now be used to reduce this problem but it will not overcome it altogether. In plugging, on the other hand, the motor direction can be reversed if necessary so that it is effective right down to standstill. For quick response systems it is therefore clearly the preferred method. A system such as that shown in *Figure 11.14* is advantageous since regeneration can be used at higher motor speeds and then the firing angle advanced to below 90° in order to plug the motor to rest.

ELECTRONIC COMMUTATOR D.C. MOTORS

The d.c. motor is by far the most popular variable-speed machine, owing to its excellent speed/torque characteristics and variable-speed capability. Its one major drawback is its mechanical commutator which places design and environmental limitations on its operation. It is, of course, feasible to use an a.c. motor drive system when a mechanical commutator is undesirable. However, as is shown later, a d.c. motor has many advantages over a.c. motors in some applications. It is in order to maintain these characteristics while overcoming the limitations of a mechanical commutator that electronic commutator (or brushless) motors are used.

In the following sections the construction of electronic commutator d.c. motors is described in more detail. These motors are not commonly used in large sizes but have gained wide acceptance in good-quality small motor drive systems. Transistors are now the switching devices. Although power transistors are described in greater detail in the next chapter their application to d.c. motors is included here.

Specification for an electronic commutator

In a d.c. motor, torque is produced by the interaction of the stator and rotor fields. Usually the stator contains salient poles (and interpoles if provided), which are energised by a field coil, and the rotor carries armature current supplied by an external d.c. source. To maintain unidirectional torque between the rotor and stator it is necessary to switch the rotor current periodically so as to keep the two fields as closely perpendicular to each other as possible. A conventional d.c. machine uses a mechanical commutator and brushes to achieve this. The commutator switches the current; this occurs at the correct instant in the rotor position by virtue of the commutator position, and the greater the number of commutator segments the smoother the torque produced.

A two-segment motor would have a torque/position characteristic as shown in *Figure 11.19*(*a*), peak torque occurring when rotor and stator fields are at 90°E and falling to zero when this angle is reduced to zero. If four segments are provided, it is possible to switch rotor current four times per revolution so that the stator field is always maintained between 45°E and 135°E ahead of the rotor field, the torque curve being as in *Figure 11.19(b)*. The mean torque is now 90% of the peak value and gives much smoother operation.

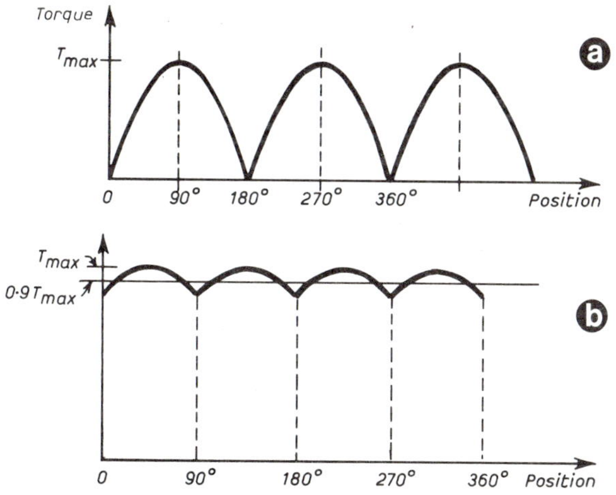

Figure 11.19. Torque/position characteristics for a d.c. motor with (a) two-segment commutator, (b) four-segment commutator

It is, of course, possible to design a d.c. motor in which the rotor has permanent-magnet poles and the stator current is switched. This, however, would require the brushes to rotate at the same speed as the rotor to maintain unidirectional torque. Although such a system is undesirable in conventional machines it can be used with advantage in electronic commutator motors, since the stationary commutator makes it easier to operate the electronic devices. *Figure 11.20* shows such an inverted machine which contains a multi-segment commutator. With rail A or rail B positive the

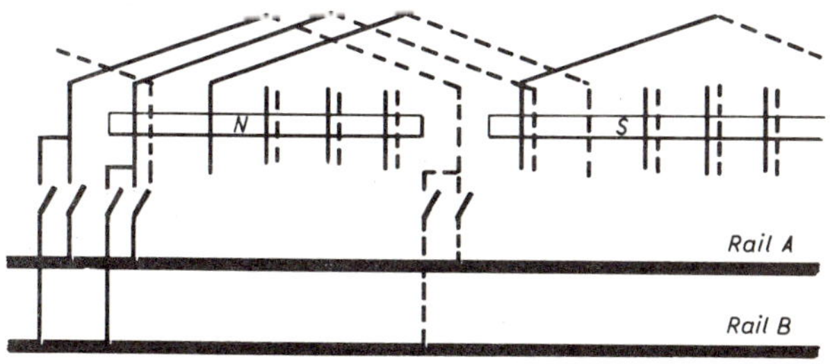

Figure 11.20. Inverted d.c. motor with multi-segment commutator

polarity of the stator flux can be readily controlled by closing the appropriate switches. As with most electronic commutator machines the rotor is salient pole in construction and the field contribution to the air-gap flux predominates over the armature contribution so that the effect of armature reaction becomes negligible.

Although the circuit of *Figure 11.20* has been used in electronic commutator motors, it uses a large number of switching devices, and can be expensive. *Figure 11.19(b)* shows that, with a little sacrifice in torque smoothness, a four-segment commutator can be used which would be much more economical. *Figures 11.21* and *11.22* illustrate two possible winding arrangements for such a machine,

Position	Switch S_1	S_2	S_3	S_4	S_5	S_6	S_7	S_8
A				X			X	
B		X			X			
A′			X					X
B′	X					X		

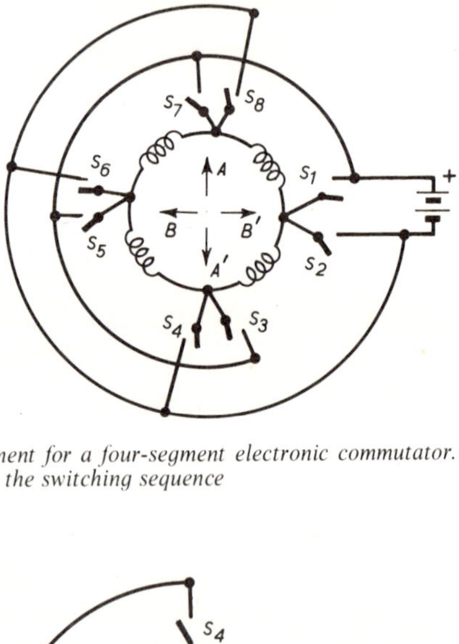

Figure 11.21. Closed-winding arrangement for a four-segment electronic commutator. The table gives the switching sequence

Position	Switch S_1	S_2	S_3	S_4
A				X
B			X	
A′		X		
B′	X			

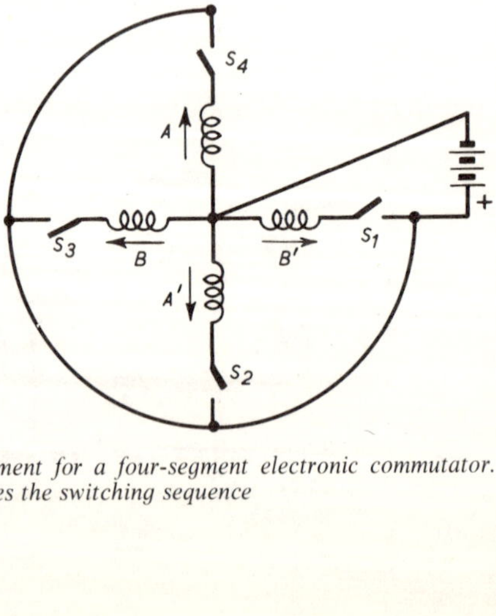

Figure 11.22. Open-winding arrangement for a four-segment electronic commutator. The table gives the switching sequence

and also show the switch operating periods. It is seen from these that although the open-winding arrangement uses only half as many switches as the closed-winding machine, its winding utilisation is only one-quarter. Therefore it would generally be physically larger and less efficient.

Figure 11.21 also shows that the switching action is very similar to that obtained by a two-phase inverter where S_4, S_7, S_3 and S_8 constitute one phase and S_2, S_5, S_1 and S_6 give the other phase. Thus essentially, as mentioned earlier, there are two functions required to be performed by the commutator in a d.c. machine. First it must sense the relative position of the stator and rotor fields. Secondly it must switch the stator field current at the appropriate instance. We will now examine these functions in detail, and see how they are performed by an electronic commutator.

Position sensors

Rotor-position sensors are required to detect the instant at which the stator current is to be switched. There should therefore be as many position signals as there are commutator segments. A four-segment motor requires switching signals at 0°E, 90°E, 180°E and 270°E. Many different types of devices may be used, those most commonly employed in electronic motors being magnetic, optical or based on the Hall effect. Only these three are considered here. In all cases it is important that the detector gives a signal regarding the position of the rotor and not its speed or direction of rotation.

Figure 11.23 shows one form of rotor-position sensor which has been used. Attached to the main rotor is an auxiliary rotor which revolves in a four-pole stator yoke. On each of the poles, which correspond to the four switching points of a four-segment motor, is wound a sensing coil which feeds the respective stator winding switch. A high-frequency oscillator supplies two coils placed at diametrically opposite ends of the yoke and arranged to produce anti-phase signals. Clearly a voltage will be induced in only those coils which are under the rotor poles, so that the system produces rotor position signals.

There are, of course, several possible modifications to this arrangement. One system uses four sense coils embedded in the main frame of the stator, the coils being supplied in parallel by a high-frequency oscillator. A series-resonant capacitor gives a sense voltage across the coils unless it is saturated by the close proximity of a rotor pole. The absence of a pick-up signal at any coil therefore indicates that the corresponding stator switch is to be operated.

Magnetic sensing devices are very robust and are not affected by dirt or dust. They require minimal auxiliary components when the main rotor poles are utilised as part of the sense system, and the signal output can normally be used to operate the stator switches directly without further amplification. However, the induced signals in the coils build up gradually; this gives slower turn on and greater dissipation in the switches, as well as possible uncertainty in the exact switching point. The presence of leakage flux also means that there is always some induced signal in the coils. Since all signals are a.c. they have usually to be rectified before being used to operate the switches.

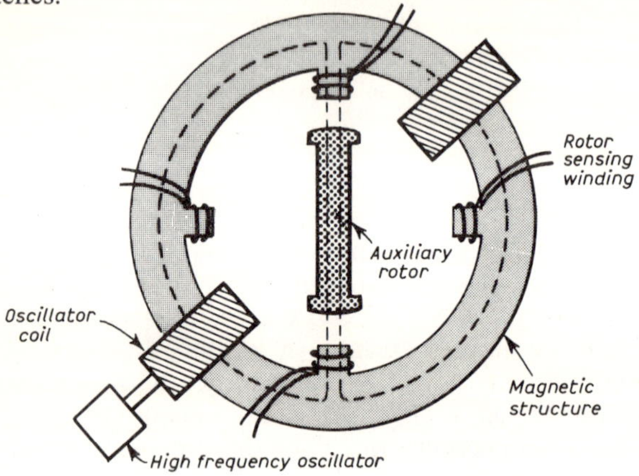

Figure 11.23. Magnetic position sensor

Optical sensors are generally less robust than magnetic systems but are also much smaller and lighter. They have therefore tended to be used in motors for aero-space applications. *Figure 11.24* shows a schematic of an optical system where a light shield with an aperture is connected to the rotor and revolves around a stationary light source. Photocells placed at the four switch points detect the commutation positions for the motor as before. Generally the output signals from the light detectors are weak and need further amplification before they can be used to operate the stator switches. However, the signals can be made to rise sharply, and since they are generally d.c. no rectification is required.

A sensor which has been extensively used in many small electronic commutator motors is the Hall effect device. It combines the robustness of magnetic sensors with the lightness of optical devices. The Hall effect is well known. A current I_c passed between two faces of a conductor placed in a transverse magnetic field B would result in a redistribution of charge carriers within the conductor and induce a voltage across the conductor in a direction perpendicular to both the current flow and the magnetic field. This voltage, known as the Hall voltage, is proportional to I_c and B.

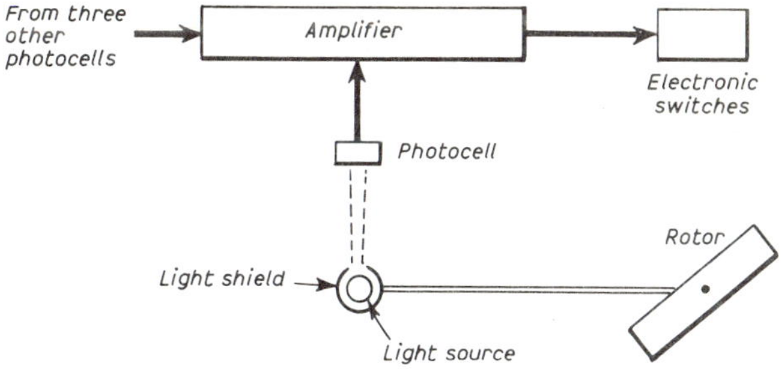

Figure 11.24. Block schematic of optical position sensor

Hall generators placed at the commutator switching positions would therefore react with the rotor field to produce the required position signals. A further advantage of the Hall detector is that the voltage reverses with the direction of the magnetic field so that only two sensors at 90°E to each other need be used for a four-segment commutator.

In addition to the advantages of small physical size, ruggedness and direction sensitivity, the output from a Hall generator is d.c. and so does not need rectification. The disadvantage of the system is that an auxiliary source is required to provide I_c and that, like a magnetic sensor, the output voltage is always present due to leakage flux, and it builds up gradually.

It is, of course, possible to use the Hall voltage to feed the stator coils directly instead of via a switch. However, the Hall generator is now required to handle much higher power levels. Presently available Hall devices are predominantly resistive, their efficiency being given by

$$\eta^2 = (V_{ho})^2/4R_h I_c^2 R_c \qquad (11.5)$$

where R_c and R_h are the control and equivalent circuit resistances respectively. This efficiency is usually of the order of 10% so that motors built on this principle would be very inefficient.

The choice between the various rotor position sensors is usually determined by the motor application. Large machines use magnetic sensors since efficiency and switching power requirements are important, although optical sensors may be used where robustness is not essential. For smaller motors Hall generators give a small and relatively cheap system. For higher efficiencies magnetic systems have been used and optical sensors are chosen where lightness is equally important.

Electronic switches

The switches used to change the direction of stator current, at a signal from the rotor position sensors, need to satisfy several requirements. Firstly they must be adequately rated to handle the full winding power. This can vary widely with the size of machine used. It is also important to establish whether the current through the switches will be unidirectional or bidirectional. A mechanical commutator can readily pass current in either direction. Electronic switches, however, are usually restricted in this aspect. Therefore special provisions must be made if bidirectional operation is required. This often arises when the same machine is to act as a motor and a generator, and also in some closed stator winding arrangements.

A mechanical switch can only be off or on, but an electronic device can often have a controlled turn on, as in a transistor. This means that, for instance, in *Figure 11.25* instead of controlling the applied voltage (V_B) to a motor to change its speed, the value of this voltage can be kept fixed but the voltage across the switch (V_s) varied.

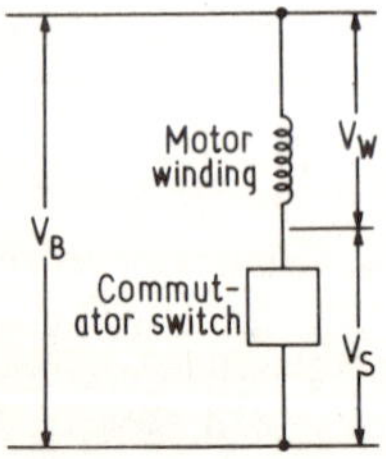

Figure 11.25. Distribution of voltage across the electronic commutator switch and the motor winding

Therefore an electronic switch can act in a switching mode, as for a mechanical commutator, or in an amplification mode, where the extent to which it is on or off is controlled. This can have several advantages.

For instance, *Figure 11.26* shows a four-segment commutator operated from Hall-effect sensors. The currents through the sensors are constant. However, if the rotor flux is assumed to be sinusoidally distributed, then the Hall voltage, being proportional to this flux, would cause the transistor to operate in an amplification mode. The stator current distribution in the four windings would therefore be such as to maintain the stator flux constant, irrespective of rotor position. Since motor torque is proportional to the product of stator

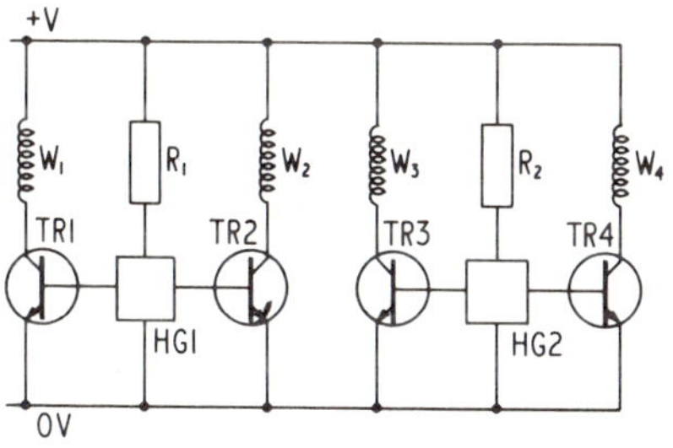

Figure 11.26. Winding arrangements on a four-segment Hall sensor electronic motor

and rotor flux, and these are now constant, such a motor is capable of extremely smooth operation. The amplification mode has one obvious disadvantage. The dissipation across the switches is much greater than if they were always either off or saturated. Thus its use is limited to relatively small motors.

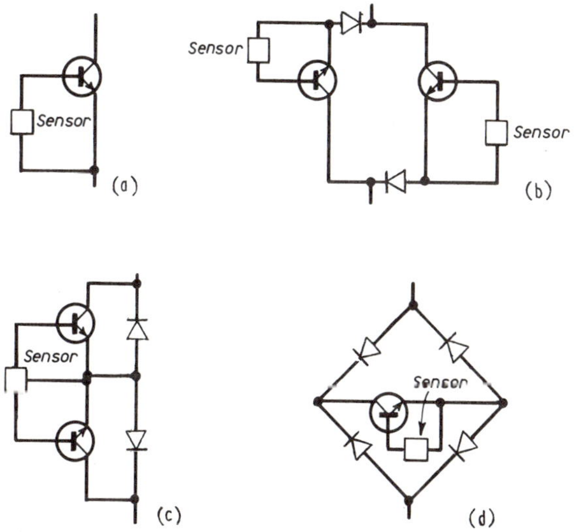

Figure 11.27. Electronic switching using transistors: (a) unidirectional operation, (b) to (d) bidirectional operation

There are two devices which have been used as switches in electronic commutator motors, namely thyristors and transistors.

Where the power to be handled is large thyristors are the obvious choice. These devices can currently switch 1 000A at 2·5kV and promise even greater power handling capability. The thyristor is not capable of amplification-mode operation, but at these high power levels the low efficiency of such a system would be unacceptable. Another characteristic of a thyristor is that it is unidirectional in operation. For bidirectional switching two devices can be connected in a parallel back-to-back arrangement, or a bidirectional power device, the triac, which is equivalent to two inverse parallel thyristors, can be used. The triac is described in Chapter 12.

Transistors are certainly much more versatile since they can be operated in an amplification or a switching mode. *Figure 11.27* shows several arrangements that may be used with a transistor. *Figure 11.27(a)* is intended for unidirectional operation, whereas *Figures 11.27(b)* to *(d)* give bidirectional capabilities. Since only one sensor is used in *(c)* and *(d)* this is generally the preferred arrangement.

The motor winding shown in *Figure 11.26* is inefficient due to poor winding utilisation. *Figure 11.28* shows an alternative arrangement

Rotor position	T_1	T_1'	T_2	T_2'	T_3	T_3'	T_4	T_4'
S_1		X	X		X			X
S_2		X	X	X	X		X	
S_3		X		X	X		X	
S_4	X	X		X	X	X	X	
S_5	X			X		X	X	
S_6	X		X	X		X	X	X
S_7	X		X			X		X
S_8	X	X	X		X	X		X

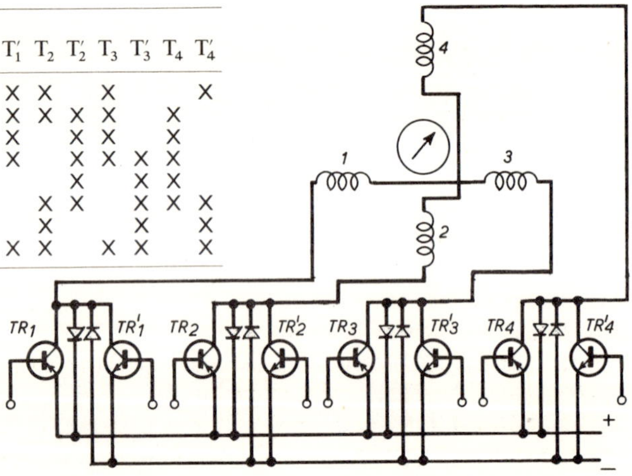

Figure 11.28. Closed-winding four-segment commutator motor. The table gives the switching sequence

which overcomes this disadvantage and also illustrates an alternative form of closed-winding. The table shows that the switches have purposely been operated so as to produce a stepped output voltage waveform. This reduces the motor voltage waveform harmonic content and gives greater efficiency.

Closed-loop control

One of the advantages of electronic commutator motors over conventional machines is that they can be readily adapted for closed-loop operation. One such system is illustrated in *Figure 11.29*, which is seen to be an adaptation of the basic motor of *Figure 11.26*. The action of the Hall generators and switching transistors TR1 to TR4 is as before. These devices supply current to the stator windings, under control of the Hall-effect position sensors, to produce a two-phase supply. The current to the generators is not now constant,

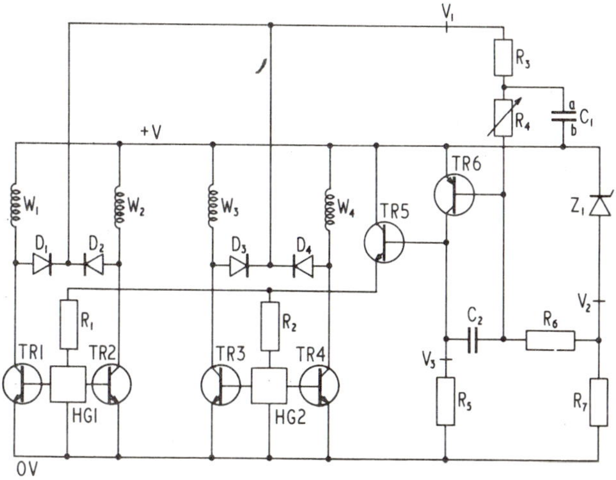

Figure 11.29. Speed-control arrangement for the motor in Figure 11.26

being under control of transistor TR5. This transistor in turn is controlled by TR6 which sums the reference voltage across Zener diode Z1 and the motor voltage feedback V_1. This voltage is obtained by rectifying the back e.m.f. from the motor via diodes D1 to D4 and smoothing in capacitor C_1. The action of the motor can be explained as follows. Suppose the machine is at standstill when the supply is applied. TR6 turns hard on driving TR5 into saturation and providing maximum control current to the Hall generators. The outputs from these provide peak base current to the switching transistors and therefore full motor voltage. As the motor speeds up, its back e.m.f. charges C_1 with plate *a* positive to a greater extent so reducing the base voltage on TR6 and therefore the Hall generator currents. At equilibrium, which can be varied by R_4, the drive to transistors TR1 to TR4 is just sufficient to provide the required motor back e.m.f. Although back e.m.f. sensing has been used in the above example, it is possible to obtain better regulation with tachogenerator feedback. Such a method is expensive and not common in small motor systems.

If the open-loop gain of the feedback system is large, as is usually the case, then the equation for the feedback loop in *Figure 11.29* for the angular velocity, ω, of the motor reduces to $\omega = R_a V_2/R_b K_2$, where $R_a = R_3 + R_4$, $R_b = R_6$ and K_2 is a motor constant. This equation shows that speed stability is determined primarily by the reference voltage V_2, the motor back e.m.f. constant K_2 and resistors R_3, R_4 and R_6. It also gives a method of selecting these to obtain a given motor speed. By suitable choice of components, speed holding capabilities of 1% or less are readily obtained over a range of loads and temperature.

Advantages of an electronic commutator

A d.c. motor has very desirable characteristics for a wide range of applications. Its one disadvantage is its mechanical commutator. This leads to the following limitations:

(1) The commutator and brushes are prone to sparking with a consequent production of radiated interference. This can be a serious hazard in airborne equipment.

(2) Sparking also reduces the life of the commutator. Conventional motors therefore need periodic servicing and brush replacement, and cannot be mounted in inaccessible positions.

(3) Usually the commutator is mounted on the rotating member (an inverted motor would require the brushes to rotate). This limits the maximum speed at which it can run and therefore the power output from a given frame size.

(4) The commutator also limits the machine design since it requires a large ratio of diameter to length of armature iron.

(5) The commutator is made from good quality copper which softens at high temperatures and limits the machine operating temperature. Dirt and moisture also attack the commutator and the machine needs to be specially protected in hazardous atmospheres.

(6) Rubbing of the brushes on the commutator, and clatter of the brushes in their holders give rise to considerable audible noise.

An electronic-commutator motor maintains all the advantages of a traditional d.c. machine, such as high starting torque and excellent efficiency at all speeds, while avoiding the disadvantages of a mechanical commutator. In addition it can also produce smoother torques in small motors. This is due to the fact that the number of commutator segments in a small conventional motor is usually limited by physical size to six or eight. Therefore the torque/position characteristic is relatively poor. An electronic commutator using four segments, on the other hand, can produce a much smoother torque by using its switches in an amplification mode.

The same performance can be obtained from an inverter-induction motor drive as from a commutatorless d.c. motor drive. What then is the difference between this system and the electronic commutator motor? From the hardware angle both motors are very similar, the induction motor drive being slightly simpler owing to the absence of the rotor-position sensor. However, there is one important difference. The frequency at which the stator switches operate in an electronic commutator motor is determined by the rotor speed, as sensed by the position sensors. This speed is in turn fixed by the winding current. For an inverter drive, on the other hand, the switch frequency is fixed by an internal oscillator in the inverter. The motor merely follows this rotating field and speed control is achieved by controlling the inverter frequency and not the supply voltage. The following further differences can be noted between the two systems.

(1) The inverter of an induction motor drive could be located at some distance from the motor. This would require only two wires connecting the driver to the motor. An electronic-commutator motor would need the sensing leads to be brought out as well; these leads are then exposed to pick-up effects.

(2) Permanent-magnet motors take a larger starting current than corresponding induction machines. This could be a serious problem since a current surge can cause demagnetisation of the permanent magnet. Smaller motors, with higher-impedance windings, are less prone to this effect.

(3) Since the inverter can operate independently of the motor in an inverter drive, several motors can be run in parallel off the same inverter.

In general, therefore, the electronic-commutator motor is best suited to applications requiring a high starting torque, good efficiency at low speeds and continuous speed variation from standstill to full speed. Owing to their poor efficiency at low speeds, inverter-induction motors are more suited for use in applications where the output is a fixed high speed, and where, by using crystal-controlled oscillators in the inverter, and a synchronous motor, the speed regulation can be made exceptionally good.

Applications

In spite of the considerable interest shown in electronic commutators over the past decade they have not been price competitive in many applications and are therefore not extensively used. Larger motors, in the region of 7·5kW (10 hp) in size, have been built as specials and run principally under experimental conditions. The thyristor has been the switching element and position sensing is achieved by optical or magnetic systems. Smaller motors, however, have recently been offered by several manufacturers. They range in sizes up to about 40W and 100 gf-cm torque. Although suitable for a variety of applications their reliability, long life and absence of interference (since there is no mechanical commutator) have found favour especially with the aerospace industry. They are also very popular in high-performance tape recorders. Here their speed-holding capabilities (achieved by speed-dependent feedback) minimises tone-pitch variation and flutter and wow. The absence of generated interference is also highly desirable in this application. Generally the small machines have transistors for the switching elements and use magnetic or Hall effect sensors.

A.C. MOTORS

As outlined in *Figure 11.2*, there are three basic types of a.c. motors (induction, synchronous and commutator) which are popularly used with thyristor controllers. There is yet another type of motor, the reluctance machine, which has not been mentioned so far since it can be considered as a special form of synchronous motor.

Reluctance motors have not been popular since they suffered from several disadvantages chiefly connected with the low power factor. Recent work has, however, succeeded in producing motors with greatly enhanced performance. Power factors as high as 0·74 can now be obtained, with efficiencies as large as 90% of the induction-motor full-load value. All reluctance motors have a certain saliency in their rotors. The stator produces a rotating field similar to an induction motor, which causes the motor to run up close to synchronism by induction-motor principles. Thereafter the salient poles lock it onto the stator field and it runs at synchronous speed, as in a synchronous motor, even though the rotor poles are not externally excited. It is this synchronous/induction action of the reluctance motor that is beginning to find widespread applications in multi-motor converter drives in the textile, steel and plastics industries. When used to drive an alternator it can also provide highly stable computer supplies.

To obtain a high power factor with good pull-in and pull-out torques it is essential to increase the ratio of the direct axis to the quadrature-axis magnetising reactance of the motor. *Figure 11.30* shows a conventional reluctance motor and indicates the paths of the direct and quadrature fluxes. To achieve a large value of direct-axis reactance, and a small value of quadrature-axis reactance, it is necessary to reduce the salient-pole arc to as low a value as possible.

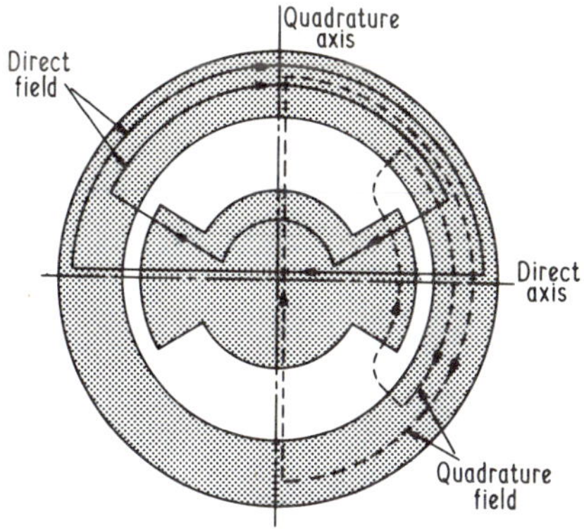

Figure 11.30. Conventional two-pole reluctance motor

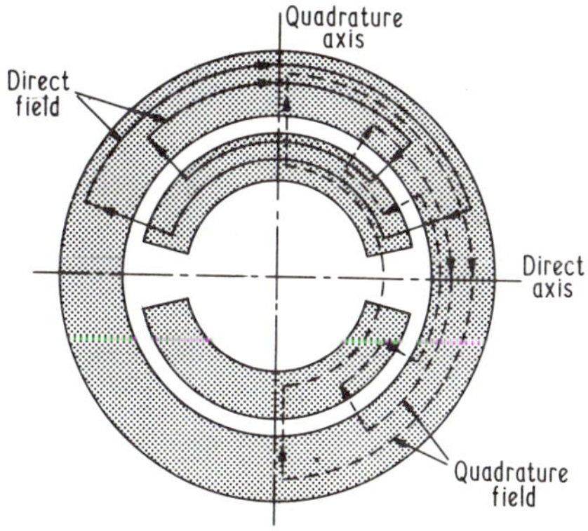

Figure 11.31. Segmental two-pole reluctance motor

Although this now increases the power factor it leads to greatly increased magnetising currents and reduces torques. A recently developed form of segmental motor which overcomes these disadvantages is shown in *Figure 11.31* for a two-pole motor. Since all the torque is experienced by the salient poles and none by the central rotor cylinder, this has been eliminated to give a motor with reduced inertia. In *Figure 11.31* only the quadrature-axis flux crosses the relatively large interpolar space so that its reactance is greatly reduced.

Instead of segmenting the rotor it is possible to attenuate the quadrature-axis flux by using flux guides or barriers, as in *Figure 11.32*. Two forms are illustrated here. In both cases it is seen that the quadrature-axis flux is decreased to a greater extent than the direct-axis flux owing to the position of the guides.

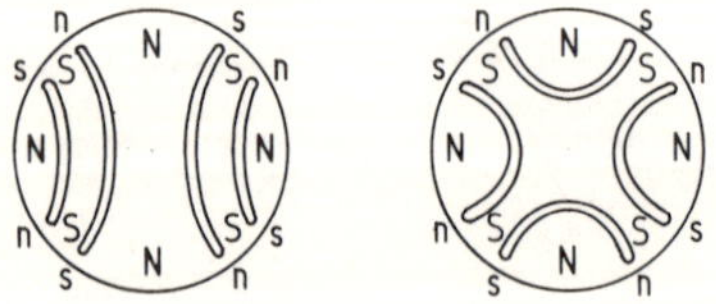

Figure 11.32. Flux barriers in reluctance motors

It has been seen by the above discussion that a reluctance motor is in effect a synchronous machine with unexcited rotor poles. Both machines, on the other hand, behave like an induction motor during run-up periods, and can be considered as special cases of the induction machine. As has been seen, the torque on the rotor of an induction motor is caused by induced currents produced by a rotating stator field. If N_1 and N_2 are the speeds of this field and of the rotor respectively, then the slip S is defined as

$$S = \frac{N_1 - N_2}{N_1} \tag{11.6}$$

The magnitude of the rotor torque is proportional to stator flux, rotor induced current and the cosine of the angle between these two phases, *i.e.*,

$$T \propto \frac{SE_1^2 R_2}{f_1(R_2^2 + S^2 X_2^2)} \tag{11.7}$$

where T = torque on the rotor
S = slip
E_1 = stator applied voltage
f_1 = stator frequency
R_2 = rotor resistance
X_2 = rotor reactance

From equation (11.7), the maximum torque T_m can be found at a slip S_m as

$$T \propto \left(\frac{E_1}{f_1}\right)^2 \tag{11.8}$$

$$S_m \propto \frac{R_2}{f_1} \tag{11.9}$$

since

$$X_2 \propto f_1$$

Substituting in equation (11.7) gives

$$T = \frac{2T_m}{S_m/S + S/S_m} \tag{11.10}$$

The input power, which crosses the air gap between stator and rotor, is primarily used in producing copper losses and torque output. These two are divided in the ratio of $S/(1-S)$ so that motor efficiency is equal to

$$\eta_M = 1-S \tag{11.11}$$

In a commutator motor, as with some types of induction motors when the motor winding is brought out to slip rings and brushes, a resistor connected in series with the rotor would result in power loss. This would mean that the output power of the machine is reduced and it must slow down to supply the same torque. A further advantage of commutator motors is that the frequency at the brushes is the supply frequency. There can, therefore, be a direct interchange of power between the a.c. lines and the rotor, as shown in *Figure 11.33*.

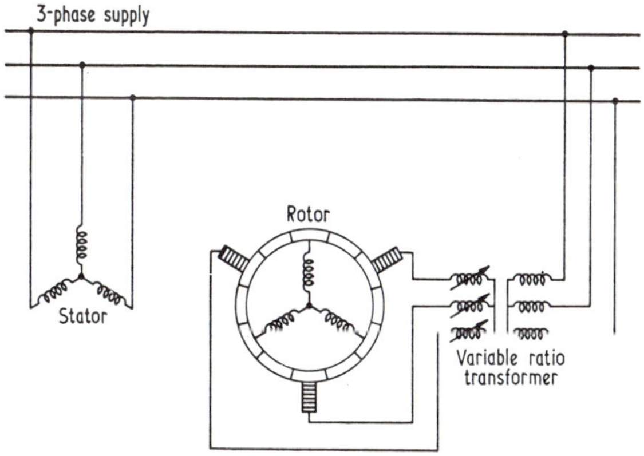

Figure 11.33. Control of an a.c. commutator machine

If power is fed from the rotor to the a.c. supply the motor will slow down. If power is supplied into the rotor its speed will increase. The a.c. commutator is similar to an induction machine with the addition of a frequency converter in the rotor to allow interchange of power between rotor and a.c. line. These machines are bigger, costlier and more difficult to design than other a.c. drive systems. They have largely been replaced by the d.c. motor and thyristor controlled rectifier driver, and an a.c. commutator machine with a thyristor controller is very rarely used. It is not intended, therefore, to consider this motor further in this chapter.

Starting

Induction motors (and other a.c. machines which run up as induction motors) draw a large starting current when connected directly to a mains supply. This is due to the fact that there is large slip between stator field and rotor speed, so that at full stator flux there is considerable induced rotor current, which is reflected to the stator windings by transformer action. To decrease the starting current one must reduce either the stator flux, by lowering the stator voltage, or the stator frequency, by a reduction of the supply frequency. The stator voltage must now also be changed in order to keep the machine flux constant. When flux reduction is required the stator voltage can be reduced by a.c. line control methods, as described in Chapter 4. Static contactors working on this principle were described in Chapter 3. Frequency control is only possible when the motor is running from an inverter or cycloconverter. The supply frequency can now be increased gradually as the motor runs up so that the peak current is limited. It will be seen later that there is now no loss of starting torque, as occurs with voltage control systems, since the motor flux is kept constant by automatic adjustment of the supply voltage. When a variable frequency supply is available for starting, synchronous and reluctance motors need not run up as induction machines since they can lock onto the stator field at low speeds, and then run up to full speed in synchronism.

Control

As has been seen, the most usual operating mode for synchronous and reluctance motors is when they are rotating at the synchronous speed of the stator field. Therefore the speed of these machines can only be varied by changing the frequency of the stator supply. The same considerations apply to an induction motor. However, owing to its construction, there are four methods which may be used to control an induction motor. These are described below.

(1) Excitation field strength control

Figure 11.34 shows the plot of equation (11.10). The following can be noted:

(1) the torque at synchronous speed, when $S = 0$, is zero

(2) at standstill, $S = 1$ and the operating point is located at $1/S_m$.

Since S_m is a function of rotor resistance and reactance, the amount by which the torque/slip curve extends along the X axis is determined by the rotor construction. If $S_m = 0{\cdot}2$ then at standstill $S/S_m = 5$ and the starting torque is 0·385 times the maximum value.

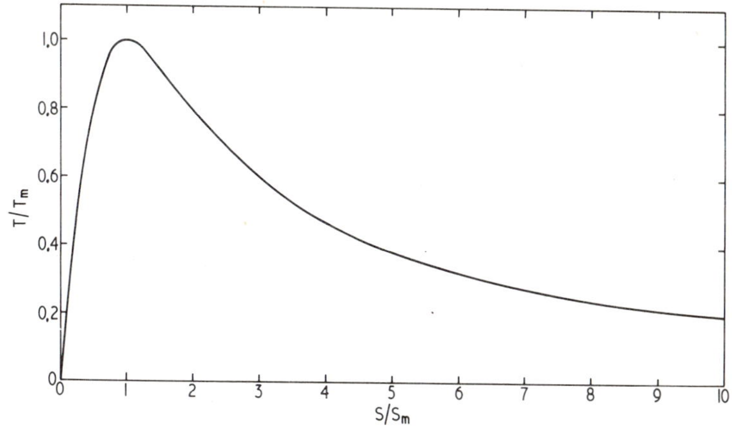

Figure 11.34. Normalised torque/slip curve

From equation (11.8), T_m is proportional to E_1^2, other parameters being fixed. Therefore, if stator voltage is halved the maximum torque is reduced by a factor of four. However, *Figure 11.34* remains unchanged since it is normalised to T_m. The starting torque will still be 0·385 T_m, but since T_m is one quarter of its original value the torque is also reduced to a quarter.

If the torque/slip (not normalised to maximum values) curve is plotted for variable voltage control, the characteristics, derived from *Figure 11.34*, would be as in *Figure 11.35*. For operation with a constant load (torque line AB), variation of the stator voltage from V_4 to V_3 will reduce the speed from E to F. Further reduction in voltage will reduce the available motor torque to below load torque and it will stop. Therefore range of speed control available is limited. Operation on points I and J is not possible since an increase in speed results in an increase in available motor torque so that the motor will rapidly run up to the positive slope part of the characteristic.

A load having a characteristic such as DC (fan loads) is more suited to stator voltage control. On any part of the curve beyond the operating point, an increase in motor speed would increase the difference between the load torque and available motor torque and so cause the motor to slow down again. Therefore, the system is inherently stable and the speed can be controlled over a very wide range.

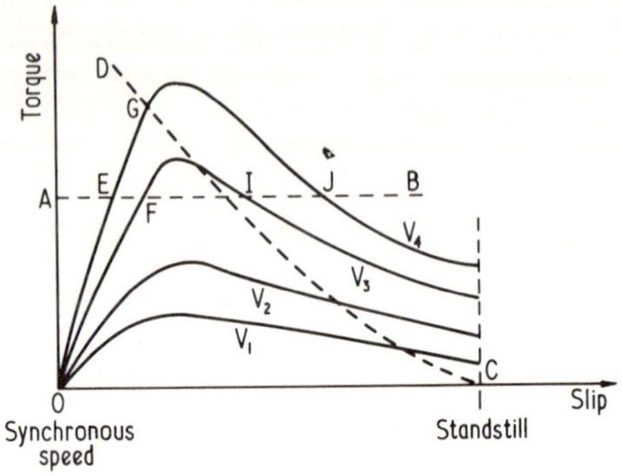

Figure 11.35. Torque/slip curves for stator-voltage variation

The main disadvantage of this system of speed control is the large slip at low speeds. From equation (11.11) efficiency is equal to (1-slip) so that for low speeds the efficiency is poor. At half speed the maximum efficiency is only 50%. Another consequence of this slip is that heat is generated in the motor, rotor loss being equal to S times the rotor input; at low speeds the heat can be excessive. Both these factors limit this method of speed control to small motors where efficiency is relatively unimportant and the heat easier to dissipate.

Voltage applied to the induction motor can be controlled by some form of line impedance such as saturable reactors. However, these are bulky and relatively slow in response. A more modern technique employs silicon-controlled rectifier in a.c. line control systems as described in Chapter 4.

(2) Excitation field speed control

An alternative control method is to vary the speed of the stator field by controlling the supply frequency. However, now the supply voltage must be changed in proportion to frequency in order to maintain motor flux constant at its most economical value.

If supply voltage E_1 is changed in proportion to the supply frequency f_1 then from equation (11.8) T_m is constant. However, from equation (11.9) S_m will vary inversely as f_1 so that as frequency is reduced the starting torque approaches the peak torque.

Equation (11.10) is used to calculate the torque for supply frequencies $f = f_1$, $f_1/2$ and $f_1/10$, where f_1 is the frequency used in *Figure 11.34*, and the results are given in *Figure 11.36*. From this it is evident that the available motor torque can be readily controlled to match the load and so give any desired operating speed. Furthermore, since the stator field speed need only be slightly greater than rotor speed (the resultant rotor current being just sufficient to overcome load torque), slip over the whole range can be very small giving high efficiency. In *Figure 11.36* slip is plotted along the horizontal

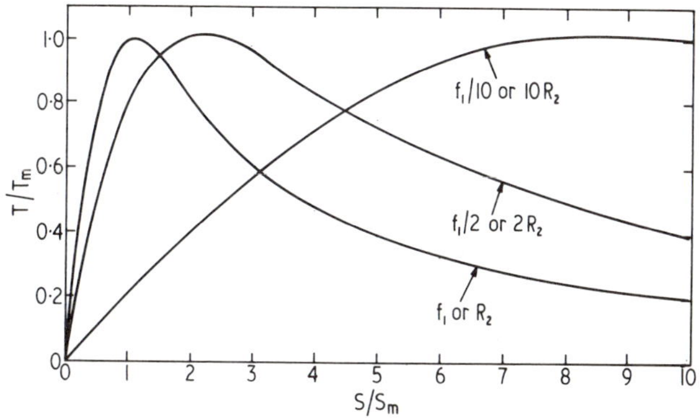

Figure 11.36. Effect of frequency (or rotor resistance) variation on the normalised curve

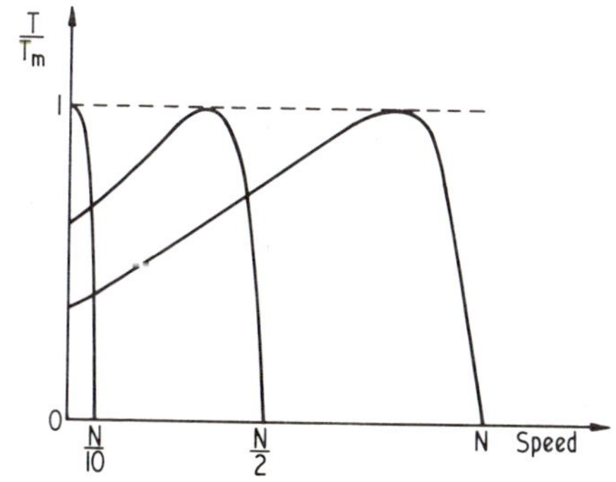

Figure 11.37. Stator frequency control

axis so that all the curves originate from the same point. However, if torque is plotted against speed as in *Figure 11.37* the graphs terminate at their respective synchronous speeds.

Owing to the high efficiency at all speeds, variable frequency control is suitable for motors of all sizes requiring infinite speed variation coupled with high performance. Frequency control can be accomplished statically by inverters or cycloconverters, as described in Chapter 7.

In Chapter 9 various forms of inverter commutation and voltage control systems were discussed. Quasi-square wave control is the

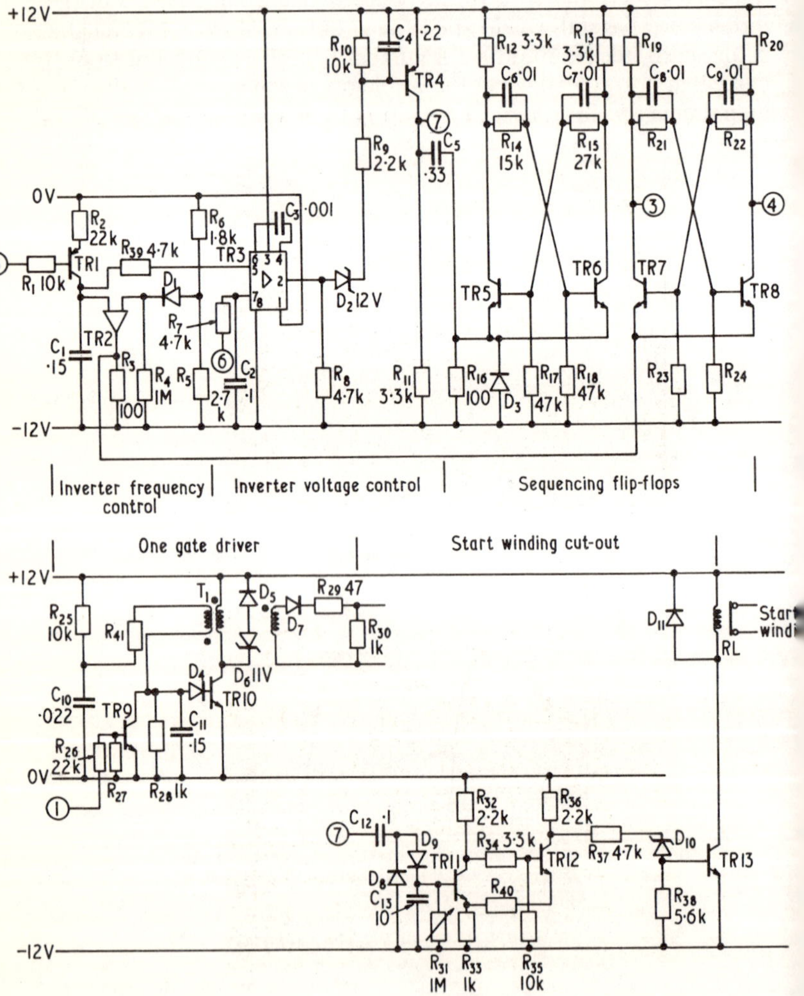

Figure 11.38. Electronic control circuit for a bridge inverter using quasi square wave control

most popular and *Figure 11.38* shows a control system which may be used to fire a single-phase version of the inverter shown in *Figure 9.26*. TR2 forms a programmable unijunction transistor oscillator (which is described in Chapter 12) whose output is used to control a flip-flop stage TR7, TR8. The sawtooth waveform at R_{39} is compared with a variable d.c. level at R_7 in comparator TR3. This produces a phase-shifted waveform which is amplified and differentiated by TR4, the signal then triggering a second flip-flop TR5, TR6. The collector of each flip-flop transistor controls a gate drive oscillator TR10, through transistor TR9, only one driver being shown in *Figure 11.38*. The inverter frequency is determined by the input at pin (5) whereas the quasi-square output by the voltage at pin (6).

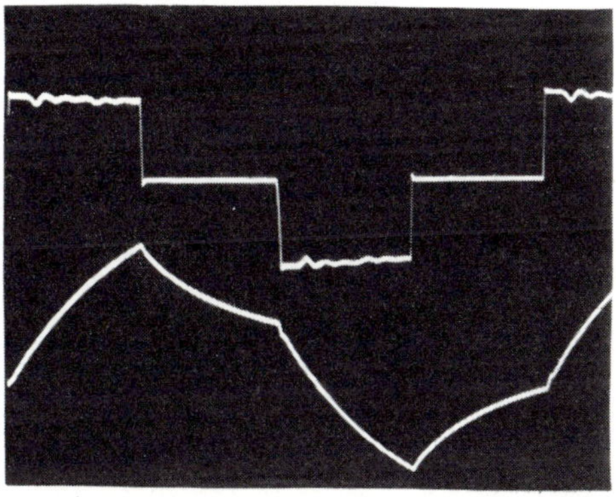

(i)

(ii)

(a)

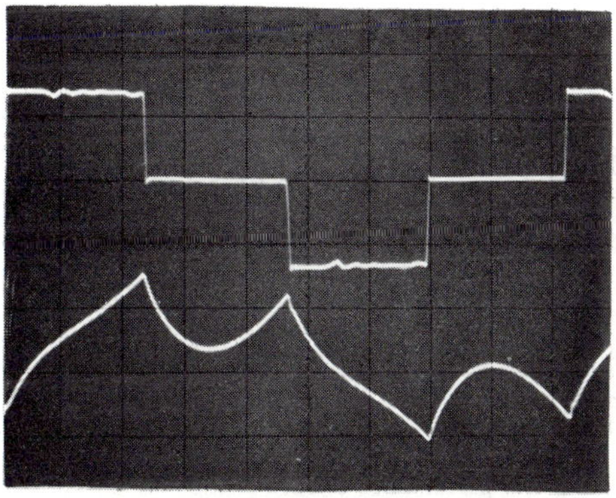

(i)

(ii)

(b)

Figure 11.39. Motor waveforms for a bridge inverter using quasi square-wave voltage control: (a) motor stalled, (b) motor on no load, (i) voltage, (ii) current

Figure 11.38 also shows a circuit for cutting out the motor start winding once the inverter frequency exceeds a preset value. The differentiated output form pin (7) is used to charge C_{13} through a diode pump circuit. When the voltage reaches the value of Schmidt trigger TR11–TR12 (set by R_{31}) transistor TR13 comes on energising the normally closed relay RL, and opening the motor start winding. *Figure 11.39* shows the motor voltage and current. The stalled

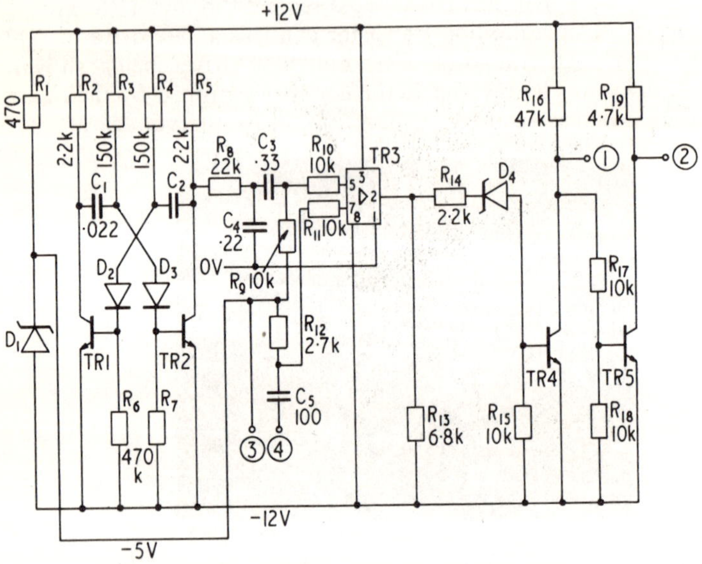

Figure 11.40. Control circuit for a bridge inverter using pulse-width modulated voltage control

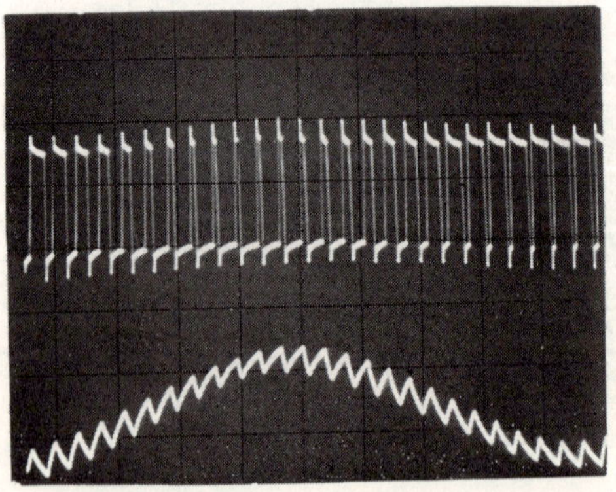

(i)

(ii)

Figure 11.41. Motor stalled waveforms for a bridge inverter using pulse-width modulation with a sine wave f_T = 555Hz, f_S = 20Hz, *(i) voltage, (ii) current*

current corresponds to that discussed in Chapter 9. However, when the motor is running, its back e.m.f. distorts this current waveform so that it is now difficult to determine equations for the ratings of the inverter devices.

Figure 11.40 illustrates a control circuit which may be used to produce a pulse-width modulated inverter output. Pins (1) and (2) drive oscillators which fire diametrically opposite pairs of bridge thyristors. The output from free-running oscillator TR1, TR2 is integrated to give an approximately triangular wave which is then compared with a reference sine wave, fed in at pins (3) and (4), in a comparator TR3. The output from the comparator, which is pulse-width modulated, drives TR4 and TR5. *Figure 11.41* shows the output voltage and current waveforms for half a cycle. The current is

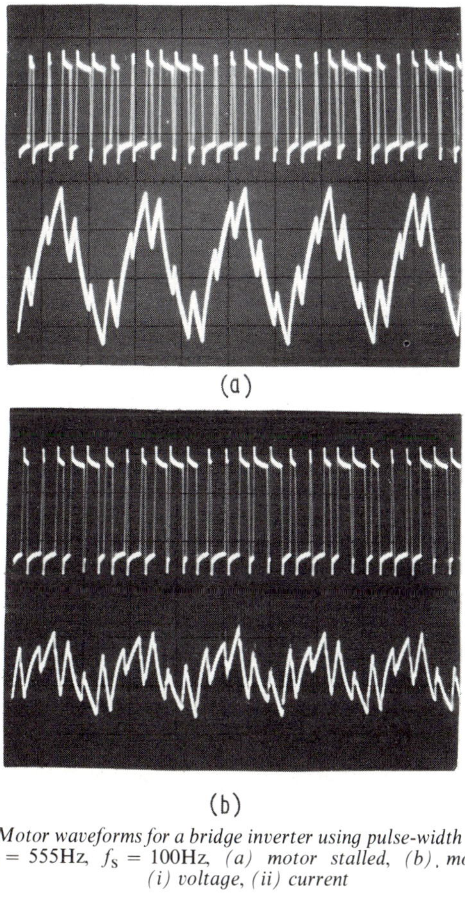

(i)

(ii)

(a)

(i)

(ii)

(b)

Figure 11.42. Motor waveforms for a bridge inverter using pulse-width modulation with a sine wave f_T = 555Hz, f_S = 100Hz, *(a) motor stalled, (b). motor on no load, (i) voltage, (ii) current* seen to be relatively sinusoidal. Since the motor inductance was low, a higher chopping frequency should be used to reduce further the high-frequency component in the current. This is more evident from *Figure 11.42* which also shows the current distortion caused by motor back e.m.f.

(3) Induced current control

In the previous sections, the value of S_m was controlled by varying f_1. From equation (11.9) it is seen that the same effect can be accomplished by changing R_2, the rotor resistance. Note, however, that S_m varies directly as R_2 but inversely as f_1. *Figure 11.36*, therefore, also represents the characteristics for variable rotor resistance control. There is one important difference between this method of speed control and that using variable frequency. In this instance, the stator field speed is constant so that at low speeds the slip energy is dissipated in the rotor resistance and the efficiency is low.

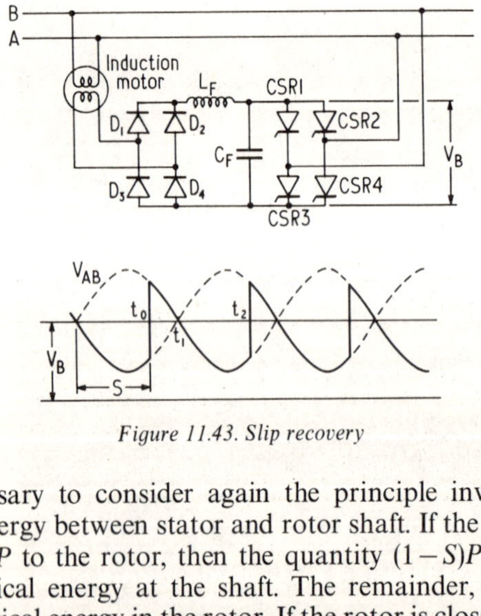

Figure 11.43. Slip recovery

It is necessary to consider again the principle involved in the transfer of energy between stator and rotor shaft. If the rotating field gives power P to the rotor, then the quantity $(1-S)P$ is converted into mechanical energy at the shaft. The remainder, SP, is in the form of electrical energy in the rotor. If the rotor is closed on itself or closed through a resistance, this energy is dissipated as heat. However, if it is possible to extract the energy from the rotor, then the effect on the speed will be the same as if this energy was dissipated. The extracted energy can now be fed back to the supply to give an overall high efficiency. It is also possible to feed energy from an external source into the rotor. The slip can now be made negative (rotor speed exceeding that of the stator field) and energy will flow from rotor to stator across the gap.

A simple slip-recovery scheme is shown in *Figure 11.43*. Rotor voltage is rectified by diodes D1 to D4 and smoothed (C_f is not normally necessary). Thyristor inverter CSR1 to CSR4 can be used to feed a controlled amount of the rotor energy back to the supply. Unlike inverters used in frequency changers, this inverter connects to an a.c. reference line AB which determines its operating frequency and also provides the energy required to turn off the thyristors. For instance, suppose CSR1 and CSR4 are fired at t_0. The thyristors turn on and remain conducting even though the supply AB reverses at t_1, energy being fed from V_B to the a.c. lines. When CSR2 and CSR3 are fired at t_2 the voltage across the previous pair reverses and turns them off. The longer the firing delays the greater the energy fed back from the rotor and the lower its speed. A.C. line commutated inverters were discussed in Chapter 6.

(4) External clutch control

In this method of control the induction motor itself runs at a fixed speed. It is coupled to the load via a clutch which acts as the variable speed system.

Eddy-current clutches are most frequently used. An electromagnet is connected to the shaft of the induction motor and is rotated at a fixed speed. This produces a rotating field whose strength can be controlled by adjusting the magnet coil current. A disc is connected to the load shaft and placed in the rotating field. Owing to normal induction motor action, the disc will rotate, the slip being adjusted by varying the magnetic field strength.

In effect, there are now two induction motors in series. The first runs at a fixed (low-slip) speed, whereas the second runs at adjustable slip. Clearly the efficiency (1 − slip) will be poor at low speeds and will be less than that of a single motor with slip control.

The advantage of using a clutch is that, since it dissipates most of the slip energy, it can be specially constructed to lose this heat more easily than if it were a motor. The method of control, however, is again limited to relatively small motors.

Summarising the above it can be seen that basically there are only two systems of speed control.

(1) where the stator field speed is fixed, the slip being variable
(2) where the stator speed is variable, slip being small and relatively constant.

The first method is inefficient. Voltage control schemes are used in small systems with fan-type loads, or provide constant torque over a limited range. To reduce dissipation in larger motors external eddy-current clutches may be used. Efficiency of variable slip systems can be increased by incorporating a form of slip-recovery scheme. However, this necessitates the use of a wound rotor induction motor, with associated slip rings and brushgear, which can make it unsuitable for certain applications (*e.g.* hazardous atmospheres).

The most promising control system is a variable-frequency drive. A cage rotor can now be used and efficiency over the whole speed range is also high. Such systems are normally the most expensive. However, if a drive is required to provide a constant torque output for wide speed variation, and the motor is located in an inaccessible position, variable-frequency control of a cage rotor machine is often the only practicable system.

As stated previously, the speed of a synchronous motor can only be controlled by means of frequency variation. Changing the rotor field does not affect the motor speed but does change the power factor that it presents to the supply. Since the field of the motor is mounted on the rotor, it is necessary to provide sliprings and brushes if the field excitation is to be supplied externally. This can be avoided by using a.c. exciters and rotating field diodes, as in brushless synchronous motors. *Figure 11.44* shows one type of motor arrangement. The armature of an a.c. exciter rotates with the synchronous

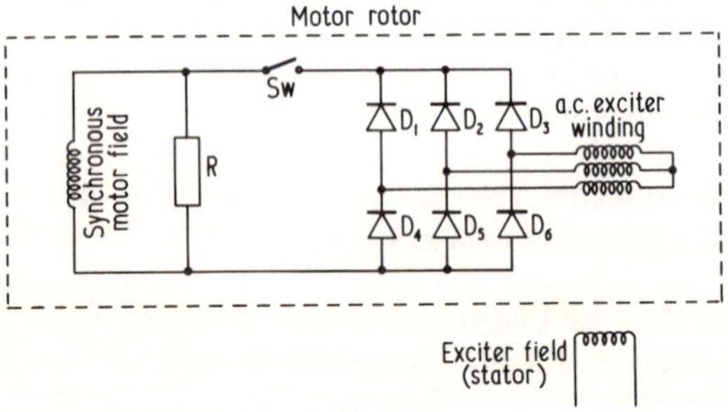

Figure 11.44. Brushless synchronous motor arrangement

motor field winding, the a.c. output from the exciter being rectified by rotating diodes D1 to D6 before being fed to the motor field. During run up, switch Sw is open and the field windings are shorted through resistor *R*. The a.c. exciter field is open circuit. When the machine nears synchronism centrifuge switch Sw closes. The exciter field is now energised so that d.c. is applied to the synchronous motor field and it locks onto the rotating stator field.

The above system suffers from two disadvantages: (1) it is not fully solid-state since switch Sw is used; (2) resistor R is connected in circuit throughout run up and synchronous operation so that the motor efficiency is reduced. *Figure 11.45* shows a better arrangement.

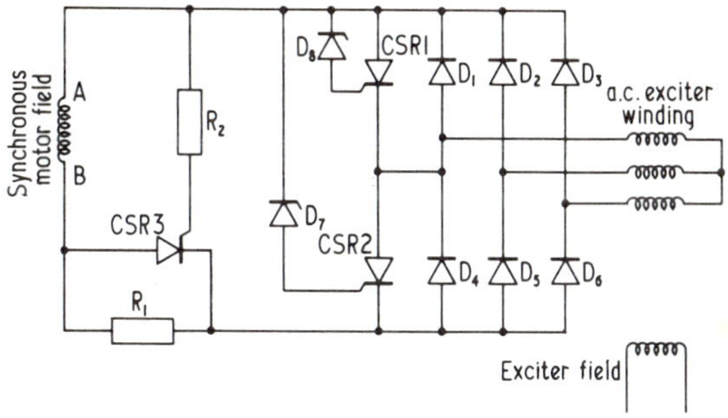

Figure 11.45. A more efficient brushless synchronous motor than that in Figure 11.44

During run up the a.c. exciter is de-energised as before. An alternating voltage is now induced in the synchronous motor field winding due to induction action, and it can be very large and destroy the field diodes unless these are protected. Assuming end B to be positive to A, current will flow through R_1 and D1 to D6. CSR3 is off since its gate is at a negative potential (equal to two diode drops) with respect to its cathode. In the next half cycle, when end A is positive to B, CSR3 is reverse-biased and still held off. The voltage induced in the synchronous motor coil now rapidly builds up. When it reaches the Zener voltage of diode D7, thyristor CSR2 is turned on followed closely by the breakover of D8 and the firing of CSR1. The motor field is now again short-circuited through CSR1, CSR2 and R_1. This prevents damage occurring to the diode bridge. At speeds close to synchronism the a.c. exciter is energised. Current flows into the motor field and CSR1 and CSR2 turn off when D1 and D4 first conduct. CSR3 is forward-biased and turns on so that R_1 is short-circuited and no longer plays any part in the exciter operation. Zener diodes D7 and D8 are chosen to ensure that their voltage is greater than the peak a.c. exciter voltage, so that CSR1 and CSR2 are held off.

With the above scheme the starting and synchronous performance of the machine is identical to a conventional motor with slip rings. Since the efficiency of the rectifier is higher the overall motor efficiency is also greater. However, there is now random synchronism so that pull-in characteristics are inferior to those obtained with slip-ring motors and slip-frequency relay synchronisation. In the majority of applications this loss of pull-in torque is not critical, and this scheme finds use in systems such as reciprocating motor drives, cement mill motors which start unclutched and synchronous condensers.

Braking

As with d.c. motors there are three methods for electrically braking a.c. machines. During plugging the direction of rotation of the motor is reversed. This is done by reversing the direction of the rotating field by interchanging two of the stator winding connections (as illustrated in Chapter 3). An induction and reluctance motor can be plugged in this manner. A synchronous motor can only be plugged if it is fitted with induction windings so that during plugging it behaves as an induction motor. Since the rotor and stator fields are moving in opposite directions, large currents are induced in the motor, resulting in heat loss. The rate of plugging can be varied by controlling the supply voltage or frequency so as to change its torque. These can be accomplished by using thyristor phase-controlled circuits or frequency changers.

In dynamic braking, the motor must act as a generator. For an induction motor to act as a generator it must be driven above the speed of the stator field. This is not possible during braking if the normal a.c. supply is maintained to the stator. However, if the stator is supplied with d.c., the excitation necessary to induce currents in the rotor is present, and at the same time the rotor exceeds the stator field speed. Operation is as in *Figure 11.6*. Since the stator poles are now stationary the rotor flux is also stationary. The system no longer operates as a transformer and load current is not present in the stator. Therefore all the mechanical input is used in generating heat in the rotor, no power being fed back to the d.c. supply. Dynamic braking in a synchronous motor is similar to that used for d.c. machines, since the rotor field is independent of the stator. All that is required is for the stator supply to be disconnected and braking resistors to be connected across the motor terminals, the rotor field still being maintained.

To regenerate in a synchronous motor the frequency of the supply must be reduced gradually so that the motor (now acting as a generator) is continually in step, the loss in rotor kinetic energy being fed back to the supply. The motor can also be disconnected from the source and its generated output (which will be a.c. whose frequency changes with the motor speed) rectified and fed into a d.c. source. An induction motor will generate so long as the rotor speed exceeds that of the stator field, provided this is not d.c. The motor inertia is converted partly into heat in the rotor, and is partly fed back to the supply. The larger the difference in rotor and stator field speeds the smaller the ratio (regenerated output)/(input power), and the greater the heat loss in the rotor. Clearly regeneration into an a.c. source for all types of a.c. motors requires a controlled frequency source.

A.C. GENERATORS

The principal objective in generator control is regulation of its output voltage. Referring to the elementary machine of *Figure 11.1* the voltage induced in coil *a-b* is directly proportional to the rate with which it cuts the stator flux, *i.e.* to the relative speed between rotor and stator field and to stator field strength. In d.c. and synchronous generators, changing the rotor-stator field speed involves control of the mechanical drive, which is inconvenient. In a synchronous machine a change in speed also produces a change in output frequency. Therefore field control is normally used, the thyristor system operating in d.c. line control or controlled rectification modes depending on whether the supply to the controller is d.c. or a.c.

Since the excitation field in an induction generator is rotating, the speed of this field can be controlled by the use of thyristor frequency changers, so as to vary the generated output. It must be remembered that, owing to the complex nature of the transformer action involved, part only of the induced rotor current is supplied to the stator output, the rest being converted into rotor heat. There is an optimum relative speed at which the output is a maximum. Excitation control may be used to control the magnitude of the output, although it is normally kept proportional to frequency, to maintain constant flux.

Synchronous generators, called alternators, are popularly used for power generation, and these are treated here in further detail.

Figure 11.46 shows several arrangements for alternator excitation. In all cases the regulator converts the input supply (if it is a.c.) into d.c., and regulates the amount of the field excitation so as to maintain the output voltage constant, at a predetermined value, irrespective of the load on the machine. *Figure 11.46*(i) illustrates the simplest arrangement which is used for smaller machines up to about 100kVA. The input to the regulator is derived from the a.c. output of the generator itself. These lines therefore perform the dual action of power supply and sensing for feedback voltage control. The regulator is normally a half-controlled single-phase thyristor bridge although half-wave circuits, using a single thyristor and a free-wheeling diode across the field, may be used. For larger machines the arrangement of *Figure 11.46*(ii) is preferred. The regulator now controls the field of an auxiliary exciter, its armature being mounted on the rotor and supplying the generator field through a rotating rectifier bridge. Both the above systems can be operated from an

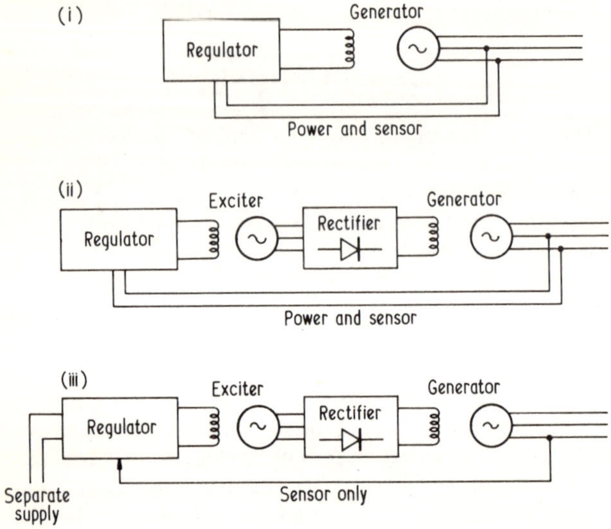

Figure 11.46. Different arrangements for alternator excitation: (i) direct self excitation, (ii) indirect self excitation, (iii) indirect separate excitation

external supply if it is available. *Figure 11.46*(iii) shows such a system for large turbo-generators. The feedback line from the generator output is now required only for voltage sensing purposes and carries no power.

The action of a thyristor regulator when used in an alternator excitation system is very similar to usual phase-control systems employed in controlled rectification. The greater the delay angle α in the firing of the thyristors during a half cycle, the lower its d.c. output, and therefore there is a fall in alternator voltage, and vice versa. There are, however, several other considerations which modify the simple control loop shown in *Figure 11.46*. For instance it is often desirable to introduce a droop in the output voltage characteristic such that the terminal voltage falls with load. This is required to enable better load sharing between parallel-connected machines. Similarly alternators must be designed so as to give a large output current under fault conditions. This is necessary to operate various circuit trip systems. With a separately excited system this occurs naturally, as in *Figure 11.47*(i) where the short-circuit current is only limited by the alternator reactance. For self excitation, a fall in terminal voltage results in a reduction of the excitation current and a further fall in the generator output. The characteristic therefore folds back as in *Figure 11.47*(ii), and sufficient current is

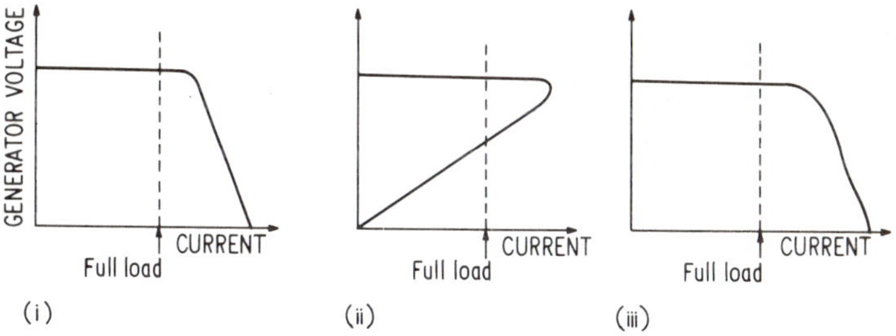

Figure 11.47. Generator characteristics up to short-circuit conditions: (i) separate excitation, (ii) self excitation, (iii) self excitation with current boost

not available to operate trip circuits. This can be overcome by feeding the alternator field, in parallel with the regulator output, with current derived from the alternator lines through a current transformer. An overload condition will now feed back field excitation and will be self-sustaining. The characteristics of the alternator are now modified as shown in *Figure 11.47*(iii).

Another problem encountered with self-excited machines, which does not occur with separate excitation, is that of starting. When the alternator first runs up to speed, its output voltage is very low, being caused primarily by residual field flux, and it will be too small to operate the thyristor gate drive system. The thyristors will therefore be held off and the machine will not be excited. To overcome this it is usual to apply continuous gate drive to the thyristors during the period that the alternator output is below a certain critical value. The easiest way of doing this is by a small normally closed relay connected between anode and gate of the thyristors. Alternatively a static control system may be used, one type being shown in *Figure 11.48*. When the a.c. line voltage (from the alternator output) is low the voltage induced in L_3 is small so that CSR1 is held off. This means that current flows via D1, the top of L_1, L_4 and the bottom of L_2. Current in L_4 induces a voltage in L_5 which fires the regulator thyristors at the start of the phase-control cycle, supplying field current. Above a certain line voltage the induced voltage in L_3 is sufficiently large to overcome the Zener voltage of D4 and fire CSR1. The currents in L_1 and L_2 are now balanced so that no voltage is induced in L_5. The gate pulse from the start-up circuit is therefore no longer

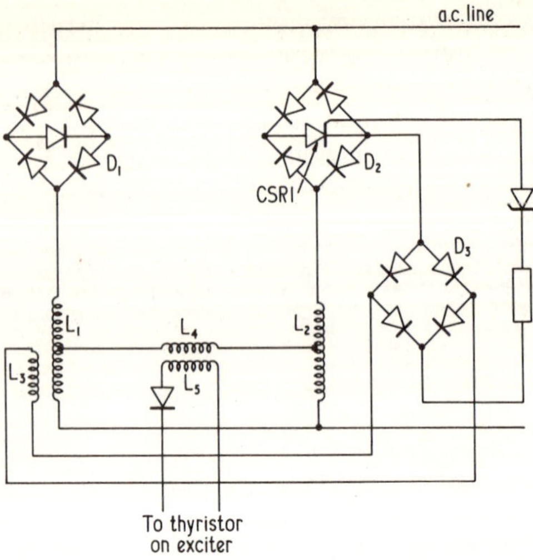

Figure 11.48. Static start-up circuit for an alternator excitation system

applied to the regulator thyristors which now operate under normal phase control.

Thyristor excitation circuits are small, robust and have a fast response due to the low system lag. When there is a fall in the output voltage of an alternator, say due to the loss across the machine sub-transient reactance when an induction motor is switched on, it is necessary to increase the field current rapidly and therefore raise the output voltage again. This calls for a field current above the normal excitation. Thyristor circuits are well suited to provide this field forcing action, and the ratio of maximum to full-load excitation field voltage, called the forcing margin, is usually three.

CONCLUSIONS

Thyristor circuits are currently in widespread use for motor control. Often they represent the only practicable method of obtaining a control system. In others they are in direct competition with traditional controllers and must then prove themselves technically or economically superior. The advantages of thyristor drives, compared to conventional methods such as Ward Leonard systems are:

(1) Quicker response. This is specially advantageous when fast acceleration or high speed accuracy is required as in reversing-mill tables.

(2) High operating efficiency due to the small volt drop across a thyristor even when it is handling large amounts of power.

(3) Less weight and lower installation costs. A thyristor control cubicle occupies a small floor area, is light and does not require any special floor surface or mounting. This compares very favourably with the aligning and mounting required for motor-generator systems.

(4) Ease of maintenance. Contrary to popular belief a thyristor system can be made very easy to maintain by incorporating diagnostic circuits during the design stage, and providing plug-in card replacements for electronics. Servicing of the controller is, of course, not required owing to the absence of any moving parts.

The disadvantages of thyristor systems are:

(1) The low overload capability which is caused primarily by the low time constant of the thyristors. This means that in most cases the control system must be rated to withstand the peak overload expected instead of the usual full-load value.

(2) Inability of a thyristor system to regenerate unless special circuit modifications are included, such as double bridge controllers. A motor-generator set on the other hand can readily pass energy in either direction.

(3) Low power factors in drives which use phase-angle control. As has been illustrated in previous chapters the load power factor is approximately proportional to the thyristor delay angle, and can be very poor at low speeds (low voltage).

(4) Generated harmonics in systems where large voltages are switched, due to the very fast operating time of the thyristor. These harmonics cause radio-frequency interference which must be suppressed.

12

Power semiconductor devices

In the previous chapters of this book the characteristics and applications of thyristors have been considered in detail. The thyristor is by far the most popular power semiconductor in use at the present time. It is, however, not the only such device. This chapter will briefly examine other types of power semiconductors. Also described will be control devices which, although modestly rated themselves, are capable of controlling the power devices.

TAKE SOME SEMICONDUCTOR

It is not intended in this brief introduction to delve into atomic physics. Instead some of the basic concepts of semiconductors are introduced for use in subsequent sections.

The structure of an atom is well known. It consists of a central nucleus and revolving electrons in various orbits. The positive charge on the nucleus is balanced by negative electronic charge. The atom is said to be stable if its outer shell is full, the electrons for the first four shells (working outwards from the nucleus) being 2, 8, 8 or 18, 8 or 32. The number of electrons in the outermost shell determines its group number. Therefore, germanium has 2, 8, 18, 4 electrons and silicon has 2, 8, 4 electrons. Both are in group IV, and although unstable on its own a crystal of germanium or silicon forms stable covalent bonds between atoms. Each shares its four electrons with neighbouring atoms and so has eight orbital electrons. However, if energy is given to the material, say in the form of heat, an electron can break away from its valency bond and cause conduction in the material. This is called intrinsic conduction. If an impurity of group V (donor) or group III (acceptor) is added to the material, there will be an abundance or shortage of electrons respectively, causing increased conduction at a given temperature. This is called extrinsic conduction. A shortage of electrons results in a *p*-type doping where holes are the majority carriers and electrons the minority carriers, whereas extra electrons results in an *n*-type material in which the carrier function is reversed.

Holes and electrons are constantly being created in a semiconductor. They disappear owing to recombination, a process which can be increased by the presence of specially created trapping centres in the crystal. These hold a carrier until an opposite polarity charge arrives for recombination.

The properties of a semiconductor can now be summarised. Firstly it has a resistivity which lies between that of a conductor and an insulator. Secondly its resistance decreases with temperature, unlike that of a conductor. The most popularly used base semiconductors are germanium and silicon. Germanium has a much lower conduction volt drop and therefore dissipates less power for a given current. However, it is very sensitive to voltage transients, and owing to its low energy gap (0·67eV) cannot work with junction temperatures above 100°C. Silicon is more frequently used in most power devices. It has a slightly higher volt drop than germanium but the high energy gap (1·106eV) means that it can work at junction temperatures of the order of 200°C.

ADD A SECOND LAYER

There are two popular techniques for forming a *p-n* junction within a semiconductor, *i.e.* diffusion and alloying. During diffusion an *n*-type silicon slice is exposed to *p*-type impurity (*e.g.* gallium) at increased temperature and pressure. This causes the impurity to diffuse to a few thousandths of an inch into the silicon over several hours. For alloying aluminium (*p*-type, *i.e.* group III element) to silicon, the aluminium is placed in contact with the *n*-type slice and the pair subjected to temperatures of 600 to 700°C. This causes the aluminium and silicon to melt and recrystallise on cooling into a *p-n* junction.

Figure 12.1 shows a *p-n* device in which diffusion of free holes into the *n* region and electrons into the *p* region gives a depletion region of reverse voltage which stops further significant diffusion. Applying an external supply with the positive connection to the *n*-type (*i.e.* reverse-biasing the diode) will cause a low reverse current to flow due to minority carriers. For forward diode bias, holes and electrons flow readily across the junction and the current is high. Shockley has shown that the forward current I_F is given by

$$I_F = I_R[\exp(qV/kT) - 1] \tag{12.1}$$

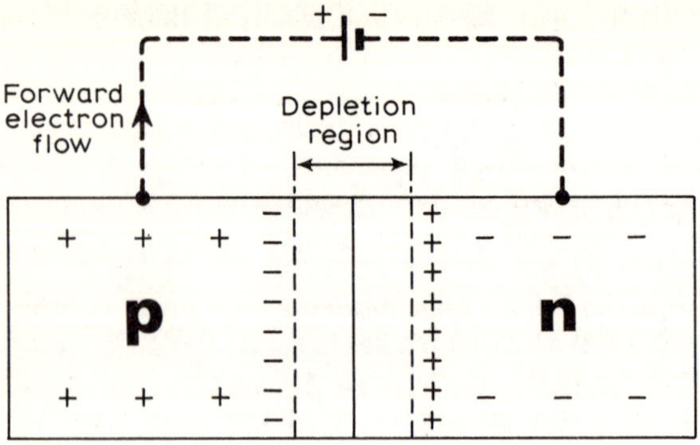

Figure 12.1. A p-n junction

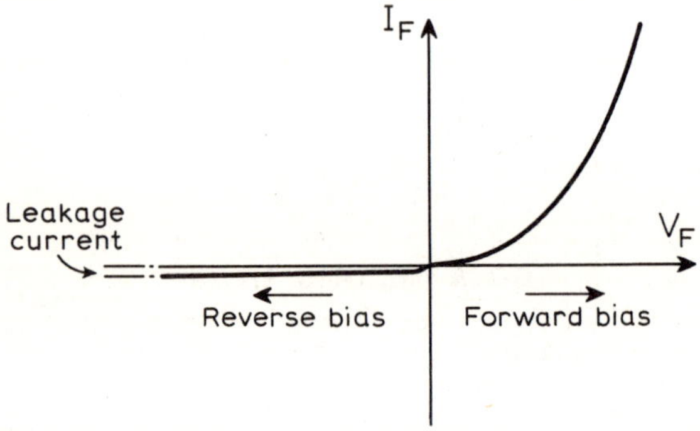

Figure 12.2. Typical diode characteristic

where I_R is the reverse junction current, k is Boltzmann's constant, q the electronic charge, T the absolute temperature and V the voltage across the junction. *Figure 12.2* shows a typical diode characteristic.

The power loss in a diode occurs for three reasons. When forward- or reverse-biased, the loss is the product of voltage and current through the device. In power devices, operating at higher frequencies a third loss due to switching becomes significantly important but is often neglected. When a diode is switched into conduction a finite time is required for the minority carriers (holes for *p-n* devices) to flow across the junction and so prime it to carry the full load current. This can result in an initial peak voltage across the device, as shown in *Figure 12.3*, which gives increased losses (voltage-current product). The forward recovery time is, however, normally small compared with the reverse recovery period described below.

The diode junction shown in *Figure 12.1* is seen to resemble a

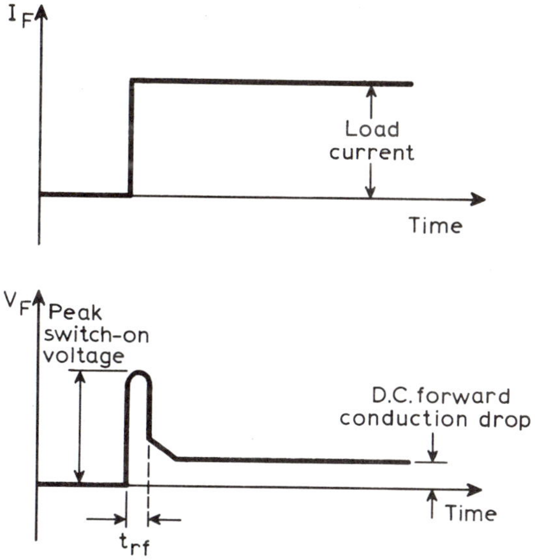

Figure 12.3. Diode switch-on characteristic

capacitor in which the two oppositely charged plates are separated by a distance representing the depletion region. This is true of most semiconductor junctions. The capacitance produced is often 'unwanted' and can have a nuisance effect on devices of varying structures. It will be referred to again later.

When a diode is switched from the forward conducting to the reverse-biased mode two changes occur. Firstly the free holes and electrons are swept away as reverse current in the external supply. This current can be large, often being limited only by the external circuit impedance. It could be described as being the surge current through the device junction capacitance, although this is not strictly true. In addition to this external current flow, holes and electrons disappear in the vicinity of the reverse biased region owing to recombination. The current decays to zero as the reverse voltage blocking capability of the diode increases as shown in *Figure 12.4*. Clearly the reverse recovery period t_{rr} must be kept as short as possible in order to limit the device dissipation. However, the decay of reverse current must not be too abrupt or it can give rise to severe voltage transients in circuit inductances ($L\,di/dt$ effects).

In circuit applications the reverse recovery current, which for power devices can be large due to the larger silicon area involved, is often overlooked, with disastrous effects. For instance, consider the chopper circuit shown in *Figure 12.5(a)*. The semiconductor

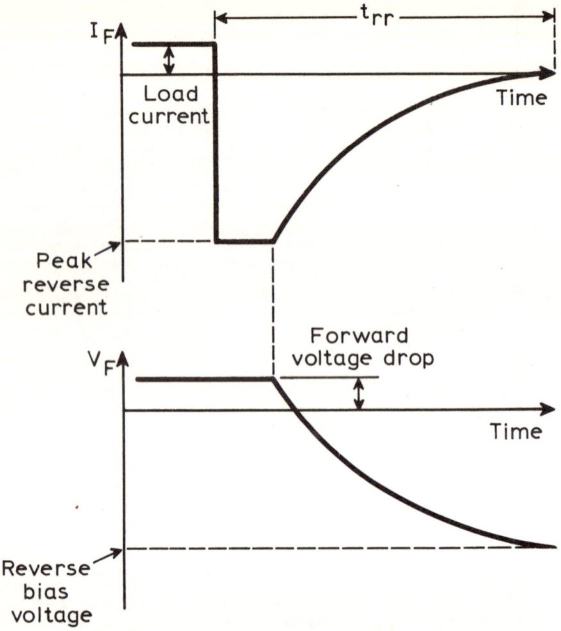

Figure 12.4. Diode switch-off characteristic

switch is opened and closed at a relatively high frequency and varying duty cycle in order to control the magnitude of the mean load voltage and current. Suppose the load is inductive. This requires a diode D, often called a free-wheeling diode, to be connected across it to prevent excessive surges in the switch when it opens. Ignoring the reverse recovery current in the diode the system would operate as follows. With the switch closed D is non-conducting and load current is supplied via the d.c. source V_B. When the switch opens the load current free-wheels in D and, provided the loss in this path is not large, the current will be substantially unchanged when the switch is again closed. The free-wheeling current will cease to flow in D and be supplied through the semiconductor switch when it is again closed, and so on. If the reverse recovery current through D is not ignored then with the load current free-wheeling through it the operation is basically as before. However, when the semiconductor switch closes this current transfers to the supply. In addition there is now a reverse recovery path as shown, and since this is of a very low impedance a surge of current can pass which could destroy the switch.

Chopper circuits are not exclusively prone to reverse current effects. *Figure 12.5(b)* shows a commonly used bridge rectifier circuit. Suppose line A is positive to B so that load current is supplied via diodes 1 and 4. When the voltage reverses and line B becomes positive the current in diodes 1 and 4 will decay to zero while that in 3 and 2 will increase to support the load. As soon as 1 and 4 turn off a

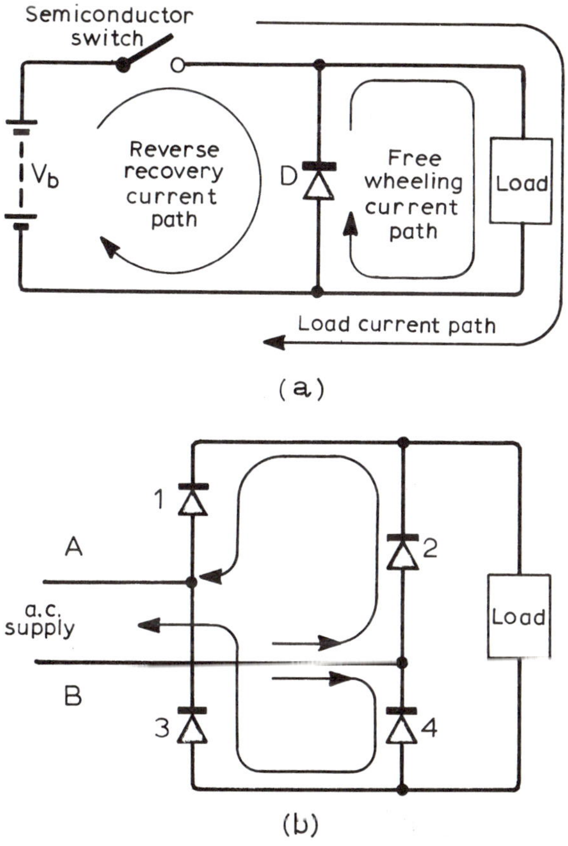

Figure 12.5. Examples of circuit reverse recovery current

reverse current path exists through them, as illustrated. This provides a short circuit across the supply lines which could destroy diodes 2 and 3. In all such applications great care must be taken to ensure that there is sufficient line impedance which would limit this short-circuit current until the reverse-biased device recovers its full blocking capability.

The encapsulation of a power diode is normally similar to those

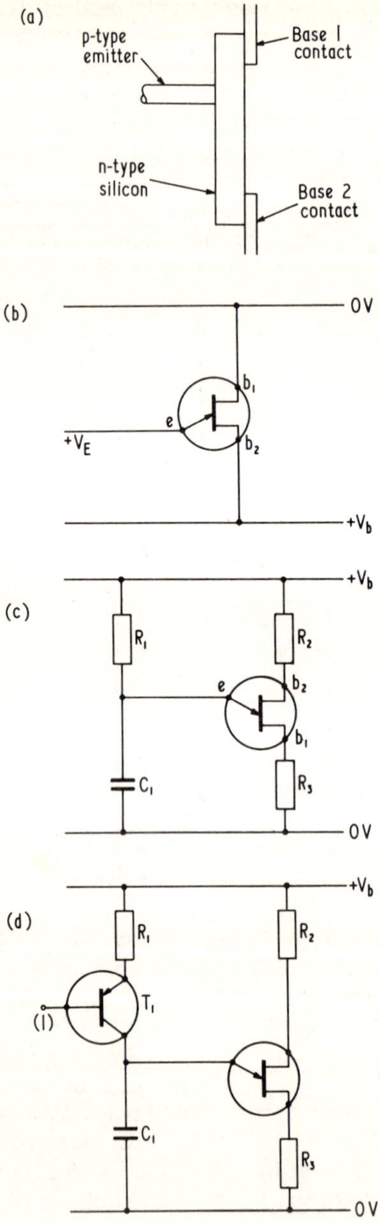

Figure 12.6. Unijunction transistor construction and applications

of other semiconductor rectifying power devices and will be discussed later.

A two-layer device which has been extensively used to control other power semiconductor devices is the unijunction transistor (UJT). Its construction is shown in *Figure 12.6(a)*, and the circuit symbol and biasing connections in *Figure 12.6(b)*. The transistor is essentially a two-layer device in which the emitter is a *p*-type rod placed asymmetrically in an *n*-type base material. Two base contacts are made to the *n* region, base 1 being closer to the emitter than base 2. This arrangement is such as to cause the *p-n* region to be reverse biased by a voltage ηV_B. V_B is the supply voltage between bases 1 and 2 whereas η is a built-in constant of the UJT, known as its intrinsic stand-off ratio, whose value is determined by the device geometry.

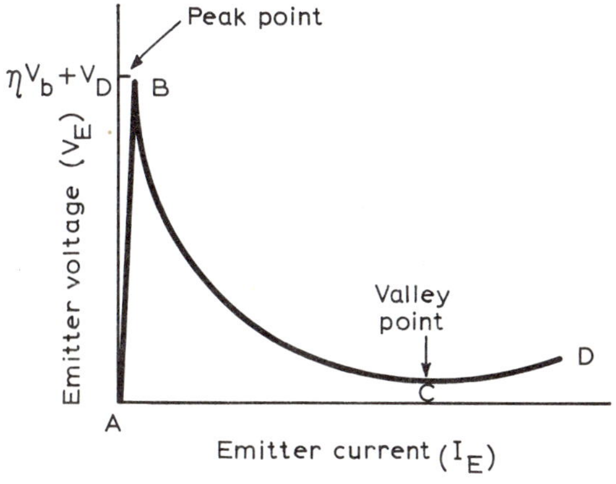

Figure 12.7. Unijunction transistor characteristic

The characteristic for a UJT is given in *Figure 12.7*. As the emitter voltage V_E increases, for a certain base voltage V_B the current increases only slightly due to the reverse *p-n* junction. Once, however, the emitter voltage reaches $\eta V_B + V_D$, where V_D is equal to the drop across the emitter-base 1 junction, this becomes forward-biased and holes are injected into the *n* material. This increase in emitter current causes the emitter-base 1 voltage to fall so that the characteristics enter a negative-resistance portion BC. The increase in conduction τ can be controlled by device doping, being given by

$$\tau = q(\mu_p p + \mu_n n) \qquad (12.2)$$

where μ_p and μ_n are the mobility of holes and electrons and p and n their concentration in the silicon. The negative region extends up to the valley point C. Beyond this region further increase in conduction is limited by hole-electron recombination so that the emitter voltage must be increased to give a greater emitter current.

Unijunction transistors are mainly used over the negative-resistance region to produce a variety of oscillator and timing circuits. Their use in firing power semiconductors has already been described. *Figure 12.6(c)* shows an elementary relaxation oscillator. Capacitor C_1 charges through R_1 towards $+V_B$ volts. Owing to the very low leakage current of the UJT when reverse-biased, the value of R_1 can be made large so that long time delays are possible. As soon as the voltage on C_1 reaches that of the peak point, its emitter-base 1 junction collapses. C_1 discharges rapidly through R_3 producing a positive spike across it. When the voltage falls to the valley point the UJT recovers and C_1 again begins to charge through R_1. The oscillator frequency can be controlled relatively easily by changing C_1 or R_1. Since R_1 operates from a d.c. source it can be connected to the emitter of a transistor, as in *Figure 12.6(d)* if required. This produces a system where the timing varies linearly with the voltage at input (1) and is commonly used in closed-loop control systems.

THEN A THIRD LAYER

Perhaps the best known semiconductor device is the transistor, which is a three-layer *p-n-p* or *n-p-n* device. *Figure 12.8* shows a *p-n-p* construction. Holes are injected across junction J1 and since the base is relatively thin these are carried across reverse-biased

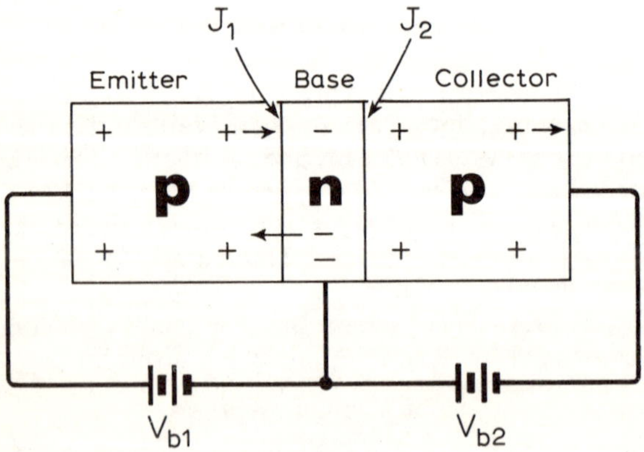

Figure 12.8. p-n-p transistor action

junction J2 and reach the collector. Those that recombine in the base region are replaced by further base electrons, which constitutes the base current. If the base layer is thin the holes lost by recombination form a small proportion of those reaching the collector—hence the device has base to collector current gain. Also the collector current is proportional to the injected base current, provided the external circuit can support this.

A power transistor is very similar to its low signal counterpart, differing only in processing techniques which are used to enhance some particular aspect of its performance. The parameters of most interest in power devices are the d.c. current gain h_{FE}, the breakover voltage BV_{CEO}, switching time and safe operating area (SOA). D.C. gain should be as high as possible since at large currents the base drive power could be appreciable, considerably reducing the system efficiency. The breakover voltage must be adequate to withstand the peak values expected in the circuit, which includes any switching transients. Switching times should be kept as low as possible since, for a transistor which is operating as a switch, it is at these instants when peak power dissipation occurs, and at high frequencies it can give considerable peak power loss.

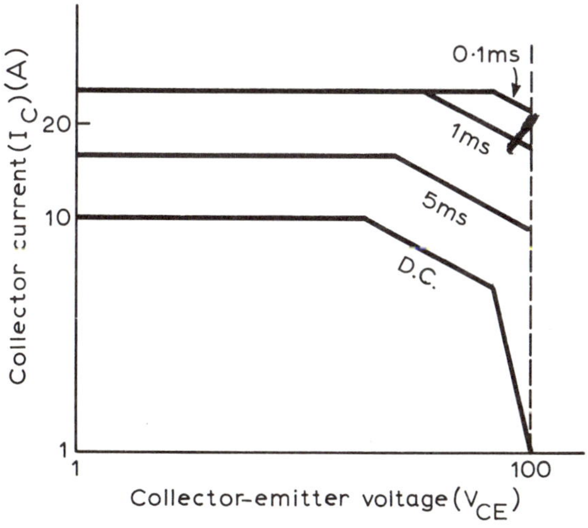

Figure 12.9. Typical safe operating curves for a transistor

A parameter which is most frequently overlooked when designing power transistor circuits is its safe operating area. This is given on transistor data sheets by sets of curves as shown in *Figure 12.9* for a 100V 10A device. As seen the transistor cannot control 10A at 100V, dissipation limiting its current carrying capability at this voltage to only 1A. Apart from d.c. the characteristics also show curves for pulsed operation. Clearly the shorter the transistor duty cycle the greater the current it can carry for the same power dissipation. The operating load line for a transistor must fall within its respective SOA. If it does not then secondary breakdown can result, destroying the device. This phenomenon of secondary breakdown can be explained by considering a transistor which is driven by a fixed base current. As the supply voltage is raised the collector current will increase, along line AB in *Figure 12.10*. However, soon, say at point

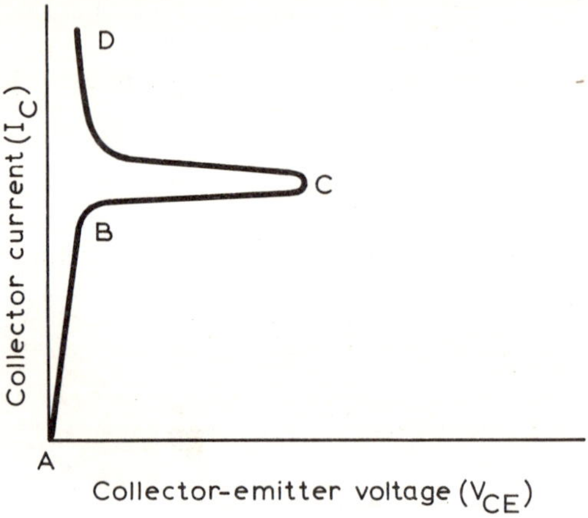

Figure 12.10. Secondary breakdown in a transistor

B, the base current will no longer be able to support the increasing collector current. The transistor will come out of saturation and any further increase in supply voltage will cause the collector voltage to rise above the previous saturation value, the current increasing only slightly due to greater leakage at the higher voltages. At C the operating point of the transistor goes outside its SOA. The electrons in the device attain sufficient energy to cause an avalanche effect in the semiconductor. Collector to emitter voltage collapses along CD while the current rapidly builds up, limited only by external circuit impedances. If the current is not kept below a safe value it can reach very high concentration levels in certain parts of the semiconductor slice. This gives rise to local hot spots and can result in melting of the semiconductor and a collector to emitter short circuit.

The design of a transistor is normally a compromise to obtain the required characteristic. For instance, if the width of the base layer or the collector resistivity is increased, then the maximum voltage and the safe operating area are increased. However, gain and switching times fall. Similarly, the process used is important. Alloying is employed only in the lower-cost germanium devices, most others being alloy-diffused, single-diffused, double-diffused or triple-diffused.

Although germanium power transistors were popular in the past they have now been largely replaced by silicon. The maximum operating temperature of germanium is about half that for silicon, and its f_T is limited to about 1MHz compared to 100MHz for silicon. Germanium has a lower collector-emitter saturation voltage, but its slower switching speed gives a dissipation comparable to silicon when operating at higher frequencies. Peak collector voltage of germanium is also low, about 50V, although diffused devices have been built up to about 300V. Silicon transistors can currently operate at 1 500V. Although *n-p-n* germanium transistors are rare, both polarity silicon devices are readily available. This is a boon in constructing transformerless audio output stages, as shown in *Figure 12.11*, and up to 200W units are currently in use. The advantages of

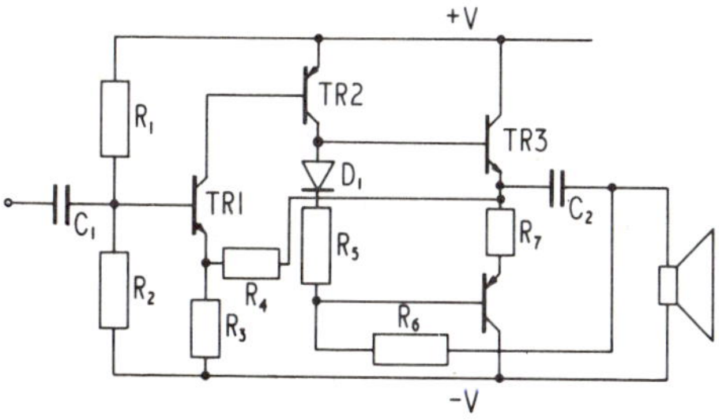

Figure 12.11. Audio amplifier with complementary output

germanium devices are the more competitive price and the lower supply voltages that are required, due to the lower saturation volt drop of germanium.

Power-transistor encapsulations are normally designed with the prime purpose of providing access to the hottest part of the dice in order to facilitate cooling. TO-3 and TO-66 with the familiar diamond shape are very popular although for high currents the devices are usually stud-mounted. Recently there has been a growth in plastic power transistors which have a metal backing plate for clamping to a heatsink. This gives a much cheaper device and advances in plastic techniques are making this a very acceptable and reliable encapsulation.

NOW TO FOUR LAYERS

Perhaps the device which has contributed most to the recent growth in power semiconductor technology is the thyristor. This has four layers, which may be formed by diffusion or alloying, as shown in *Figure 12.12*. Normally lead G2 is not used. When the thyristor is

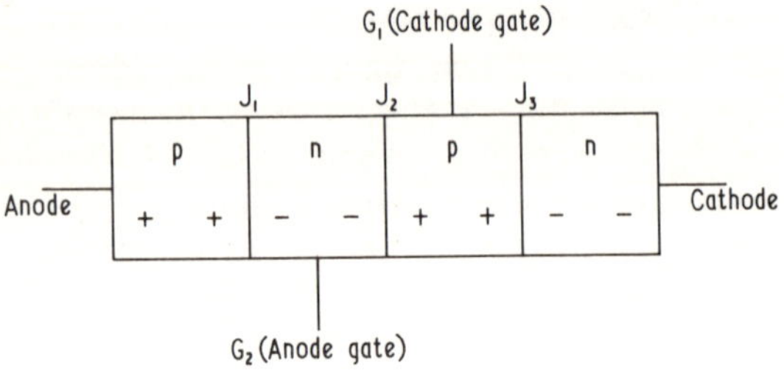

Figure 12.12. Four-layer p-n-p-n device

connected with anode positive with respect to cathode, junctions J1 and J3 are forward-biased while J2 is reverse-biased and supports only a small leakage current. There are several ways in which the thyristor can be turned on. The voltage across the device can be raised such that the energy of minority carriers crossing J2 is sufficient to turn it on, or heat can be used to excite the electrons across this junction. Alternatively a rapid voltage rate of change across anode and cathode would cause current flow through the device capacitance and trigger the thyristor. The recognised method, however, is to bias G1 positive or G2 (if provided) negative, the injected current acting as base drive in a transistor and turning the thyristor on.

Several modifications can be made to the basic four-layer structure of a thyristor to enhance certain of its properties. As for a transistor, however, this normally gives a deterioration of some other parameter. For instance, during turn off if the anode to cathode potential is reversed the charges at junctions J1 and J3 are rapidly swept away in the form of reverse current. However, junction J2 recovers relatively slowly by a process of recombination. This can be speeded up by adding certain impurities, notably gold, during diffusion of the layers, which reduces carrier lifetime. However, gold increases the forward volt drop of the device and also its leakage current, so limiting the maximum voltage it can block. Therefore high-voltage devices cannot be obtained with fast turn-off times. A similar trade-off exists when designing thyristors with fast turn on. This requires the inner *n* region to be relatively thin. However, this region supports the reverse voltage across the device and should be thick to do so. Therefore high-voltage thyristors have higher switching losses and lower *di/dt* capability than comparable lower voltage rated devices.

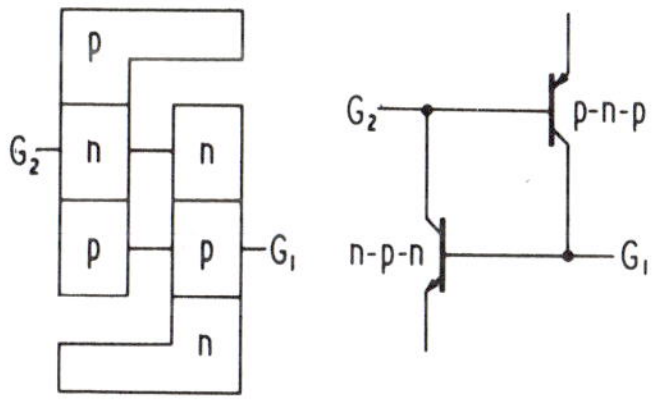

Figure 12.13. Two-transistor analogy of a thyristor

Figure 12.13 gives the two-transistor analogy of a thyristor. This shows that although a thyristor is normally turned on by a positive pulse on terminal G1, it can also be initiated into conduction by negative drive on gate G2. Due to the geometry of the device more current is now required, and this terminal is not usually brought out on higher-power thyristors. However, in some instances, notably when operating in logic circuits, there is advantage in having this additional trigger facility. Devices which have both gate leads accessible are called silicon controlled switches (SCS).

Examining *Figure 12.13* again it is evident that, with the transistors conducting, a negative potential on G1 could turn the device off if it diverted all the base drive away from the *n-p-n* stage. This therefore gives a three-terminal device which can be turned on with positive gate potential and turned off with negative potential. It is known as a gate turn-off switch (GTO). Since the turn-off signal is applied at junction J2 it is more efficient than normal reverse-biased methods. There are many conflicting requirements in the design of GTO devices. For instance, to have sufficient turn-off gain (ratio of anode current to gate current required to turn the device off) the gain of the *p-n-p* device should be low. This means that the *n-p-n* transistor must have high gain to maintain the overall gain greater than unity during switch on. The value of α_{pnp} can be decreased by reducing emitter efficiency and increasing the width of the *n*-base. Both these modifications, however, increase the forward voltage drop of the device. To increase α_{npn}, on the other hand, requires an increase in emitter efficiency and a reduction in base width. Apart from conflicting with the above requirements this makes the device sensitive to spurious gate signals and reduces its breakdown voltage. Generally a GTO needs greater gate drive than a thyristor, has a higher latching current and forward voltage drop, and is poorer in current overload and *dv/dt* capabilities. However, it can be turned off in a much shorter time than a thyristor.

There are several low-power four-layer devices which have been used to fire thyristors. A Shockley diode is a two terminal thyristor (no gate lead) which breaks over when the anode voltage exceeds a certain maximum value, normally about 100V. These devices are not very stable with regard to their breakdown voltage and are rarely used.

The silicon unilateral switch (SUS) can be considered as a thyristor with anode gate, and a Zener between gate and cathode on one dice, as in *Figure 12.14*. Although the gate is brought out it need not be

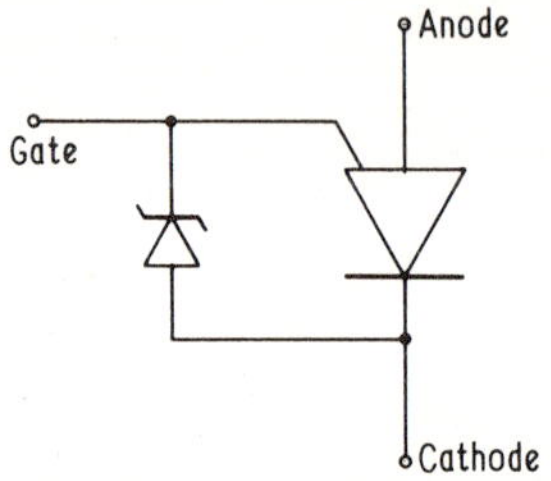

Figure 12.14. Equivalent circuit of a silicon unilateral switch

used. The device will turn on when the anode voltage exceeds that of the Zener. The characteristics are as for a thyristor, the appropriate value of gate current being determined by the Zener. Normally these devices operate at about 5 to 10V. A silicon bilateral switch (SBS) is two SUSs connected back to back on a dice. It breaks over in quadrants I and III as for a triac (*Figure 12.19*).

Two further four-layer trigger devices are shown in *Figure 12.15*. These are the complementary unijunction transistor (CUJT) and the programmable unijunction transistor (PUT). In a CUJT with base 1 biased positive to base 2 by V_B the two transistors will turn on when the emitter voltage goes more negative than $\eta V_B + V_D$. The value of η is now more tightly controlled than for a conventional UJT. The PUT is similar in operation to an ordinary unijunction transistor with the exception that the emitter voltage at which the device triggers is determined by the base resistors R_2 and R_1. Therefore the value of η can be programmed as required. By choice of suitable compensation components this device can be made less susceptible to temperature drifts than a conventional UJT.

Figure 12.16 shows the PUT when used in a relaxation oscillator circuit similar to that of *Figure 12.6(c)*. External resistors R_3 and R_4 determine the value of η for the device and timing circuit R_1 C_1 can

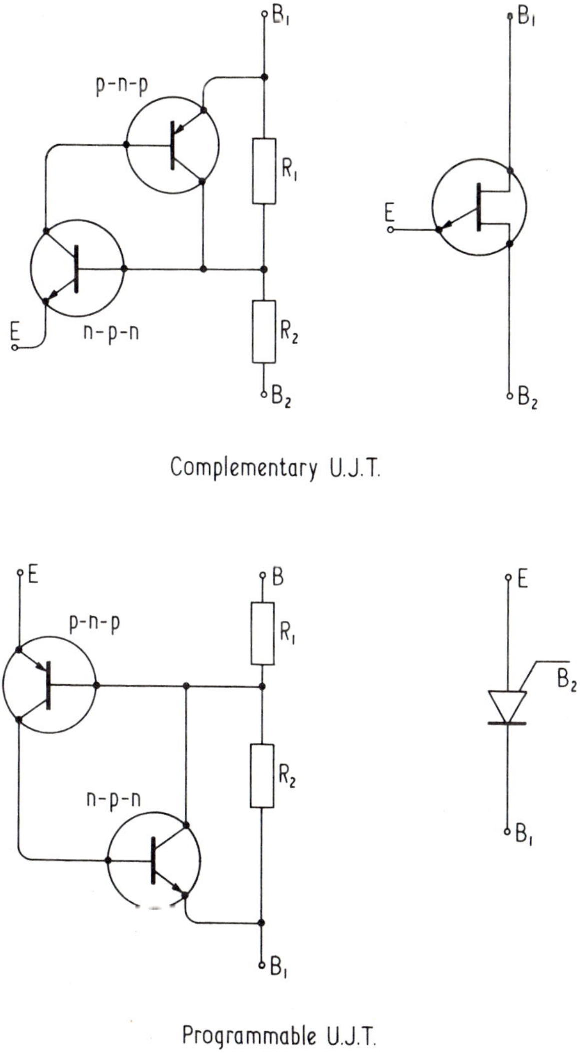

Figure 12.15. Four-layer unijunction transistors

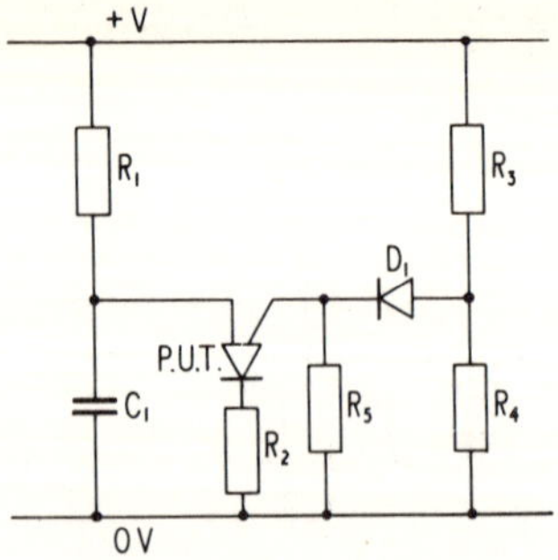

Figure 12.16. PUT oscillator

then be modified to fix the oscillator frequency. Diode D1 compensates for the temperature variation in V_D and so enables this circuit to maintain much greater frequency stability than conventional UJT systems.

FINALLY THE FIFTH LAYER

A four-layer power device conducts essentially in one direction only. Two of these can, of course, be connected back-to-back as in *Figure 12.17* to give a bidirectional device. Examining this diagram it is clear that in essence there are five distinct layers n_1 p_1 n_2 p_2 n_3, and

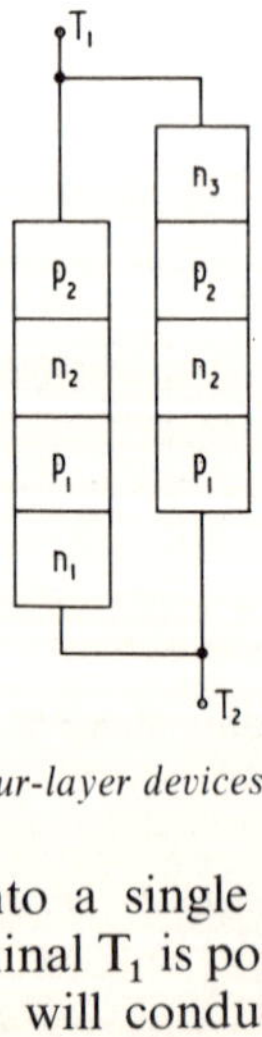

Figure 12.17. Two four-layer devices connected back to back

these can be combined into a single structure. This is shown in *Figure 12.18(a)*. When terminal T_1 is positive to T_2 by a large enough voltage, layers p_2 n_2 p_1 n_1 will conduct, by normal thyristor type action in which the anode voltage exceeds the rated breakover value. For reverse polarity layers p_1 n_2 p_2 n_3 will carry the current so that irrespective of the voltage polarity a *p-n-p-n* structure is presented to the external system.

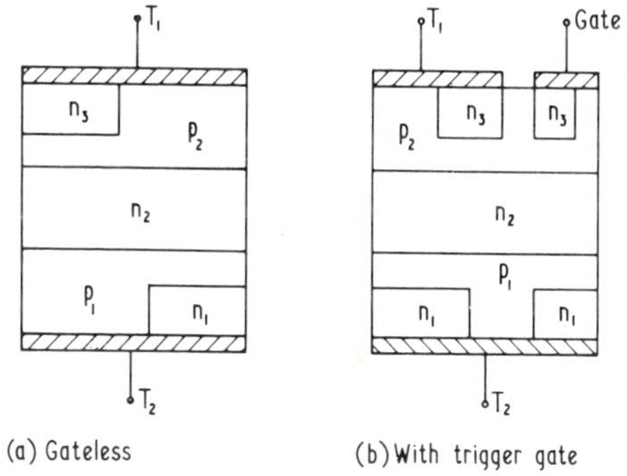

Figure 12.18. Five-layer semiconductor switch

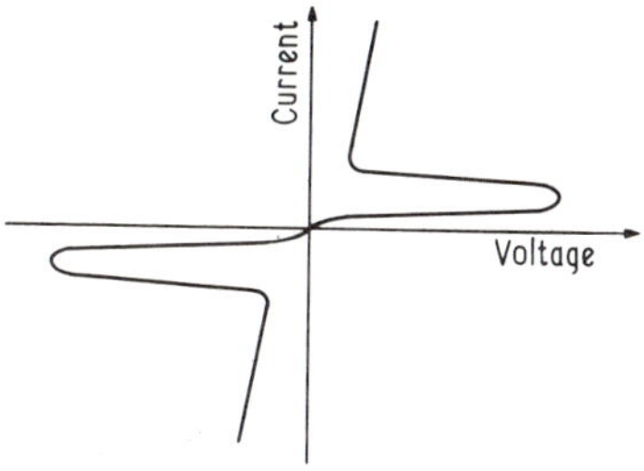

Figure 12.19. Five-layer switch characteristic. Gate current (when available) causes the device to switch at lower anode voltages, as with a thyristor

Five-layer gateless devices can be designed for a variety of breakover (or trigger) voltages and current capabilities. When the voltage is about 30V and the peak current below 1A, the device is called a diac. It is now popularly used to fire other semiconductor power switches. For higher voltages and currents (with power levels well into the kilowatt range) the device is a sidac. *Figure 12.19* shows the characteristics for five-layer structures, the current and voltage scales being determined by the type of device.

It is, of course, possible to trigger a five-layer arrangement by relatively low-power signals, as was done for thyristors. *Figure 12.18(b)* shows an arrangement of the semiconductor layers for such a device. It is evident that there are still only five layers although these have been sectionalised. A triggerable five-layer device is called a triac. It can conduct in quadrants I or III (see *Figure 12.19*) and may be fired by positive gate current (into the device) or negative gate current (out of the device), the gate signal being applied between the gate and T_1 terminals. The operation of the triac can be explained as follows. With T_2 positive to T_1 (quadrant I mode) the structure will block voltages below its breakover value. If the gate is now made positive to T_1 (mode I^+) the device is in essence a thyristor with layers $p_1\ n_2\ p_2\ n_3$ giving the *p-n-p-n* arrangement and gate terminal at P_2. The triac will therefore turn on. If T_2 is positive to T_1 and gate current is negative (mode I^-), region $p_2\ n_3$ becomes forward biased between T_1 and gate. This causes emission of electrons into junction n_2 which takes the place of gate current in a thyristor. Layers $p_1\ n_2$ $p_2\ n_3$ are again switched on and provide load current. During mode III operation (terminal T_1 positive to T_2) layers $p_2\ n_2\ p_1\ n_1$ will be active in handling the load energy. For mode III^- the p_2 (terminal T_1) and n_3 (gate) region is again forward biased and causes emission of electrons into the main body of the device and eventual turn on. Similarly for mode III^+ layers p_2 (gate) and n_3 (terminal T_1) are forward biased and cause the emission required to trigger the triac.

A triac is not equally sensitive to gate signals in all its four operating modes, *i.e.* I^+, I^-, III^+, III^-. Generally it is least sensitive to III^+ signals. This is principally due to device geometry. It is possible to accentuate this effect further so that the triac is insensitive to gate signals in this mode. This will mean that with a negative gate signal the triac conducts in both directions, whereas with positive signals it is a rectifier, like a thyristor. Such a device has been called a logic triac. It finds use in certain special applications where the output mode is controlled by the polarity of the input logic level.

Owing to its inherent simplicity of operation a triac is rapidly gaining acceptance in applications where two back-to-back thyristors were previously used. Only one device is now required, reducing the cost and drive circuit complexity. However, a triac suffers from the disadvantage of being sensitive to dv/dt effects on inductive loads. This is illustrated by the waveforms shown in *Figure 12.20*. At t_0 the triac is conducting load current in the reverse direction to the supply voltage polarity. This is due to the inductive energy in the load causing a regenerative period between t_0 and t_1. When the load current falls to zero at t_1 the triac goes off. Although the load voltage is now zero the voltage across the triac rises rapidly to the instantaneous a.c. supply at this point. This large dv/dt effect can cause the triac to be triggered into conduction in the absence of a gate signal. A similar commutation period occurs at t_4. A back-to-back connected pair will also be subjected to the waveforms shown in *Figure 12.20*, but now only one device is conducting at any time and the non-conducting thyristor has a full half cycle to recover before being subjected to this rapid voltage change. It is therefore more capable of withstanding the transient voltage.

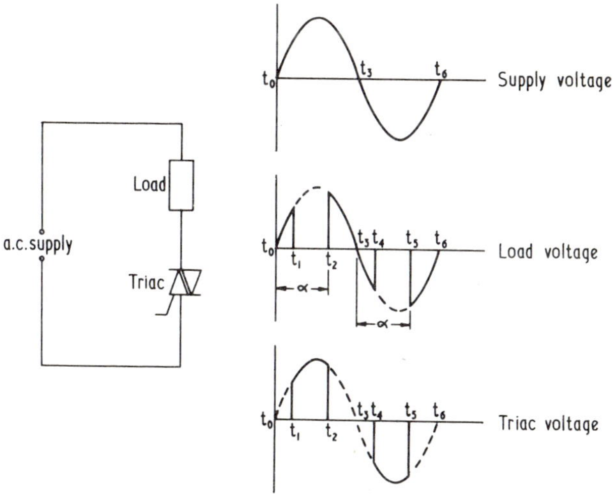

Figure 12.20. dv/dt effects in a triac a.c. phase controller

It is customary in triac circuits which are operating into inductive loads to connect an *R-C* voltage suppressor across the device, as was done for a thyristor. This improves the turn-off capability of the triac. However, triacs are seldom used to control very inductive loads. They are best suited to controlling resistive elements such as heaters, and are ideally matched to the PA424 zero voltage controller described in Chapter 4.

A CHOICE OF SYSTEMS

A wide range of different semiconductor devices has been surveyed so far. All of these are used either directly to regulate power, or to control another power semiconductor which is capable of handling the expected voltage and current levels of the load. The choice between these devices is often an economic one, and the circuit designer must be prepared to abandon a routine solution if an alternative system looks more promising. The list of devices described have by no means been exhaustive. Several modifications can be made to the basic structures to obtain slightly differing characteristics. For instance, all semiconductors are sensitive to light. This has led to a range of photo-sensitive transistors, diodes and thyristors which can be caused to conduct by incident light. The encapsulation is now specially modified to provide a clear area for light to enter the semiconductor. These photo-sensitive devices are finding extensive use in power systems. *Figure 12.21*, for instance,

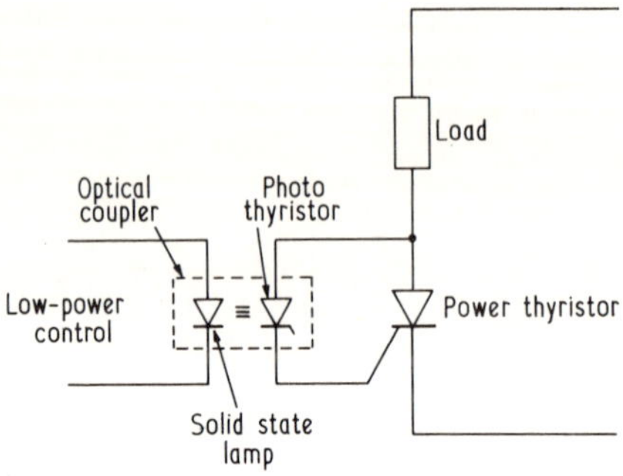

Figure 12.21. Firing a thyristor by an auxiliary photo-thyristor

shows a control circuit for a power thyristor using an auxiliary low-power photo-thyristor. The photo-thyristor is required to withstand the peak voltage of the power device, but it only carries its gate current for the time necessary to turn it on. The low-power control circuit is isolated from the power system by air or vacuum. The trigger signal passes a pulse of current through the light source, which in this instance is a solid-state light-emitting diode. This switches on the photo-thyristor and provides gate current through the main thyristor. Apart from excellent isolation this circuit also has the advantage that the control system only provides a low signal (for a light-emitting diode this could be 5mA at 2V), whereas the gate signal, which for large thyristors can be of the order of 1A at 3V, is derived from the load supply. This greatly simplifies the design of the gate driver circuit.

For straightforward rectifier application a diode is often the obvious choice of devices. These are now available in power ratings approaching 5 000A and several kilovolts—and threaten to get even bigger. A diode, however, cannot regulate the energy it passes and when this is required, a thyristor, transistor, gate turn-off switch or triac must be used. Power transistors resemble low-power transistors in principle and therefore have the advantage that they are familiar to a wide range of people. They are expanding in power capabilities and are currently available up to 250A and 1·5kV, although not from the same device. In the high-power end, however, transistors are more expensive than other semiconductors, although they are the only devices which are capable of operating in a linear mode and are therefore unrivalled in certain applications. Apart from gate turn-off switches a transistor is also the only device which can be turned off by a low-level signal so that it finds greater use in d.c. systems.

A gate turn-off switch was intended as a rival to both thyristors and transistors, occupying a position in between their two power ranges. However, it must be emphasised that in spite of several promising starts this device is still very much in the development stage and currently available only in a few sizes. Since a GTO is designed to be turned off by a gate signal there is no advantage to its use in a.c. systems where natural commutation occurs. A thyristor would now be more efficient and cheaper. When connected to a d.c. source, however, it can be turned on and off by short duration current pulses on its gate. It is therefore more efficient than a transistor, which required continuous base drive over its conduction period, and a thyristor which must operate with elaborate capacitor-inductor commutation circuitry. A gate turn-off switch is also much faster than a thyristor during turn off (by a factor of about 20) but lags by an equal amount behind the switching speed of average power transistors. Generally a GTO should be cheaper than a comparable transistor in power levels above 2kW and should find a successful market in this region for lower-frequency applications.

For a.c. systems requiring bidirectional conduction, the triac is often the best choice. It has grown in popularity relatively recently as prices have fallen. It is more convenient to use than two back-to-back thyristors since it has only one gate. However, a triac is very prone to failure by dv/dt effects and is not used on very inductive loads. It is available in a wide range of power levels and is currently second to the thyristor in the controllable semiconductor power stakes. Since a sidac needs no gate connections it is cheaper than a comparably rated triac although the trigger circuitry needed is more involved. *Figure 12.22* illustrates two firing techniques. For both systems the pulse coupled to the device via the transformer momentarily raises the voltage to greater than its breakover value. The device then remains conducting until the supply voltage reverses. Parallel pulse techniques need more drive energy but have the advantage that the transformer is not in the main load current path.

The thyristor is used in a wide range of a.c. applications, although in the low-power end it is strongly challenged by the triac. However, for high powers it reigns supreme. Single devices are available to regulate upwards of a megawatt of power, and this level is rapidly rising. For d.c. systems thyristors require relatively elaborate commutation circuitry, and for low-power applications transistors or GTOs are preferable. However, since thyristors were one of the first power semiconductor devices with a control capability, considerable knowledge has been amassed on commutation techniques, so that this aspect is no longer a technical problem but still remains a very important economical consideration.

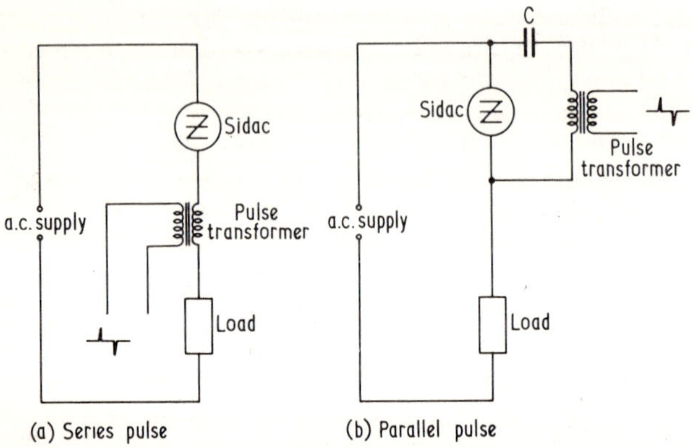

Figure 12.22. Firing circuit for a sidac

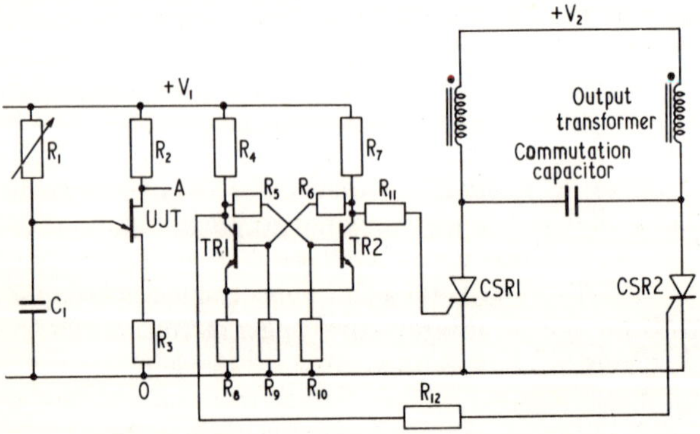

Figure 12.23. Push-pull thyristor inverter driven from a synchronised UJT oscillator

In the trigger devices the UJT, CUJT and PUT are all used in similar circumstances, when long time delays or low-voltage oscillators are required. Since they provide a sharp trigger pulse and can readily be synchronised to an external system they are very popular for thyristor drivers. *Figure 12.23* shows a push-pull inverter which is triggered by a flip-flop circuit produced by transistors TR1 and TR2. The unijunction transistor is now used to control the frequency of this bistable circuit, and the system frequency is set by timing circuits R_1 C_1. Synchronising to any external system is obtained by coupling negative pulses from it to point A on the unijunction transistor. The SUS and SBS can also be used to provide delayed pulses, and although not as popular as unijunction devices they can be operated in very similar circuits. Owing to their thyristor-type characteristics they will not turn off once fired, if the pulse current exceeds their holding current. This puts a limit on the minimum value of R_1 in *Figure 12.23*.

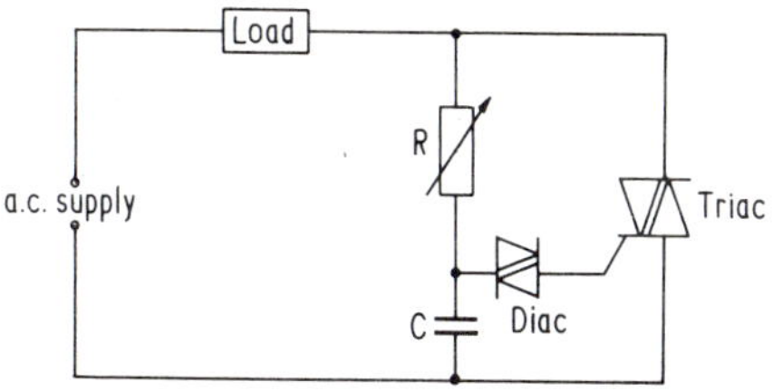

Figure 12.24. Simple diac-triac phase-control system

Diacs are generally the cheapest of trigger devices. However, since they breakover at a relatively high voltage (30V) they cannot be used in low-voltage systems. They are bidirectional in operation and when used with triacs they give some of the most elegant of control circuits. *Figure 12.24*, for instance, shows a full-wave phase-control system which has found extensive use for heater controllers, light dimmers and motor speed variation. It is perhaps the simplest of circuits especially since devices are currently available with triac and trigger diode in a single case. Without doubt these two five-layer devices are complementary and will increase each other's popularity in power systems.

Bibliography

The bibliography for Chapter 2 onwards is limited to years 1967 onwards. Earlier papers are normally quoted in these later references. The material published during this period has been selected so as to cover a wide range of topics, as well as to provide further reading material for the relative beginner and experienced user. Chapter 1 lists a few currently available books on thyristors.

CHAPTER 1

Silicon Controlled Rectifier Designer's Handbook, Editor R. MURRAY, Westinghouse Electric Corporation, 1963, 392 pages

A. LYTEL *Electronic Motor Control*, Foulsham-Sams, 1964, 224 pages
F. E. GENTRY, F. W. GUTZWILLER *et al.*, *Semiconductor Controlled Rectifiers*, Prentice-Hall Inc., 1964, 383 pages
B. D. BEDFORD and R. G. HOFT *Principles of Inverter Circuits*, John Wiley & Sons, 1964, 413 pages

A. LYTEL *ABC's of Silicon Controlled Rectifiers*, Foulsham-Sams, 1965, 128 pages
A. GRIFFIN and R. S. RAMSHAW *The Thyristor and its Applications*, Chapman and Hall Ltd., 1965, 175 pages

A. LYTEL *Silicon Controlled Rectifiers*, Foulsham-Sams, 1967, 128 pages
SCR Manual, Editor F. W. GUTZWILLER, General Electric, 1967, 513 pages

Semiconductor Devices in Power Engineering, Edited by J. SEYMOUR, Sir Isaac Pitman & Sons Ltd., 1968, 172 pages

A. KUSKO *Solid State D.C. Motor Drives*, M.I.T. Press, 1969, 122 pages

R. M. DAVIS *Power Diode and Thyristor Circuits*, Cambridge University Press, 1971, 265 pages
B. R. PELLY *Thyristor Phase-Controlled Converters and Cycloconverters*, Wiley-Interscience, 1971, 434 pages

CHAPTER 2

J. C. HEY 'Rating and Application of SCR's at Higher Power Frequencies; *IEEE Record of Technical Program*, Dallas (1967), p. 951

M. SAEGUSA 'The *di/dt* Ratings of Thyristors', *Review of the Electrical Communication Laboratory*, **15**, No. 5–6, May–June (1967), p. 343

I. LUNDSTRÖM 'Temperature Rise in Thyristors During Turn On', *Int. J. Electronics*, (1967), **23,** No. 1, p. 69

E. S. YANG 'Turn-Off Characteristics of pnpn Devices', *Solid State Electronics*, **10,** (1967), p. 927

E. S. YANG 'On the Variation of Small-Signal Alpha's of a pnpn Device with Current', *ibid*, p. 641

H. F. HAIGHT 'Testing Procedures and Test Equipment for Controlled Rectifiers', *Solid State Technology*, Nov. (1967), p. 19

D. A. KLOPFER 'How Effective are Selenium Devices as Surge Suppressors?' *Electro-Technology*, Sept. (1967), p. 38

F. B. GOLDEN 'Power Semiconductor I^2t Ratings', 1968 IEEE International Convention, New York, March (1968), p. 147

P. C. JACOBS 'Fuse Characteristics Essential to the Protection of Thyristors', *ibid*, p. 150

K. LIPMAN 'The Need for Current-Limiting Fuse in Circuits Containing Power Semiconductors', *ibid*, p. 151

J. MERRETT 'Voltage Transient and *dv/dt* Suppression in Thyristor Bridges', *Mullard Technical Communications*, March (1968), No. 92, p. 40

G. KOHL 'A Mesa-Like Edge Contour for Si High Voltage Thyristors', *Solid-State Electronics*, (1968), **11,** p. 501

W. KAPALLO 'On-Characteristics of Planar SCR's With Regard to the Use of These Devices in Monolithic Circuits', *ibid*, p. 437

R. G. DALE '*di/dt* Rating is not Adequate to Determine Thyristor Turn-On Failures', 3rd Annual IGA Group Meeting, Chicago, 29 Sept.–30 Oct. (1968), p. 567

G. MELLGREN 'The Use of Bilaterally-Cooled Thyristors in Large Converters for Industrial Motor Drives and Traction Equipment', *ibid*, p. 583

R. SVENSSON 'High-Power Thyristors', *ASEA Journal*, (1968), **41,** No. 1, p. 3

V. A. KUZ'MIN 'The Switching Transient of a Thyristor', *Radio Engineering and Electronic Physics*, **13,** No. 4 (1968), p. 616

D. I. GRAY 'This SCR is Not For Burning', *Electronics*, Sept. (1968), p. 96

F. E. GENTRY 'The Amplifying Gate Thyristor', IEEE Int. Electron Devices Meeting, Washington DC, Oct. (1968)

F. B. GOLDEN 'Fuse Co-Ordination With Power Semiconductors', IEEE International Convention N.Y., March (1968)

H. F. HAIGHT 'Industrially Useful Circuits and Techniques Employed in Test Equipment for Controlled Rectifiers', *Solid State Technology*, Feb. (1968), p. 35

C. G. DEWEY '200 kV Solid State Valve Proves Feasible for HVDC', *Electrical World*, 29 April (1968), p. 34

'Thyristor Stacks for HVDC', *Electrical Review*, 10 May (1968), p. 685
J. J. AGHASSI 'Here's a Good Switch: Radiation-Resistant Thyristors', *Electronics*, 1 April (1968), p. 65

M. PERUGINI 'Forum on SCR's', *Electronic Products*, Pt. 1, March (1969), p. 18; Pt. 2, 15 March (1969), p. 44
D. W. BORST 'Inverter Thyristors—What Affects Them—How to Characterize Them', *Electronic Components*, Oct. (1969), p. 1177
R. A. KOKOSA 'A High Voltage, High Temperature *p-n-p-n* Inverter Switch', *IEEE International Conv. Digest*, March (1969), p. 204
R. DANCY 'Which Thyristor', *Electronic Equipment News*, Nov. (1969), p. 60
R. G. DANCY 'The ACE Thyristor', *Electrical Review*, 28 Feb. (1969), p. 305
F. H. C. PYE 'The Cooling of Semiconductors', *Electronic Components*, Dec. (1969), p. 1457
G. SCHUBERT 'Behaviour of Power Thyristors at Medium Frequencies', *Brown Boveri Review*, No. 9 (1969), p. 466
F. B. GOLDEN 'Clearing Time, the Forgotten Fuse Parameter', *The Electronic Engineer*, **28,** No. 7 July (1969)
D. E. PICCONE 'Are You Confused by High *di/dt* SCR Ratings?', *The Electronic Engineer*, **28,** No. 1 January (1969)
P. A. WALTERS 'Step-Like Shifts in the Voltage-Current Characteristics of Power Thyristors and Their Effect on Transient Thermal Impedance Determinations', *IEEE Trans*, **ED-16,** May (1969), p. 497
D. N. HARSTAD (SCR Model Simplifies Computer Programs', *Electronic Design*, **22,** 25 Oct. (1969), p. 92
W. RITCHEY 'Use—Don't Abuse the SCR', *Electronic Design*, **5,** 1 March (1969), p. 76
W. RITCHEY 'There Are Many Ways To Fire an SCR', *Electronic Design*, **6,** 15 March (1969). p. 234
R. BRADLEY 'High Power Thyristor Design for Performance and Reliability', *IEE Conference Pub.,* No. 53 Pt. 1, May (1969), p. 22
M. OTSUKA 'A New Edge Contour for Si High Voltage Thyristors', *ibid*, p. 32
H. OKA 'High Power High Voltage Thyristors and Their Applications', *ibid*, p. 47
P. A. BYLUND 'Double-Sided Cooled Thyristors in Converters for Industrial Motor Drives and Traction', *ibid*, p. 61
T. MATSUMOTO 'Chemical Liquid Cooling and Construction for Power Thyristors', *ibid*, p. 79
P. G. NEWBERY 'The Correct Protection of Power Thyristors by High Speed H.R.C. Fuse-Links', *ibid*, p. 125
J. Y. HAYDEN 'Liquid-Filled Fuses for the Protection of Thyristors', *ibid*, p. 132
J. J. L. WEAVER 'Development of a Thyristor Valve for H.V.D.C. Transmission', *ibid*, p. 339
K. M. JONES 'The Prospects of Using Thyristor 'Strings' in Distribution System Development', *ibid*, p. 422
E. OHNO 'Thyristor Strings for High Voltage Applications', *ibid*, p. 406

W. D. GILMOUR 'Thyristors and Associated Devices', *Wireless World*, March (1969), p. 120

J. B. RICE 'Design of Snubber Circuits for Thyristor Converters' IEEE 4th Annual IGA Group Meeting, Detroit, 12–16 Oct. (1969)

E. ANWANDER 'Thyristor Converter Valve for 100kV d.c. Bridge Voltage', *Brown Boveri Review*, No. 2 (1969), p. 79

D. COOPER 'Development, Characterization and Application of a New Geometry Inverter High Power Switch', *Electronic Components*, Feb. (1970), p. 191

T. MATSUZAWA 'Some Electrical Characteristics of Reverse Conducting Thyristor', *IEEE Trans*, **ED-17,** No. 9 Sept. (1970), p. 816

R. A. KOKOSA 'A High-Voltage, High-Temperature Reverse Conducting Thyristor', *ibid*, p. 667

J. MERRETT 'Practical Transient Suppression Circuits for Thyristor Power-Control Systems', *Mullard Technical Communications*, March (1970) No. 104, p. 82

J. H. PARKER 'A Look Inside the Thyristor', *Students Quarterly Journal (IEE)*, Sept. (1970), p. 147

K. LUDWIG 'New Thyristor and Diode Assemblies for Power Electronics Systems', *Siemens Rev*., (1970), p. 252

A. M. MARTIN 'Thyristor Development—the Next Decade', *Electrical Times*, 30 April (1970), p. 12F

P. S. TAMBE 'H.R.C. Fuse-Link', *Electrical Equipment*, Jan. (1970), p. 28

H. RIEGER 'Protecting Thyristor Converters Against Overvoltages from the a.c. System', *Siemens Review*, No. 9 (1970), p. 494

E. T. SCHONHOLZER 'Fuse Protection for Power Thyristors', IEEE 5th Annual IGA Group Meeting 5–8 Oct. (1970), p. 455

P. C. JACOBS 'Application of Fuses with Power Semiconductors in Direct Current Circuits', *ibid*, p. 487

E. REIMERS 'A Low Cost Thyristor Fuse', *ibid*, p. 509

F. DURNYA 'Transient Power Losses in Epitaxial Thyristors,' *ibid*, p. 525

H. A. STEINBUEGGE 'Circuits and Procedures for Measuring SCR Turn-On Switching Losses', *ibid*, p. 555

F. B. GOLDEN 'Liquid Cooling of Power Thyristors', *ibid*, p. 573

G. D. EVAN-HART 'Staggered-Finger Heat Dissipators', *Electronic Components*, Sept. (1970), p. 1001

E. JEFFS 'High-Voltage Thyristors Stimulate a Growing Wealth of Applications', *Design Electronics*, May (1971), p. 61

M. J. ROBINSON 'Greater Efficiency for Heat Dissipation', *Electron Equip. News (G.B.)*, **13,** No. 1 April (1971), p. 14

K. G. ASHAR 'High Speed Thyristor' *IBM Tech. Disclosure Bull., (USA)*, **13,** No. 10 March (1971), p. 2826

J. FEENMAN 'The Versatility of the HRC Fuse in Protecting Semiconductor Equipments', *Electrical Review*, **188,** No. 3, 15 January (1971), p. 85

K. OAKMAN 'Regenerative Gate Thyristors', *Electronic Components*, **12,** No. 4, 19 March (1971), p. 241

'Twin-Thyristor Pair Needs no Trigger Circuit', *Electronics*, 24 May (1971), p. 21E
R. KELLER 'BStN15, a New Disk Thyristor with a Heat Sink (Thyblock)', *Siemens Rev. (Germany)*, **38,** No. 7, July (1971), p. 330
F. F. MAZDA 'Using the Thyristor', *Electronic Components*, 17 Sept. (1971), p. 947
D. L. W. HOLLOWAY 'Fast Switching Thyristor for Inverter', *Elect. Eng.*, **48,** No. 8, Aug. (1971), p. 30
R. J. BASSETT 'An Investigation into the Mode of Switching of Inverse Gate Thyristors', *Int. J. Electron.*, **31,** No. 5, Nov. (1971), p. 453
P. S. RADERECHT 'A Review of the 'Short-Emitter' Principle as Applied to p-n-p-n Silicon Controlled Rectifiers', *Int. J. Electron.*, **31,** No. 6, Dec. (1971), p. 541
H. W. TURNER 'Cartridge Fuse Developments', *Elec. Rev.*, **189,** No. 19, 5 Nov. (1971), p. 663
Y. SATO 'All Diffused High Power Thyristors', *Fuji Elec. Rev.*, **17,** No. 5 (1971), p. 262
R. REIMERS 'An Approach to the Testing of High Performance Thyristors', IEEE 6th Annual IGA Group Meeting, Cleveland, Ohio, USA, 18–21 Oct. (1971), p. 769
J. D. BALENOVIČH 'Thyristor High Frequency Rating by Concurrent Testing and Computer Simulation', *ibid*, p. 753
A. P. CONNOLLY 'PWM Characterization of SCR's', *ibid*, p. 777
R. W. WENDELBURG 'What's Missing in SCR's and Power Diodes?', IEEE 6th Annual Meeting of IGA Group, Cleveland, Ohio, USA, 18–21 Oct. (1971), p. 897
J. T. MITCHELL 'Thyristor Data Sheets—A Users Viewpoint', *ibid*, p. 901
R. R. GILBERT 'Thyristor Parameter Correlation Using Monte Carlo Techniques', *ibid*, p. 905
W. McMURRAY 'Optimum Snubbers for Power Semiconductors', *ibid*, p. 885

E. JACKS 'H.R.C. Fuses for Earth Fault Protection of Medium Voltage Industrial Systems', IEE Conference on International Medium Voltage Earthing Practices, London, England, 21–23, March (1972), p. 176
'Overcurrent Protection of High-Power Semiconductor Devices', *Elect. Times*, **161,** No. 7, 17 Feb. (1972), p. 31
G. A. OSBORN 'Thyristors 1972—A State-of-the-Art Review', *ibid*, p. 125

CHAPTER 3

R. BEAUDRY 'Soft Starting a.c. Motors', *Machine Design*, 25 May (1967), p. 171
A. HARRIS 'Semiconductor Switching of Low-Power Circuits', *Electronics World*, June (1967), p. 33
B. HURST 'Thyristor Starting Techniques for Induction Motors', *Industrial Electronics*, Part 1, Jan. (1967), p. 25. Part 2, Feb. (1967), p. 74

L. ZENTGRAF 'Modular System of Thyristor Switches', *Siemens Review*, No. 10 (1967), p. 395

F. ROTHWARF 'Silicon-Controlled Rectifier Circuitry for Producing Pulsed Magnetic Fields', *Review of Scientific Instrum.*, **38,** No. 9, Sept. (1967), p. 1241

W. D. RABURN 'SCR Switching with Ionizing Radiation', *IEEE Trans.*, **NS-14,** No. 6, Dec. (1967), p. 187

'Solid State Contactors', *Electrical Review*, 12 April (1968), p. 549

A. M. DARBIE 'Protection Circuits for Solid-State Power Supplies', *Electronics World*, April (1968), p. 49

A. EMANUEL-EIGELES 'Thyristorized Starting Device for a Wound-Rotor Induction Motor', *Control*, Oct. (1968), p. 864

C. F. BATTERSBY 'Present Techniques in Gate Firing', *IEE Conf. Pub.*, No. 53, Pt. 1 (1969), p. 146

M. E. ROBERTS 'A Thyristor Assisted Mechanical On-Load Tap Changer', *ibid*, p. 185

H. GOSHIMA 'Non-Break A.C. Power Source Switching Equipment', *ibid*, p. 193

B. FELTBOWER 'Thyristor Contactors', *ibid*, p. 448

E. Y. ROCHER 'Response Time of Thyristors: Theoretical Study and Application to Electronic Switching Networks', *IBM J. Res. Develop.*, July (1969), p. 447

A. A. MAHMOUD 'Power Conditioning Utilizing High-Speed Power Source Transfer Switching', IEEE 5th Annual IGA Group Meeting, 5–8 Oct. (1970), p. 679

A. A. MAHMOUD 'A Solid-State Asynchronous Static Switch', *ibid*, p. 707

F. B. GOLDEN 'Thyristor Switching Losses and their Measurement', Conf. on Power Electronics, Budapest, Hungary (Budapest: Magyar Elektrotechnikai Egyesulet), 25–27 Nov. (1970)

M. DANKA 'Thyristor Auxiliary Equipment of a Main-Line Electric Traction Vehicle', *ibid*

F. F. MAZDA 'Static Contactors', *Electronic Components*, August (1970), p. 889

F. F. MAZDA 'The Thyristor as a Switch', *Electronic Components*, 23 Oct. (1970), p. 1153

P. HARRISON 'Small Contactors with Electronic Preamplifiers', *Brown Boveri Rev.*, No. 7 (1971), p. 317

G. POTISK 'Static Protection Equipment for Busbars', *ibid*, p. 265

L. JACKSON 'Developments in Distance Protection', *Electrical Review*, 29 Oct. (1971), p. 627

R. WANNER 'Protection of Large Generators', *ibid*, p. 257

J. BONIGER 'Solid-State Overcurrent-Time Relays', *ibid*, p. 312

A. SEDGWICK 'The Hybrid Solution', *Electronic Components*, 17 Dec. (1971), p. 1273

R. M. MARSTON '24 Easy-to-Build Burglar Alarms', *Radio-Electronics*, Aug. (1971), p. 46
M. KEOHANE 'Pulse-Modular Thyristor BTW35 as a Radar Magnetron Driver', *Mullard Tech. Comm.*, No. 111, July (1971), p. 27
B. L. W. HOLLOWAY 'A High Power, Fast Switching Thyristor for Inverter and Pulse Applications', *SERT J. (G.B.)*, **5,** No. 2 (1971), p. 36
J. M. GARRETT 'The Evolution of a High-Power Fast-Switching Thyristor', *Elect. Eng. (Australia)*, **48,** No. 5, May (1971), p. 33.
S. B. THAKUR 'Effect of N^+ Cathode Impurity Concentration on the Switching of all Diffused Thyristors', Symposium on Electronics, (Indian Inst. Technol.) Madras, India 8–10 March (1971)
B. V. CORDINGLEY 'Improving the Turn-On Performance of High-Power Thyristors: Some Recent Advances', *J. Sci. & Technol., (GB)*, **38,** No. 1 (1971), p. 2
I. I. PONNER 'Design Criteria for a Static Three-Phase Switch with Three Delta-Connected Thyristors', *IEEE Trans. Ind. Electron. & Contr. Instrum.*, **IECI-18,** No. 4, Nov. (1971), p. 161
P. N. AVASTHY 'Photo-Thyristorised Controller', *IEE-IERE Proc. India*, **9,** No. 4, July-Aug. (1971), p. 151
D. E. BAKER 'Solid-State Remote Power Controllers for Electrical Systems', *Westinghouse Eng.*, **31,** No. 5, Sept. (1971), p. 135
G. R. HOGAN 'Force Commutated Static Switching in Uninterruptable a.c. Power Systems', *IEEE Internat. Electrical & Electronics Conf.*, Toronto, Ontario, Canada, 4–6 Oct. (1971), p. 78
A. O. MARCKEL 'Motor Start Switch', *IBM Technical Disclosure Bull.*, **14,** No. 6, Nov. (1971), p. 1942

M. W. WILLIAMS 'SCR Reset for Integrator Provides High Speed', *Electronics*, **45,** No. 1, 3 Jan. (1972), p. 84

CHAPTER 4

W. SHEPHERD 'Thyristors in the Control of Electrical Systems', *Control*, Part 1, Jan. (1967), p. 28, Part 2, Feb. (1967), p. 83, Part 3, March (1967), p. 130
E. S. McVEY 'Design Information for a Switched a.c. Regulator', *IEEE Trans.*, **IECI-14,** No. 2, Dec. (1967), p. 51
E. S. McVEY 'Design Data for a Multiple Output d.c. Regulator System', *IEEE Trans.*, **IGA-3,** No. 4, July/Aug. (1967), p. 337.
P. C. SEN 'Thyristor Controlled Single Phase Variable Inductor', *IEEE Trans.*, **MAG-3,** Sept. (1967), p. 240
P. C. SEN 'Three Phase Variable Inductor Using Thyristors', IEEE IGA Group Meeting, Pittsburg, Pa. 2–5 Oct. (1967)
S. B. DEWAN 'Analysis of a Filter Circuit for Non-Linear Loads', *IEEE Trans.*, **MAG-3,** March (1967), p. 34
W. SHEPHERD 'Thyristor-Driven Induction Motor', *Electrical Times*, 13 July (1967), p. 57

G. N. SARIDIS 'A Linear SCR Control Amplifier', *IEEE Trans.*, **IGA-4,** No. 6, Nov./Dec. (1968), p. 619

K. N. STANTON 'Instrumentation for Thyristor Control', *ibid*, p. 638

B. W. LINGARD 'Steady State Performance and Analysis of the Series Resistance-Capacitance Circuit with Control by Adjustable Thyristor Triggering', *ibid*, p. 644

P. C. SEN 'Smooth Control of Reactive Power Using Static Device', *IEEE Trans.*, **MAG-4,** No. 3, Sept. (1968), p. 325

R. THOMPSON 'A Thyristor Alternating-Voltage Regulator', *IEEE Trans*, **IGA-4,** No. 2, March/April (1968), p. 162

A. EMANUEL-EIGELES 'A.C. Voltage Regulation by Means of Chopper Circuits', *Electronics Letters*, 26 Jan. (1968), p. 26

T. W. HAHN 'Electronic Refrigeration Defrost Control—Report of Performance, *IEEE Trans.*, **IGA-4,** No. 1, Jan./Feb. (1968), p. 114

W. J. DUNCAN 'A Simple Thyristor Furnace Controller', *Journal of Scientific Instruments*, Series 2, 1, (1968), p. 58

M. H. JACOBS 'Solid-State Switching', *Electrical Times*, 15 Feb. (1968), p. 253

M. S. ERLICKI 'New Aspects of Power Factor Improvements', *IEEE Trans.*, **IGA-4,** July/Aug. (1968), Pt. 1, p. 441, Pt. 2, p. 447

R. B. COWDWLL 'New Dimensions in Shielding', *IEEE Trans.*, **EMC-10,** No. 1, March (1968)

J. E. BRIDGES 'Magnetic Field Pickup by Flexible Braid Coaxial Cables', *ibid*, p. 130

J. H. BULL 'Voltage Spikes in LV Distribution Systems and their Effects on the Use of Electronic Control Equipment', *ERA Report*, No. 5254 (1968)

C. A. MEADOWS 'A New Method of Measuring the Attenuation of Interference Suppressors', Ministry of Technology (RAE) Technical Report 68134, May (1968)

L. O. REXRODE 'Noiseproof Digital Electronic Control System', *Control Engineering*, Nov. (1968), p. 66

R. SCHALLER 'Radio Interference Suppression of Silicon Controlled Rectifiers', *Siemens Electronic Components Bull.*, **3,** No. 1 (1968), p. 23

D. HUMPHRIES 'Thyristor Control of Fan Cooling and Heating Systems', *Mullard Technical Comm.*, No. 92, March (1968), p. 65

J. MUNGENAST 'Voltage Control of Induction Motors', *Machine Design*, 25 April (1968), p. 167

B. W. LINGARD 'Steady-State Performance and Analysis of the Series Resistance-Capacitance Circuit with Control by Adjustable Thyristor Triggering', *IEEE Trans.*, **IGA-4,** No. 6, Nov./Dec. (1968), p. 644

W. A. SCHMIDT 'Solid State Automatic Charger Control', *Electronic Components*, July (1969), p. 835

S. W. GRAY 'Thermostatic Fan Control for Central Heating', *Mullard Tech, Comm.*, No. 102, Nov. (1969), p. 46

P. ATKINSON 'A Linear Temperature Control System Using Thyristors', *The Radio and Electronic Engineer*, **38,** No. 1, July (1969), p. 23

J. ANDERSON 'Numerical Power Control', *Instrum. and Control Syst.*, Dec. (1969), p. 71
G. RATTCLIFF 'A.C. Power Control Incorporating Zero-Voltage Switching', *Electronic Components*, May (1969), p. 580
S. B. DEWAN 'A d.c. Controlled Variable Reactor Using Thyristors', *IEEE Trans.*, **MAG-5,** Sept. (1969), p. 342
R. E. GOERS, 'Quiet-Wiring Zone', *IEEE Trans.*, **IGA-5,** No. 3, May/June (1969), p. 273
J. P. DEMETRY 'Achieving Filter Effectiveness and Insuring Reliability', IEEE 11th G-EMC Symposium Record (1969), p. 290
R. R. FERBER 'The Shielding of Electromagnetic Pulses by Use of Magnetic Materials', *ibid*, p. 73
H. W. DENNY 'Some RF Characteristics of Bonding Systems', *IEEE Trans.*, **EMC-11,** No. 1, Feb. (1969), p. 15
L. F. BABCOCK 'Direct Graphical Determination of Optimum Shielding and Suppression', *ibid*, p. 9
P. A. DENES 'Tiny Filters Block the Path of Radio Frequency Interference', *Electronics*, August (1969), p. 58
R. W. BUNTENBACH 'Ground Circuits', *Instruments and Control Systems*, **42,** Nov. (1969), p. 135
R. L. HICKS 'Transient Interference on Station Control Wiring', *Ontario Hydro Res. Quart.*, **21,** No. 1, (1969), p. 13
W. GRIMMOND 'Mains Supply for Data Processing Systems', *IEE Conference Pub.*, No. 60, Dec. (1969), p. 155
T. E. SKOPAL 'Stop Noise Problems Before They Start', *Electronic Design*, January (1969), p. 90
F. G. WILLARD 'Prevention and Treatment of Noise in Control Signals', *IEEE Trans.*, **IGA-5,** No. 3, May/June (1969), p. 266
W. C. SCHMIDT 'Electrical Noise Reduction in Industrial Control Systems, *ibid*, p. 259
R. J. MOHR 'Interference Coupling—Attack it Early', *Electronic Design News*, 1 July (1969)
J. MERRETT 'The Design of Lower Power Control Systems with Particular Reference to Transients and to Experience Gained with d.c. Motor Speed Control', *IEE Conf. Pub.*, No. 53, Pt. 1, May (1969), p. 161
S. LLOYD 'A Thyristor a.c. Regulator with Sinusoidal Output', *ibid*, p. 168
E. M. POLLARD 'A 20 MW Thyristor a.c. Switch for Induction Heating Power Control and Protection', *ibid*, p. 177

S. B. DEWAN 'Input Filter Design with Static Power Converters', *IEEE Trans.*, **IGA-6,** No. 4, July/Aug. (1970), p. 378
W. BRUNNLER 'Low-Power Silicon Thyristors BSt, BO1, and BO2' (Part 2), *Siemens Electronic Components Bulletin*, **5,** No. 1 (1970), p. 22
A. J. BAGGOTT 'Data Transmission Over Distribution Systems', *Electrical Times*, 12 March (1970), p. 55
P. J. GALLAGHER 'Trigger Circuit Reduces Supply Waveform Distortion', *Electronic Engineering*, Nov. (1970), p. 67
N. R. GRUENBERG 'SCR Controlled 1350 kW and 2000 kW Electric

Booster Glass Melting Systems', IEEE 5th Annual IGA Group Meeting, 5–8 Oct. (1970), p. 769

J. G. WEIT 'Noise in Signal Systems Associated with Static Converters; *ibid*, p. 799

R. S. SEGSWORTH 'Power Supply Systems for Induction Furnaces', *ibid*

J. L. STEENBURGH 'Electromagnetic Interference Worry You?', *Electronic Design*, **15,** March (1970), p. 199

M. HUNYAR 'Induction Motor Supplied by Back-to-Back Connected Thyristors', Conf. on Power Electronics, Budapest, Hungary, 25–27 Nov. (1970) (Budapest: Magyar Elektrotechnikai Egyesulet)

T. MLINKO 'Triggering Problems of Back-to-Back Connected Thyristor Supply to Induction Motor Drives', *ibid*

J. A. BARON 'Harmonic Generation by HVDC Converters', Univ. Waterloo Ontario, Canada, Thesis (1970)

F. F. MAZDA 'A.C. Line Control', *Electronic Components*, 20 Nov. (1970), p. 1307

M. RAMAMOORTY 'Single Phase to 3-Phase Converter', *Proc. IEE*, **118,** No. 9, Sept. (1971), p. 1288

G. H. THOMPSON 'Intergral-Cycle Controller with Variable Initial Switch-On Angle', *Design Electronics*, Dec. (1971), p. 43

A. R. OWEN 'Cycle Syncopation a.c. Thyristor Control', *Electrical Review*, 1 Jan. (1971), p. 15

J. BUDEK 'Burst-Firing Techniques Using Triacs', *Design Electronics*, Feb. (1971), p. 27

D. R. GRAFHAM 'Thyristor Control of a.c. Power', *Elect. Rev.*, **188,** Pt. 1 No. 20, 14 May (1971), p. 663. Pt. 2 No. 21, 21 May (1971), p. 694, Pt. 3 No. 22, 28 May (1971), p. 731

T. HIMEI 'Steady-State Characteristics of Thyristor Control Circuit with a Series *RLC* Load', *Elec. Eng. Jap. (U.S.A.)*, **90,** No. 1–2, Jan.-Feb. (1971)

J. LEMMRICH 'Induction Motor Controlled by Stator Voltage Regulation', *Ingenieur (Netherlands)*, **83,** No. 20, 21 May (1971), p. 57

S. IVNER 'The Influence of Thyristor Converters on Supply Networks', *ASEA J. (Sweden)*, **44,** No. 2 (1971), p. 37

L. N. HULLEY 'Varying Power Input by Integral Cycle Firing of Thyristors', *Contr. & Instrum.*, **3,** No. 3, March (1971), p. 25

T. MATSUHASHI 'Thyristor Phase Control with Less Third Harmonics', *Elect. Eng. Jap.*, **90,** No. 4, July-Aug. (1971)

W. McMURRAY 'The Thyristor Electronic Transformer: A Power Converter Using a High Frequency Link', *IEEE Trans. Ind. & Gen. Appl.*, **IGA-7,** No. 4, July-Aug. (1971), p. 451

K. HARADA 'On the Magnetic Firing Circuits for Thyristors', *Mem. Fac. Eng. Kyushu Univ.*, **31,** No. 2, Sept. (1971), p. 17

J. LISSER 'Mains Distortion by Thyristor Equipment', *Holectechniek*, **1,** No. 2, Oct. (1971), p. 75

A. A. COLLIE 'Solid-State Control of Domestic Appliances', *Electron. & Power*, 18 Jan. (1972), p. 19

R. BEAN 'Sources of Noise in Domestic Light Dimmers', *Electrical Times*, 13 April (1972), p. 45

CHAPTER 5

K. HEINTZE 'Electronic d.c. Regulator for the Speed-Control of Motive Power Units with Overhead Current Supply', *Bulletin I.R.C.A. Cybernetics and Electronics on the Railways*, March (1967), p. 122

N. SATO 'Improvement of SCR Chopper Circuit', *Elect. Eng. Japan*, **87,** No. 2, Feb. (1967), p. 75

N. LINGREN 'Electric Cars—Hope Springs Eternal', *IEEE Spectrum*, April (1967), p. 48

M. ALTMAN 'The Electric Automobile—Its Future', *IEEE National Convention Record*, **15**(7) (1967), p. 89

E. OHNO 'High-Voltage Multiple Phase Thyristor Chopper for Traction Motor Control', *IEEE Trans.*, **MAG-3,** No. 3, Sept. (1967), p. 232

J. W. MARCH 'Motor for Electric Cars', Presented at Colloquium on Electric Cars 21st April, (1967). The Institution of Electrical Engineers.

I. S. PAYNE 'Control Systems for Electric Cars', *ibid*, p. 6/1

I. I. PONNER 'Resonance Commutated Thyristor d.c. Chopper Design Criteria', *IEEE Trans.*, **IEC-15,** No. 1, Nov. (1968), p. 28

B. H. SMITH 'A Simple Bilateral Variable-Ratio d.c. Pulse Converter', *ibid*, p. 1

B. H. SMITH 'Commutation in a Bilateral Variable-Ratio d.c. Pulse Converter', *ibid*, p. 6

R. J. EADIE 'Electronic Speed Control for d.c. Motors', *Electronic Engineering*, Jan (1968), p. 10

C. E. BAND 'Chopper Control on a 1500V d.c. Traction System', *Bulletin I.R.C.A. Cybernetics and Electronics on the Railways*, Dec. (1968), p. 473

FRITZ MARTEN 'Open-Loop and Closed-Loop Control of Traction Vehicles', *Siemens Review*, **35,** No. 4 April (1968), p. 160

H. IRIE 'Thyristor Chopper for Separately Excited d.c. Motor Control', *Elec. Eng. Japan*, **88,** No. 4 (1968), p. 9

W. J. EVANZIA 'Battery-Powered Cars: Fact or Fantasy?', *Electronic World*, May (1968), p. 39

D. GURWICZ 'Controls and Chargers Associated with the Proposed Electric Car', Presented at the Second Annual Colloquium on Electric Road Passenger Vehicles Oct. (1968), The Institution of Electrical Engineers

R. A. ZECCOLA 'Thyristor SCR Chopper Control System for Transportation Equipment', IEEE Conf. Rec. 3rd Ann. IGA Group Chicago, Ill. 29 Sept.–30 Oct. (1968), p. 471

C. E. BAND 'Development of and Operational Experience with a High Powered DC Chopper for 1500 volt d.c. Railway Equipment', *IEE Conf. Pub.*, No. 53, Pt. 1, May (1969), p. 277

W. C. JAUQUET 'D.C. Choppers for Railway Applications', *ibid*, p. 289

T. KITAOKA 'Thyristor d.c. Choppers and High Voltage Inverters for d.c. Traction', *ibid*, p. 263

W. J. YATES 'Selecting the Correct Capacitor for Use in Thyristor Circuits', *ibid*, p. 140

DAVID COOPER 'D.C. to d.c. Conversion', *Instruments and Control Systems*', **42,** Feb. (1969), p. 83

J. P. PANWELS 'The Electronic Chopper' Monthly bulletin of International Railway Congress Association, March (1969), p. 88

M. HAGENMAIER 'Thyristor-Controlled Battery Locomotive for Underground Mine Operation', *Siemens Review*, No. 7 (1969), p. 276

H. LOCKER 'A Trolleybus with Electronic d.c. Chopper Control', *Brown Boveri Review*, Jan (1969), p. 50

'Pulse Width Modulation for d.c. Motor Speed Control', *Design Electronics*, April (1969), p. 24

'Speed Control of Series Motors', *Electrical Equipment*, June (1969), p. 31

R. M. DAVIS 'Thyristor Control of a Multiload System with d.c. Supply', *Proc. IEE*, **116,** No. 5, May (1969), p. 801

Z. ZABAR 'Guidelines on Adaption of Thyristorized Switch for d.c. Motor Speed Control', *IEEE Trans.*, **IECI–17,** No. 1, Feb. (1970), p. 10

G. SHEARD 'High-Speed Pulse Width Modulated Thyristor Amplifier for d.c. Motor Control', *Int. J. Control*, **12,** No. 1, (1970), p. 49

F. F. MAZDA 'D.C. Line Control', *Electronic Components*, Dec. (1970), p. 1369

'Thyristor Choppers in Electric Traction', *ACEC Revue (Belgium)*, No. 2 (1970), p. 45

M. KOHUT 'A d.c. Chopper Controlled Electric Track', Conference on power electronics, Budapest, Hungary, 25–27 Nov. (1970) (Budapest Magyar Elektrotechrikai Egyesulet)

P. MAGYAR 'Testing of d.c. Thyristor Chopper Drives', *ibid*

S. BARTOS 'Theoretical and Practical Problems of d.c. Choppers', *ibid*

F. RABA 'Czechoslovakian Thyristor Drive for Trolley Buses' *ibid*

L. KYR 'Thyristor Chopper Control for Trolleybus Traction Drives', *ibid*

H. LOCKYER 'The d.c. Chopper Used on Trolleybuses with Full Electronic Control', *Brown Boveri Rev.*, **10** (1970), p. 419

P. KNAPP 'Solid-State Regulating Units for Motoring and Braking d.c. Traction Vehicles', *Brown Boveri Rev.*, **6/7** (1970), p. 252

J. V. BYRNE 'Compatible Controller-Motor System for Battery Electric Vehicle', *Proc. IEE*, Feb. (1970), p. 369

B. BERMAN 'Battery Powered Regenerative SCR Drive', *ibid*, p. 657

T. JINZENJI 'Thyristor Chopper-Control Devices for d.c. Electric Railcars', *Toshiba Review*, May–June (1970), p. 5

R. PARIMELALAGAN 'Steady-State Investigations of a Chopper-Fed d.c. Motor with Separate Excitation', *IEEE Trans.*, **IGA-7,** No. 1, Jan./Feb. (1971), p. 101

X. VOGEL 'The Solid-State d.c. Regulating Unit and its Components', *Brown Boveri Rev.*, (1971), p. 521

D. B. CORBYN 'D.C. Power Control for Aluminium and Electrolytic Loads', *Proc. IEE*, **115,** No. 11, Nov. (1968), p. 1693

F. F. MAZDA 'An Electric Vehicle Controller', *Electronic Components*, 19 March (1971), p. 235

C. E. BAND 'Some Control Systems on Railway Vehicles', IEE Colloquium on Electronics on trains, London, 15 Jan. (1971)

P. BOWLER 'A d.c. Motor Servo Controlled by a Pulse Width Modulated Inverter', *Direct Current*, **2,** No. 1, Feb. (1971), p. 45

D. COOPER 'State of the Art in Power Conversion' *Electron. Equip. News*, **13,** No. 2, 5 May (1971), p. 68

W. E. RIPPEL 'Dual SCR Chopper as a Motor Controller for an Electric Car', *IEEE Trans.*, **VT-20,** No. 2, May (1971), p. 23

J. K. HALL 'Improvement of Turn-Off Performance of Series-Connected Thyristors by Magnetic Components', *IEEE Trans.*, **MAG-7,** No. 2, June (1971), p. 297

'SNCB Adopts the Series Chopper', *Ry. Gaz. Int. (GB)*, **127,** No. 5, May (1971), p. 85

A. E. SLOAN 'The Thyristor Chopper and its Use in Motor Control for the Mechanical Handling Industry', Proc. Joint Conf. on the Electronic Control of Mechanical Handling, Nottingham, England, 6–8 July (1971) (London: IERE), p. 399

J. J. MORRISON 'Electronic Control of Battery Electric Vehicles', *ibid*, p. 45

'Stockholm Orders First Chopper Train', *Ry. Gaz. Int. (GB)*, **127,** No. 7, July (1971), p. 281

F. FRENECH 'A Thyristor (SCR) Phase Controlled Propulsion System with Linear, Regenerative Characteristics for Electrically Powered Vehicles', IEEE Vehicular Technology 22nd Annual Conference, Detroit, Mich. USA., 7–8 Dec. (1971)

L. SIMPSON 'SCR Speed Control for Electric Drills', *Electron. Aust.*, **33,** No. 2, May (1971), p. 48

T. KITAOKA 'Chopper Control Equipment for Hanskin Electric Railway', *Mitsubishi Denki Eng.*, No. 29, June (1971), p. 13

G. SCHOLTIS 'Thyristor Control for d.c. Power Equipments', *Mod. Ry.*, **28,** No. 277, Oct. (1971), p. 152

V. von BUCHSTAB 'SCR Lift Truck Drives Dominate with Efficiency and Versatility', *Mater. Manage. & Distrib.*, **16,** No. 10, Sept. (1971), p. 33

C. E. BAND 'Netherlands Railways EM '64 Chopper Coaches', *Ingenieur*, **83,** No. 45, 12 Nov. (1971), p. E137

I. S. YEFREMOV 'Thyristor-Impulse Converters with Common Commutator for d.c. Rolling Stock', *Elec. Technol. USSR*, **2** (1971), p. 103

H. FICKENSCHER 'A Hybrid Integrated Pulse-Width Modulator', IEEE 6th Annual Meeting of the IGA Group, Cleveland, Ohio, USA, 18–21 Oct. (1971), p. 815

J. J. MORRISON 'Electronic Control of Battery Electric Vehicles', *Radio & Electron. Eng.*, **42,** No. 2, Feb. (1972), p. 91

B. BERMAN 'Battery Powered Regenerative SCR Drive', *IEEE Trans. Ind. Appl.*, **IA-8,** No. 2, Mar.–Apr. (1972), p. 190

B. BERMAN 'All Solid-State Method for Implementing a Traction Drive Control', *ibid*, p. 195

CHAPTER 6

Y. WALLACH 'Power Dissipation in Rectifiers', *Instrument Review*, December (1967), p. 497

Y. TANNO 'Rectifier Utilizing Single-Phase to Polyphase Converter', *Elect. Eng. Japan*, **87,** No. 6, June (1967), p. 43

D. A. MUELLER 'Selecting Solid State a.c/d.c Power Supplies', *Automation*, December (1967), p. 67

B. FELTBOWER 'Thyristor Assemblies as Output Stages', *IEE Conf. Pub.*, No. 37, Servocomponents, (1967), p. 145

E. S. McVEY 'The Design of a Simple Single Phase Silicon Controlled Rectifier Regulator', *IEEE Trans.*, **IECI-14,** No. 1, April (1967), p. 8

E. A. PARRISH 'A Theoretical Model for Single-Phase Silicon Controlled Rectifier Systems', *IEEE Trans.*, **AC-12,** No. 5, Oct. (1967), p. 577

B. J. KABRIEL 'High-Speed 3-Phase Thyristor Converter with Several Unusual Features', *Proc. IEE*, **114,** No. 1, Jan. (1967), p. 102

J. D. AINSWORTH 'Harmonic Instability Between Controlled Static Converters and a.c. Networks', *Proc. IEE*, **114,** No. 7, July (1967), p. 949

L. L. FRERIS 'Effects of Interaction Among Groups in a Multigroup a.c.—d.c. Converter', *ibid*, p. 965

'Low Cost Control Circuits', *Industrial Electronics*, Aug. 1967, p. 347

J. D. AINSWORTH 'The Phase-Locked Oscillator—a New Control System for Controlled Static Converters', *IEEE Trans.*, **PAS-87,** No. 3, March (1968), p. 859

N. G. HINGORANI 'A New Constant Excitation Angle Control for a.c./d.c./a.c. Static Converters', *ibid*, p. 866

R. WELLS 'Operating Data for Power Rectifiers', *Electrical Times*, Part 1 18 April, p. 623. Part 2 25 April, p. 670. Part 3 30 May, p. 901 (1968)

D. L. DUFF 'Semiconverter Rectifiers go High Power', *IEEE Trans.*, **IGA-4,** No. 2, March/April (1968), p. 185

C. G. DEWEY 'Development of Experimental 20-kV, 36-MW Solid-State Converters for HVDC Systems', *IEEE Trans.*, **PAS-87,** No. 4, April (1968), p. 1058

J. REEVE 'A Technical Assessment of Artificial Commutation of HVDC Converters with Series Capacitors', *IEEE Trans.*, **PAS-87,** No. 10, October (1968), p. 1830

J. B. RICE 'Commutation *dv/dt* Effects in Thyristor Three-Phase Bridge Converters', *IEEE Trans.*, **IGA-4,** No. 6, Nov./Fec. (1968), p. 665

I. K. DORTORT 'Phase Shifting of Harmonics in a.c. Circuits of Rectifiers', *ibid*, p. 655

P. MUCHNICK 'A.C. Regulated Power Supplies', *Electronics World*, April (1968), p. 54

K. TACHIBANA 'Harmonic Currents in Catenary Systems from Chopper Control', *ibid*, p. 203

L. MALESANI 'Improved Delay Circuit for Thyristor Linear Power Control', *Electronic Engineering*, Jan. (1969), p. 84

R. A. STEVENS 'Three-Phase Point-on-a-Cycle Switch', *Electronic Engineering*, Jan. (1969), p. 82

W. N. CHEUNG 'Pulse Delay Circuit Suitable for Triggering Thyristors', *Electronic Engineering*, Feb. (1969), p. 211

T. GILSIG 'Artificial Commutation of Converters Through Injection Technique', *IEEE Trans.*, **PAS-88,** No. 7, July (1969), p. 1052

K. WINKLER 'Abnormal Operating Conditions in Controlled Single-Phase Rectifier Circuits and Preventive Measures', *Brown Boveri Rev.*, **11/12** (1969)

H. F. STORM 'Solid-State Power Electronics in the U.S.A.', *IEEE Spectrum*, Oct. (1969), p. 49

A. R. BAILEY 'Thyristor-Stabilized Power Supplies', *Wireless World*, August (1959), p. 388

G. HEYNER 'Technical and Economic Advantages of Thyristor HVDC Converters', *Electrical Review*, Dec. (1969), p. 907

J. B. RICE 'A System for Recording Thyristor and Rectifier Current Waveforms', *IEEE Trans.*, **IGA-5,** No. 6, Nov./Dec. (1969), p. 779

L. L. FRERIS 'The Performance of a Converter with Anode Reactance', *IEEE Trans.*, **PAS-88,** No. 10, Oct. (1969), p. 1537

D. J. KING 'Thyristor Power Supplies for Colour TV Receivers', *Mullard Tech. Comm.*, No. 102, Nov. (1969), p. 29

J. ARRILLAGA 'Fault Development Control in a.c.—d.c. Converters', *Proc. IEE*, **116,** No. 7, July (1969), p. 1201

A. YAIR 'Bridge Rectifiers with Double and Multiple Supply', *Proc. IEE*, **116,** No. 5, May (1969), p. 811

I. HOSONO 'Construction, Control and Reliability of Converters', *IEE Conf. Pub.*, No. 53, Pt 1 May (1969), p. 154

H. MORI 'D.C. 200V 330kA Thyristor Rectifier for Electrolysis', *ibid*, p. 225

I. JURI 'Thyristoformer for 100kA', *ibid*, p. 234

A. GAVRILOVIC 'Application of Thyristor Converters to HVDC Transmission', *ibid*, p. 328

G. E. GARDNER 'Application of Thyristors to HVDC Transmission', *ibid*, p. 347

G. L. BREWER 'A Technical and Economic Appraisal of Thyristor Equipment for HVDC for Reinforcing a.c. Distribution Systems', *ibid*, p. 354

H. KONDO 'Development of High Voltage Thyristor Converter for HVDC Transmission at Electrotechnical Laboratory', *ibid*, p. 398

J. REEVE 'Accelerated Recovery from Commutation Faults in Bridge Connected a.c./d.c. Converters', *ibid*, p. 518

J. L. PAINE 'Determination of d.c. Bus Fault Currents for Thyristor Converters', IEEE 5th IGA Group Meeting 5–8 Oct. (1970), p. 637

T. E. DUNN 'System Compatibility: Static Power Supplies and Power Distribution', *ibid*, p. 643

E. MULLER 'Gate Control Sets for Low-Power Thyristors', *Siemens Rev.*, **37,** No. 9 (1970), p. 490

F. T. BENNELL 'Making the Best Use of Thyristors', *Electrical Times*, 30 April (1970), p. F15

T. HORIGOME 'A 100-kV Thyristor Converter for High-Voltage d.c. Transmission', *IEEE Trans.*, **ED-17,** No. 9, Sept. (1970), p. 809

D. B. GIESNER 'Operating Modes of the 3-Phase Bridge Converter', *Internat. J. Elect. Engng. Educ.*, **8,** (1970), p. 373

O. LACINA 'Computer Methods Used in Thyristor Converter Design', Conference on Power Electronics, Budapest, Hungary, 25–27 Nov. (1970) (Budapest: Magyar Elektrotechnikai Egyesulet)

M. KORALUN 'Thyristor Voltage Control Circuit for Power Supply of Electro-Filters', *ibid*

K. HORVATH 'Modern High-Voltage Rectifiers for Electro-Filters', *ibid*

M. VALEK 'Three-Phase Thyristor Bridge Circuit Neutral-Point Switches', *ibid*

V. A. NAIDIS 'Analysis of Thyristor Converter Power Circuits', *Elect. Technol. USSR (GB)*, **4,** (1970)

A. A. ARZUMANOV 'High-Power Stabilized Thyristor Rectifier', *Instrum. & Exp. Tech. (USA)*, No. 4, July-Aug. (1970), p. 1157

F. F. MAZDA 'Controlled Rectification', *Electronic Components*, 22 Jan. (1971), p. 106

W. N. N. CHEUNG 'Realisation of Converter Control Using Sample-and-Delay Method', *Proc. IEE*, **118,** No. 5, May (1971), p. 701

X. VOGEL 'Thyserts—Static Converter Modules for High Powers', *Brown Boveri Rev.*, **11** (1971), p. 507

K. SCHMUCK 'Reaction Effects on the Supply System Caused by Hexapulse Static Converters with Sequential Control', *ibid*, p. 514

D. JORDAN 'Rectitrol—A Battery Charger', *ibid*, p. 532

S. IVNER 'The Influence of Thyristor Converters on Supply Networks,' *ASEA Journal*, **44,** No 2 (1971), p. 37

R. WELLS 'Applying Phase Control to Bridge-Connected Converters', *Electrical Times*, 12/19 Feb. (1971), p. 83

L. L. FRERIS 'Modes of Operation of Bridge-Connected Converters', *Electrical Times*, Part 1, 1 Oct. (1971), p. 39, Part 2, 15 Oct. (1971), p. 43

M. G. J. FRY 'Selecting Stepless Voltage Control Equipment', *Electrical Review*, 13 Aug. (1971), p. 221

D. B. GIESNER 'Behaviour of HVDC Links Under Balanced a.c. Fault Conditions', *Proc. IEE*, **118,** No. 3/4, March/April (1971), p. 591

L. L. FRERIS 'Multigroup Converters with Series a.c. Connection', *Proc. IEE*, **118,** No. 9, Sept (1971), p. 1247

J. S. C. HTSUI 'Method of Digital Computation of Thyristor Switching Circuits,' *Proc. IEE*, **118,** No. 8, Aug. (1971)

R. WELLS 'Waveforms of Thyristors in Power Rectifiers', *Electrical Times*, **159,** No. 1, 1 Jan. (1971), p. 27

'Canadian HVDC Link will have Solid State Valves', *Electrical Review*, **188,** No. 4, 22 Jan. (1971), p. 119

B. BJORKLUND 'Railway Vehicles with Thyristor Converters and Silicon Rectifiers', *ASEA J. (Sweden)*, **44,** No. 2 (1971), p. 35

H. TAKABAYASHI 'Prototype 10kV, 100A Thyristor Converter System', *Hitachi Rev. (Japan)*, **20,** No. 1, (1971), p. 138

'Compensating Reactive Power the Static Way', *Electrical Review*, **188,** No. 26, 25 June (1971), p. 866

M. KUNTNER 'Thyristor Auxiliary Rectifiers for Munich Rapid-Transit System', *Siemens Rev.*, **38,** No. 7, July (1971), p. 333

S. IWANAMI 'Thyristor Converter for Electric Railway Substation', *Mitsubishi Denki Eng. (Japan)*, No. 28, March (1971), p. 30

R. RAMAKUMAR 'A Study of Rectification Systems Used with a High-Reactance Three-Phase Power Source', Proc. of Frontiers of Power Technology Conference, Stillwater, Okla., USA, 30 Sept.–1 Oct. (1971) (Stillwater: Okla. State Univ., 1971)

I. BOKSJO 'A Thyristor Converter for HVDC in Regular Operation', *Direct Current & Power Electron.*, **2,** No. 2, May–June (1971), p. 58

T. GILSIG 'Test Circuits for High Voltage d.c. Thyristor Valves', IEEE Conf. Rec. of Internat. Symposium on High Power Testing, Portland, Ore, USA, 21–23 July (1971), p. 147

J. J. WAWZONEK 'A Naturally Commutated Thyristor Inverter for a.c. Motor Drives', IEEE Proc. of National Aerospace Electronics Conf., Dayton, Ohio, USA, 17–19 May (1971), p. 137

B. A. ROWE 'HVDC Transmission—An Application of (Power) Semiconductors', *Ingenieur*, **83,** No. 41, 15 Oct. (1971), p. 109

M. R. WESCOTT 'Full-Wave Rectifier with Constant-Current Load', *Proc. IEE*, **118,** No. 10, Oct. (1971), p. 1501

G. KARADY 'The Calculation of Turn-on Overvoltages in a High Voltage d.c. Thyristor Valve', *IEEE Trans. Power App. & Syst.*, **PAS 90,** No. 6, Nov.–Dec. (1971), p. 2802

J. K. HAGGERTY 'Application Factors for Thyristor Converter d.c. Motor Drives', *IEEE Trans. Ind. Gen. Appl.*, **IGA-7,** No. 6, Nov.–Dec. (1971), p. 712

T. TSUBOI 'Control of a.c. Electric Locomotives with a Thyristor System and Regenerative Braking', *Hitachi Rev.*, **20,** No. 9 (1971), p. 364

V. R. VAJRAMUSHTI 'A.C.-D.C. Conversion in Electric Traction', *IEE-IERE Proc. India*, **9,** No. 5, Sept.–Oct. (1971), p. 186

R. P. De PUY 'Digital Control for Large Electromechanical Rectifier Systems', *IEEE 6th Annual Meeting of the IGA Group, Cleveland, Ohio, USA*, 18–21 Oct. (1971), p. 799

A. STROMBERG 'Modern Electroplating Rectifiers', *Asea J.*, **45,** No. 1, (1972), p. 18

R. E. F. BUGG 'Thyristor Power Supplies for Television Receivers', *Mullard Technical Communications*, No. 114, Apr. (1972), p. 129

D. R. CULLIFORD 'Stabilised Power Supplies at the Rutherford Laboratory', *Electronics & Power*, April (1972), p. 133

J. M. LAVALLEE 'Thyristor Trigger Circuit for a Reversible d.c. Motor Drive with Regenerative Braking', *Mullard Tech. Commun.*, **12,** No. 113, Jan. (1972), p. 70.

CHAPTER 7

R. A. HAMILTON 'Thyristor Adjustable Frequency Power Supplies for Hot Strip Mill Run-Out Tables', *IEEE Trans.*, **IGA-3,** No. 2, March/April (1967), p. 168

W. R. LIGHT 'A Synchronous Tap Changer Step-Up Cycloconverter,' *IEEE Trans.*, **IGA-3,** No. 3, May/June (1967), p. 244

R. S. BLAND 'Factors Affecting the Operation of a Phase-Controlled Cycloconverter', *Proc. IEE*, **114,** No. 12, Dec. (1967), p. 1908

T. H. BARTON 'A 5-kw Law-Frequency Power Amplifier of Improved Frequency Response', *IEEE Trans.*, **IECI-14,** No. 1, April (1967), p. 33

F. SPIEGELBERG 'Thyristor-Controlled Drives for Rotary Printing Presses', *Brown Boveri Rev.*, May/June (1967)

F. SPIEGELBERG 'Rectifier-Fed Drives for Printing Presses and Paper-Processing Machines', *ibid*

J. C. H. BONE 'Variable Frequency Drives', *Electrical Times*, 25 July (1968), p. 37

J. MACTAGGART 'Analogue-Computer Study of a Converter Jump Phenomenon', *Proc. IEE*, **116,** No. 8, August (1968), p. 1173

M. YAMAGUCHI 'Characteristics of Graetz-Convection Type Cycloconverter', *Elect. Eng. Japan*, **88,** No. 10 (1968), p. 28

L. J. LAWSON 'The Practical Cycloconverter', *IEEE Trans.*, **IGA-4,** No. 2, March/April (1968), p. 144

S. B. DEWAN 'A.C. to a.c. Frequency Converters for Induction Heating and Melting', *IEE Conf. Pub.*, No. 53, Pt. 1, May (1969), p. 440

J. E. JENKINS 'A Modified Cycloconverter for Use with High Frequency Sources', *ibid*, p. 313

D. SCHOEN 'A.C. Adjustable Speed Drives', *Instruments and Control Systems*, Sept. (1969), p. 165

S. B. DEWAN 'Harmonic Analysis of a.c. to a.c. Frequency Converter', *IEEE Trans.*, **IGA-5,** No. 1, Jan/Feb. (1969), p. 29

A. E. RICH 'Concepts of Gearless Ball Mill Drives', *ibid*, p. 13

M. MEYER 'Recent Developments in the Field of Variable-Speed Drives', *Siemens Rev.*, Dec. (1969)

J. LANGER 'Static Frequency Changer Supply System for Synchronous Motors Driving Tube Mills', *Brown Boveri Review*, No. 3 (1970), p. 112.

H. STEMMLER 'Drive System and Electronic Control Equipment of the Gearless Tube Mill', *ibid*, p. 120

E. BLAUENSTEIN 'The First Gearless Drive for a Tube Mill., *ibid*, p. 96

S. B. DEWAN 'A Method for Harmonic Analysis of Cycloconverters', *IEEE Trans.*, **IGA-6,** No. 5, Sept./Oct. (1970), p. 455

T. M. HAMBLIN 'Cycloconverter Control Circuits' IEEE 5th Annual IGA' Group Meeting 5–8 Oct. (1970), p. 559

T. TSUCHIYA 'Basic Characteristics of Cycloconverter Type Commutatorless Motors', *IEEE Trans.*, **IGA-6,** No. 4, July/Aug. (1970), p. 349

J. NOPIRAKOWSKI 'Inverters and Cycloconverters and Their Industrial

Use' Conference on Power Electronics, Budapest, Hungary, 25–27 Nov. (1970) (Budapest: Magyar Elektrotechrikai Egyesulet)
G. J. BERG 'Three-Phase Static Frequency Changer', *ibid*

N. L. SCHMITZ 'The Cycloconverter Driven Doubly-Fed Induction Motor', *IEEE Trans.*, **PAS-90,** No. 2, March/April (1971), p. 495
W. F. LONG 'Cycloconverter Control of the Doubly Fed Induction Motor', *IEEE Trans.*, **IGA-7,** No. 1, Jan./Feb. (1971), p. 95
F. F. MAZDA 'Frequency Converters', *Electronic Components*, 16 April (1971), p. 392
I. TAKAHASHI 'Triggering Schemes and Waveform Analysis of Cycloconverters', *Elec. Eng. Jap.*, **90,** No. 2, Mar.–Apr. (1971), p. 95
M. RAMAMOORTY 'Thyristor Controlled Frequency Doubler', *J. Inst. Eng. (India) Elec. Eng. Div.*, **51,** No. 10, Pt. EL5, June (1971), p. 241
R. A. SOMMER 'Solid State Frequency Converters, a Survey', IEEE Conf. Rec. of 10th Biennial Conf. on Electric Process Heating in Industry, Milwaukee, Wis. USA, 20–21 Apr. (1971), p. 5
W. E. FRANK 'Applying the Induction Heating Static Frequency Converter', *ibid*, p. 10
W. McMURRAY 'A Study of Unsymmetrical Firing for Phase-Controlled Converters', IEEE 6th IGA Group Meeting, Cleveland, Ohio, USA, 18–21 Oct. (1971), p. 313

P. G. HOLMES 'A Dynamic Model of a Closed-Loop Cycloconverter-Fed Induction Motor System', Conf. Proc of the 7th Universities Power Engineering Conf., Bradford, Yorkshire, England, 5–7 Jan. (1972) (Bradford, Yorkshire: Bradford Univ.), p. 129

CHAPTER 8

J. W. MOYTO 'Computer Aided Analysis of Thyristor Current Ratings', IEEE 2nd Annual IGA Group Meeting, Oct. (1967), p. 1
Y. TAKEDA 'Analysis and Design of Morgan Circuit with Resistive Loads', *Elect. Eng. Japan*, July (1967), p. 59

O. A. KOSSOV 'Comparative Analysis of Chopper Voltage Regulators with LC Filter', *IEEE Trans.*, **MAG-1,** No. 4, Dec. (1968)
E. OHNO 'Analysis of Multiphase Thyristor d.c. Choppers for Traction Motor Control', *Elect. Eng. Japan*, **88,** No. 3 (1968), p. 1.
S. HAYASHI 'Analysis of a 3-Phase Static Inverter Circuit with Rectifiers', *Elect. Eng. Japan*, **88,** No. 12 (1968), p. 79
C. D. BECK 'Motor Drive Inverter Ratings', *IEEE Trans.*, **IGA-4,** No. 6, Nov./Dec. (1968), p. 589
T. J. TAKEUCHI 'Anslysis of Single-Phase Bridge-Type SCR Inverter Circuits and Control of Single-Phase Induction Motor by Means of Such Inverters', *Elect. Eng. Japan*, **88,** No. 1 (1968), p. 54
C. D. BECK 'Motor Drive Inverter Rating', *Automation*, July (1968), p. 82

J. W. MOTTO Jr. 'Computer Aided Thyristor Overload Current Ratings for Motor Drive and Other Applications Requiring Repetitive Overloads', IEEE 3rd Annual IGA, 29 Sept.–30 Oct. (1968), p. 593

N. YAMADA 'Determination of Ratings for Semiconductor Assemblies for Variable Load with Periodicity', *IEE Conf. Pub.*, No. 53, Pt. 1, May (1969), p. 101

A. K. PODKOLINSKI 'A Design Approach to Rating of Thyristors in S.C.R. Controlled Battery Driven Trucks', *ibid*, p. 108

A. J. BLUNDELL 'The Effect of On-State Temperature Coefficient on Thyristor Junction Temperature Calculations', *ibid*, p. 53

B. U. NEWSAM 'Transient Thermal Resistance—its Measurement and Use in Rating of Thyristors', *ibid*, p. 85

T. MARUHASHI 'Stability and Approximate Analysis of Series Inverter Without Gate Control', IEEE 4th Annual IGA Group Meeting, Detroit, 12–16 Oct. (1969), p. 445

K. R. JARDAN 'General Analysis of Three-Phase Inverters', *IEEE Trans.*, **IGA-5,** No. 6, Nov./Dec. (1969), p. 672

S. B. DEWAN 'Optimum Design of Input—Commutated Inverter for a.c. Motor Control', *ibid*, p. 699

E. REIMERS 'Design Analysis of Multi-Phase d.c. Chopper Motor Drive' IEEE 5th Annual IGA Group Meeting, 5–8 Oct. (1970), p. 587

Y. HIRANE 'Analysis of Parallel-Type, Two-Phase Converters', *Elect. Eng. Japan*, **90,** No. 3, May–June (1970)

H. H. HO 'Generalised Principle of Operation of 3-Phase Thyristor Inverters', *Proc. IEE*, July (1970), p. 1333

H. OKAMOTO 'High Precision Analysis of McMurray Inverters with Inductive Loads', *Elect. Eng. Japan*, **90,** No. 3, May–June (1970)

F. F. MAZDA 'The Design of Choppers and Inverters', *Electronic Components*, 14 May (1971), p. 509

R. G. HOFT 'Digital Computer Simulation of Thyristor Choppers and Inverters', *Simulation*, **17,** No. 2, Aug. (1971), p. 71

E. REIMERS 'Design Analysis of Multiphase d.c. Chopper Motor Drives', *IEEE Trans. Ind. Appl.*, **IA-8,** No. 2, Mar.–Apr. (1972), p. 136

CHAPTER 9

Y. YU 'Static d.c. to Sinusoidal a.c. Inverter Using Techniques of High Frequency Pulse-Width Modulation', *IEEE Trans.*, **MAG-3,** No. 3, Sept. (1967), p. 250

S. B. DEWAN 'Thyristor Converters Using Saturable Reactors', *ibid*, p. 224

B. MOKRYTZKI 'Pulse Width Modulated Inverters for a.c. Motor Drives', *IEEE Trans.*, **IGA-3,** No. 6, Nov. (1967), p. 493

R. J. RAVAS 'Staggered Phase Carrier Cancellation—a New Circuit Concept for Lightweight Static Inverters', Supplement to *IEEE Trans.*, **AES-3,** No. 6, Nov. (1967), p. 433

B. MOKRYTZKI 'Solid State a.c. Motor Drives', IEEE 13th Annual Pulp & Paper Conf., 3–5 May (1967), p. 16

M. STEMMLER 'The Use of Static Converters to Vary the Speed of Three-Phase Drives Without Losses', *Brown Boveri Review*, May/June (1967), p. 217

V. SCHEIDER 'Thyristor Static Frequency Converters with d.c. Link for the Speed Control of Sectional Motor Drives', *Siemens Rev.*, No. 11 (1967), p. 410

C. F. KELLERS 'A Megawatt Solid-State Inverter to Power the Cornell 10 GeV Electron Synchrotron', *IEEE Trans. on Nuclear Science*, June (1967), p. 496

M. NAKAOKA 'Thyristor Sinsusoidal Inverter with Dual-Mode Voltage Regulating System', *Tech. Rept. Osaka University (Japan)*, Oct. (1967), p. 469

N. MAPHAM 'An SCR Inverter with Good Regulation and Sine-Wave Output', *IEEE Trans.*, **IGA,** March/April (1967), p. 176

C. RIDGERS 'Voltage Stabilized Sinusoidal Inverters Using Transistors', *Radio and Electronic Engineer*, Aug. (1967), p. 109

J. E. MURRAY 'Analog Simulation Simplifies Static Power-Inverter Design', *Electro-Technology*, Dec. (1967), p. 36

R. L. MENDES 'Firing an Adjustable Frequency Three-Phase Inverter', *Electronic Engineering*, Jan. (1967), p. 41

D. C. GRIFFITH 'Standby Power for Critical Loads', *Automation*, Sept. (1967), p. 86

P. F. PITTMAN 'Staggered Phase Technique Shrinks Power Conditions', *Electro-Technology*, June (1968), p. 55

B. MOKRYTZKI 'The Controlled Slip Static Inverter Drive', *IEEE Trans.*, **IGA-4,** No. 3, May/June (1968), p. 312

H. KAWAMOTO 'Regulated d.c.—d.c. Converter', *Japan Telecom. Review*, **10,** No. 4 (1968), p. 242

A. E. CRAWFORD 'Ultrasonic Generator Using Thyristors', *Electronic Engineering,* Feb. (1968), p. 66

H. Y. WONG 'D.C. In = 3-Phase Out in High Power Static Inverter', *Electronic Engineer*, Sept. (1968), p. 40

E. F. CHANDLER 'The Wide Speed Range Inverter', IEEE Conf. Record 3rd Annual IGA Group Meeting Chicago, 29 Sept.–30 Oct. (1968), p. 839

R. K. JARDAN 'Modes of Operation of 3-Phase Inverters, *ibid*, p. 517

M. R. BRANDA 'Inverter Instrumentation—What Should it Tell You?', *ibid*, p. 847

K. M. WATKINS 'Operational and Economic Considerations of Precision High-Power Static Inverter Systems', *ibid*, p. 811

A. KUSKO 'Application of Static Uninterruptible Power Systems to Computer Loads', *ibid*, p. 635

O. N. BRATKOVA 'High Frequency Power Sources Based on Thyristors for Series Micro Welding', *Weld Prod.*, **15,** No. 7 (1968), p. 66

S. D. T. ROBERTSON 'A Variable Low-Frequency Inverter Using Thyristors', *IEEE Trans.*, **IGA**, Sept./Oct. (1968), p. 501

K. Y. G. LI 'New 3-Phase Inverter Circuit', *Proc. IEE,* **115,** No. 11, Nov. (1968), p. 1677

C. D. BECK 'Motor Drive Inverter Ratings', *IEEE Trans.,* **IGA-4,** Nov./Dec. (1968), p. 589

W. R. KNIGHT 'A Case for Static Inverters', *Industrial Electronics,* Part 1 Oct. (1968), p. 406, Part 2 Nov. (1968), p. 470

H. GOSIMA 'Non-Break a.c. Power Source Switching Equipment', *Review of Elect. Communications Laboratory,* July/Aug. (1968), p. 564

J. D. FARBER 'Static Inverter Stand-By a.c. Power for Generating Station Controls', *IEEE Trans.,* **PAS,** May (1968), p. 1270

A. J. HUMPHREY 'Inverter Commutation Circuits', *IEEE Trans.,* **IGA,** Jan./Feb. (1968), p. 104

Y. YU 'The Influence of Circuit Parameters on the Harmonics of a PWM d.c.-to-Sinusoidal Inverter', Proc. 8th Annual IEEE Region 3 Convention, Huntsville, Ala., U.S.A., 19–21 Nov. (1969), p. 232

M. A. KNIGHT 'One-kVA, Three-Phase d.c.-a.c. Inverter with Digital Control', *IEEE Trans.,* **AES-5,** No. 6, (1969), p. 989

S. B. DEWAN 'Voltage Control in Pulse Width Modulated Inverters with Clamped and Independent Commutations', IEEE 4th Annual IGA Group Meetings, 12–20 Oct. (1969), p. 621

I. B. LANDAY 'Wide-Range Speed Control of 3-Phase Squirrel-Cage Induction Motors Using Static Frequency Converters', *IEEE Trans.,* **IGA,** Jan./Feb. (1969), p. 53

K. HEUMANN 'Development of Inverters with Forced Commutation for a.c. Motor Speed Control up to the Megawatt Range', *ibid,* p. 61

F. EKELUND 'Thyristor Power Converters for the Supply of Telephone Exchanges', *IEE Conf. Pub.,* No. 53, Pt. 1, May (1969), p. 200

R. A. HAMMOND 'Industrial Variable Speed Drives Using Capacitor Commutated Thyristors', *ibid,* p. 305

K. H. WILLIAMSON 'Simple 3-Phase a.c. Motor Control System for Motors Below 5 HP', *ibid,* p. 320

S. D. T. ROBERTSON 'A Tuned Circuit Commutated Inverter', *ibid,* p. 297

M. J. LEACH 'Experience with a 200 kVA Variable Frequency Converter for Mill Roller Table Motors', *ibid,* p. 465

S. WATABE 'A 3-Phase 250 kVA no Break Power Supply with Current Limiting Filter', *ibid,* p. 216

R. A. HAMILTON 'High Power Thyristor Inverters for Essential Service', *ibid,* p. 505

H. R. CAMENZIND 'See-Saw Circuit gives Sine-Wave Power', *Orbit,* Oct. (1969), p. 44

D. WALLSTEIN 'Medium-Frequency Static Frequency Changers of the Thysert Range for Inductive Melting and Heating', *Brown Boveri Review,* **11/12** (1969), p. 581

K. M. WATKINS 'The Application of High Power Inverter Systems', *IEEE Trans.,* **IGA-5,** No. 5, Sept./Oct. (1969), p. 588

A. E. RELATION 'Uninterruptable Power for Critical Loads', *ibid,* p. 582

D. FEMLING 'Solid State Inverter', *Inst. Cont. Syst.,* March (1969), p. 133

L. T. REES 'Capacitor Charging from Batteries with d.c.-d.c. Converters', *IEEE Trans.*, **IGA-5,** No. 5, Sept./Oct. (1969), p. 600

G. W. ERNSBERGER 'Inverter-Converter Parallel Operation', Report NASA-CR-1224, Westinghouse Electric Corp., Lima., Ohio, Jan. (1969), p. 107, Contract NAS3-2792

F. GOURASH 'Modularization of High-Power Inverters and Converters', IEEE 4th Intersociety Energy Conversion Engineering Conf., Washington, 22–26 Sept. (1969), p. 830

C. P. R. WRAITH 'Static Power Systems for Computer Continuity', *Electrical Times*, 27 Feb. (1969)

R. HAINES 'Static Inverters for Emergency Power', *Electrical Times*, Feb. (1969)

W. KIRYLUK 'Static Inverter Applications', *Electrical Equipment*, August (1969), p. 20

G. FLYNN 'Static Inverters—How to Tame a.c. Power', *Electronic Products*, June (1969), p. 105

C. W. LORICK 'Inverters Supply Uniform and Constant Power for Automated Boiler Controls', *Electric Light and Power*, Oct. (1969), p. 82.

R. P. VERES 'New Inverter Supplies for High Horsepower Drives', *IEEE Trans.*, **IGA-6,** No. 2, March/April (1970)

S. P. JACKSON 'Multiple Pulse Modulation in Static Inverters Reduces Selected Output Harmonics and Provides Smooth Adjustment of Fundamentals', *IEEE Trans.*, **IGA-6,** July/Aug. (1970), p. 357

S. P. JACKSON 'Varying the Output Voltage of Static Inverters with Simple Modulation', *IEEE Trans.*, **MAG-6,** March (1970), p. 59

E. F. CHANDLER 'Wide Speed Range Inverter', *IEEE Trans.*, **IGA-6,** No. 1, Jan./Feb. (1970), p. 19

E. F. CHANDLER 'Wide Speed Range Inverter', *ibid*, p. 19

R. E. BEDFORD 'Analysis and Performance of a Three-Phase Ring Inverter', *IEEE Trans.*, **IGA-6,** No. 5, Sept./Oct. (1970), p. 488

W. B. GUGGI 'Sine Wave Inverter System', IEEE Conf. Record 1970 5th Ann. IGA Group Meeting, 5–8 Oct. (1970), p. 517

R. YOUN '360 Hz Complementary Impulse Commutated Inverter', *ibid*, p. 531

L. J. PENKOWSKI 'Fundamentals of a Pulse Width Modulated Power Circuit', *ibid*, p. 669

R. D. ADAMS 'Several Modulation Techniques PWM Inverter', *ibid*, p. 687

R. H. CUNNINGHAM 'Application Considerations of Static Inverters as Used in Critical Power Systems', Proc. 20th Internat. ISA Iron and Steel Instrumentation Symposium, Pittsburgh Pa., USA, 23–25 March (1970), p. 82

D. A. POPOV 'A Comparative Evaluation of Static Inverters Using Transistors Operating as Switches', *Elec. Technol. USSR (GB)*, **2,** (1970), p. 141

S. B. DEWAN 'Output Voltage in three-Phase Pulse-Width Modulated Inverters', *IEEE Trans.*, **IGA-6,** No. 6, Nov.–Dec. (1970), p. 570

A. KERNICK 'Static Inverter for Aerospace Applications Provides High-Quality a.c. Power Despite Disruptive Influences', *Westinghouse Eng. (USA)*, **30,** No. 6, Nov. (1970), p. 174

M. RAMAMOORTY 'Variable Frequency d.c. to a.c. Parallel Inverter Circuit Using Silicon Controlled Rectifiers', *J. Inst. Eng. (India) Elec. Eng. Div.*, **51,** No. 2, Pt. EL 1, Oct. (1970), p. 6

R. YOUN *Complementary Impulse Commutated Inverter*, Univ. Missouri, Columbia, USA, Thesis (1970), 112 pages

A. SHIRASHOJI 'Recent High Voltage Static Inverter for Rolling Stock', *Mitsubishi Denhi Eng. (Japan)*, No. 26, Sept. (1970), p. 17

S. NYUDO 'D.C. to d.c. Power Converter', *Mem. Def. Acad (Japan)*, **10,** No. 1, June (1970), p. 187

K. R. JARDAN 'Inherent Impedance and Unbalanced Load Conditions of Three-Phase Inverters', Conf. on Power Electronics, Budapest, Hungary, 25–27 Nov. 1970 (Budapest: Magyar Elektrotechnikai Egyesulet)

I. RACZ 'High Frequency Thyristor Power Supply', *ibid*

A. YAIR 'New 3-Phase Inverter with Three Thyristors', *Proc. IEE*, **118,** No. 7, July (1971), p. 901

F. F. MAZDA 'Inversion', *Electronic Components*, 25 June (1971), p. 705

M. PEARD 'Frequency Converter', *IBM Tech. Disclosure Bull. (USA)*, **13,** No. 12, May (1971), p. 3727

'Pulse Width Modulation for Multi-Motor Drives', *Electrical Review*, **189,** No. 3, 16 July (1971), p. 87

'Germans Build 2500 hp Prototype Diesel with Three-Phase Motors', *Ry. Gaz. Int. (GB)*, **127,** No. 5, May (1971), p. 185

S. IIDA 'Characteristics of Single-Phase Bridge-Type SCR Inverter Circuit Having Series Diodes', *Elect. Eng. Jap.*, **90,** No. 4, July–Aug. (1971)

A. BELLINI 'Induction Machine Frequency Control: Three-Phase Bridge Inverter Behavior and Performance', *IEEE Trans. Ind. & Gen. Appl.*, **IGA-7,** No. 4, July–Aug. (1971), p. 488

R. M. HERBERT 'No-Break and Stand-by Supplies', *Elec. Times*, **160,** No. 17, 22 Oct. (1971), p. 35

J. A. RADE 'Variable Frequency Drives as Process-Line Helpers', *Eng. Dig.*, **32,** No. 7, July (1971), p. 65

T. KUME 'Thyristor d.c. Static Inverter', IEEE 6th IGA Group Meeting, Cleveland, Ohio, USA, 18–21 Oct. (1971), p. 299

J. B. FORSYTHE 'A Pulse Width Modulated Three Phase Complementary Commutated Inverter', *ibid*, p. 321

S. B. DEWAN 'Harmonic Analysis of a Synchronized Pulse Width Modulated Three Phase Inverters', *ibid*, p. 327

J. T. SALIMI 'An Inverter for Traction Applications', *ibid*, p. 393

L. H. WALKER 'Inverter for VPS with Subcycle Fault Clearing Capability', IEEE 6th Annual Meeting of the IGA Group, Cleveland, Ohio, USA, 18–21 Oct. (1971), p. 361

W. B. GUGGI 'Practical Design Considerations for Sine Wave Inverters', *ibid*, p. 869

D. A. COWARD 'A Technique for Sychronizing a Parallel Static-Inverter System', *IEEE Trans. Ind. Electron & Contr. Instrum.*, **IECI-19,** No. 1, Feb. (1972), p. 21

K. THORBORG 'New Type of Three-Phase Thyristor Inverter', *Asea J.* (Sweden), **45,** No. 1 (1972), p. 9

CHAPTER 10

T. WEISZ 'Disregard Load Impedance in SCR Inverter Design', *Electronic Design,* May (1967), p. 68

S. B. DEWAN 'Practical Considerations in the Design of Commutation Circuits for Choppers and Inverters', IEEE 4th Annual IGA Group Meeting, Detroit, 12–16 Oct. (1969), p. 469
V. WOUK 'High Efficiency, High Power Load Insensitive d.c. Chopper for Electronic Automobile Speed Control', *ibid,* p. 393
P. M. ESPELAGE 'A Wide-Range Static Inverter Suitable for a.c. Induction Motor Drives', *IEEE Trans.,* **IGA-5,** No. 4, July/Aug. (1969), p. 438
H. KALIS 'D.C. Chopper with High Switching Reliability and Without Limitation of the Adjustable Mark Space Ratio', *IEE Conf. Pub.,* No. 53, Pt. 1, May (1969), p. 208
A. R. BAILEY 'The Use of High Frequency Choppers for Traction Purposes', *ibid,* p. 271

F. F. MAZDA 'High Frequency Thyristor Choppers', *Electronic Engineering,* Feb. (1970), p. 34
F. F. MAZDA 'Voltage Boosting Reduces Thyristor Commutation Components', *Electronic Engineering,* Oct. (1970), p. 57
E. L. KELSEY 'Development and Evaluation of a High-Efficiency Power Inverter', Report NASA-TN-D-5881, 36 pp, NASA Langley Res. Centre, Langley Station Va. USA, June (1970)

S. HAYASHI 'High-Frequency Parallel Inverter Circuits Consisting of Thyristor Elements', *Elect. Eng. Japan,* **90,** No. 1–2, Jan./Feb. (1971)
F. F. MAZDA 'Special Modifications to Choppers and Inverters', *Electronic Components,* 23 July (1971), p. 803
J. P. LANDIS 'A Static Frequency Converter for 10 kHz Power', IEEE Record of 10th Biennial Conf. on Electric Process Heating in Industry, Milwaukee, Wis., USA, 20–21 Apr. (1971), p. 71
I. HORVAT 'A 180 kW 8–11 kHz Thyristor Frequency Converter for Induction Heating', IEEE 6th Annual Meeting of the IGA Group, Cleveland, Ohio, USA, 18–21 Oct. (1971), p. 837

CHAPTER 11

W. DITTRICK 'Electronic Motor DMc 3, a New Miniature Brushless d.c. Motor', *Siemens Rev.,* **34** (1967), No. 3, p. 97
A. E. BURTON 'A Brushless d.c. Motor', *The Engineer,* Dec. (1967), p. 829

P. G. BHAGWAT 'Speed Control of d.c. Shunt Motor Using Thyristor Bridge', *Control*, May (1967), p. 222

E. OHNO 'The Thyristor Commutatorless Motor', *IEEE Trans.*, **MAG-3,** No. 3, Sept. (1967), p. 236

D. A. PAICE 'Voltage Control for Small Induction Motors', *Control Engineering*, Oct. (1967), p. 73

A. WALRAVEN 'Controlling the Speed of Small Induction Motors by Means of Thyristors', *Philips Technical Review*, **28,** No. 1 (1967), p. 1

W. SHEPHERD 'An Experimental Closed-Loop Variable-Speed Drive Incorporating a Thyristor Driven Induction Motor', *IEEE International Convention Record*, Pt. 13 (1967), p. 82

G. W. WILCOCK 'Transient and Steady State Analysis of Inverter Supplied Induction Motors', Royal Aircraft Establishment Technical Report 67252, Oct. (1967)

B. M. BIRD 'Slip-Energy-Recovery Schemes and the Use of Pole-Change Windings', *Proc. IEE*, **114,** No. 10, (1967), p. 1436

W. K. O'NEIL '20 kVA Static Frequency Changer for Direct Drive Induction Servo-Motor', *IEE Conf. Pub.*, No. 37, (1967), p. 226

G. A. SWEENEY 'Static Adjustable Frequency Drives', *Automation*, May (1967), p. 90

S. NONAKA 'A Salient-Pole Single-Phase Synchronous Motor Driven by a Square-Wave Inverter', *Elect. Eng. Japan*, Oct. (1967), p. 6

W. MERHOF 'Brushless Excitation of Synchronous Machines by Rotating Semiconductors', *Brown Boveri Review*, Sept. (1967), p. 539

D. C. KING 'The Use of Thyristors in the Static Excitation of a.c. Generators', *Student's Quarterly Journal, (IEE)*, Sept. (1967)

W. F. WRIGHT '60 MW Set with Brushless Excitation', *Electrical Times*, **152,** No. 19 (1967), p. 731

K. R. McCLYMONT 'Experience with High-Speed Rectifier Excitation Systems', IEEE Winter Power Meeting, New York, 29 Jan.–3 Feb. (1967)

I. R. SMITH 'Brushless and Self-Excited Single-Phase Alternator', *Proc. IEE*, **114,** No. 8, Aug. (1967), p. 1089

L. M. PANIS 'Rectifier Excitation of Large Turbo-Generators', *Brown Boveri Review*, **54,** No. 2/3, (1967), p. 69

W. S. BROWN 'A Lightweight Motor for an Electric Car', *NEREM Record*, (1968), p. 126

M. YAMAGUTI 'A Three-Phase Shunt-Commutatorless Motor', *Elect. Eng. Japan*, **88,** No. 2 (1968), p. 38

J. M. FRANKEL 'Brushless d.c. Torque Motors', *Instruments and Control Systems*, Oct. (1968), p. 115

C. ST. J. LAMB 'Ring-Type Stationary Armature d.c. Motor', *Electrical Engineer*, April (1968), p. 8

W. W. YATES 'Brushless d.c. Motors', *Electrical Engineer*, 10 Feb. (1968)

P. L. ALGER 'Speed Control of Wound Rotor Motors with SCR's and Saturistors', *IEEE Trans.*, **IGA-4,** No. 5, Sept./Oct. (1968), p. 477

B. J. CHALMERS 'Induction-Motor Losses Due to Non-Sinusoidal Supply Waveforms', *Proc. IEE*, Dec. (1968), p. 1777

E. M. SABBAGH 'Characteristics of an Adjustable Speed Polyphase Induction Machine', *IEEE Trans.*, **PAS-87,** No. 3, March (1968), p. 613
E. A. KLINGSHIRN 'Polyphase Induction Motor Performance and Losses on a Non-sinusoidal Voltage Source', *ibid*, p. 624
P. C. KRAUSE 'Comparison of Computer and Test Results of a Static a.c. Drive System', *IEEE Trans.*, **IGA-4,** No. 6, Nov./Dec. (1968), p. 583
J. C. H. BONE 'Variable Frequency Drives', *Electrical Review*, 18 Oct. (1968), p. 564
R. D. THORTON 'High Frequency Motors for Electric Propulsion', IEEE Intersociety Energy Conversion Engng. Conf. Boulder, Colorado, 13–17 Aug. (1968), p. 797
J. SHANKS 'Variable-Speed a.c. Motors for Mine Fans and Converters', *LSE Engng. Bulletin*, Sept. (1968), p. 23
D. A. PAICE 'Speed Control of Large Induction Motors by Thyristor Converters', IEEE 3rd Annual IGA Group, Chicago, Ill., 29 Sept.–30 Oct. (1968), p. 753
P. N. MILJANIC 'The Through-Pass Inverter and its Application to the Speed Control of Wound Rotor Induction Motors', *IEEE Trans.*, **PAS-87,** No. 1, Jan. (1968), p. 234
B. J. CHALMERS 'Steady-State Asynchronous Characteristics of Salient Pole Motors with Rectifiers in the Field Circuit', *Proc. IEE*, **115,** No. 7, July (1968), p. 987
P. A. B. WOOLDRIDGE 'Considerations Affecting the Design Philosophy of Solid-State Exciters', *IEEE Trans.*, **PAS-87,** No. 5, May (1968), p. 1288
P. L. DANDENO 'Effect of High Speed Rectifier Excitation Systems on Generator Stability Limits', *IEEE Trans.*, **PAS-87,** Jan. (1968), p. 190
D. MIDGLEY 'Alternator Self-Excitation', *Electrical Times*, 7 Nov. (1968), p. 640
D. MIDGLEY 'Electronic Control of Alternator Excitation', *Electronics Letters*, **4,** No. 6, 22 March (1968), p. 101
D. C. GILCHRIST 'Thyristor Automatic Voltage Regulators', *G.E.C. Journal*, **35,** No. 1 (1968), p. 3
I. R. SMITH 'Brushless and Self-Excited 3-Phase Synchronous Machines', *Proc. IEE*, **115,** No. 11, Nov. (1968), p. 1655

I. OHNO 'Study of Photocommutator d.c. Motor', *Elect. Eng. Japan*, **89,** No. 1 (1969), p. 51
C. D. MANNING 'Optoelectronic Control of Brushless Alternators', *Proc. IEE*, **116,** No. 7, July (1969), p. 1222
R. D. THORTON 'The Electrical Car Race', *Proc. IEEE Electrical & Electronics Engng. Resources Roundup,* 16–18 April (1969), p. 121
W. RADZIWILL 'A Highly Efficient Small Brushless d.c. Motor', *Philips Technical Review*, **30,** No. 1 (1969), p. 7
A. LUJIC 'Controlling Brushless d.c. Motors', *Machine Design*, 30 Oct. (1969), p. 113
D. ROSENMAYR 'Thyristor Elements of Variable-Speed Lifting-Gear Drives with Induction Motors', *Brown Boveri Review*, **56,** No. 2, Feb. (1969), p. 72

E. C. BARNES 'Performance and Characteristics of Induction Motors for Solid State Variable Frequency Drives', IEEE 4th Annual IGA Group Meeting Detroit, 12–16 Oct. (1969), p. 653
G. W. McLEAN 'Performance and Design of Induction Motors with Square Wave Excitation', *Proc. IEE*, **116,** Aug. (1969), p. 1405
B. J. CHALMERS 'Design of Induction Motors Supplied by Thyristors', *Electrical Review*, 11 April (1969)
S. D. T. ROBERTSON 'Torque Pulsations in Induction Motors with Inverter Drives', IEEE 16th Annual Petroleum and Chemical Industry Conf., Los Angeles, Calif., USA., 15–17 Sept. (1969)
A. LUDBROOK 'Basic Economics of Thyristor Adjustable Speed Drive Systems', IEEE 4th Annual IGA Group Meeting (1969), p. 591
I. D. LANDAU 'Wide Range Speed Control of 3-Phase Squirrel-Cage Induction Motors Using Static Frequency Converters', *ibid*, p. 91
R. L. RISBERG 'A Wide Speed Range Inverter Fed Induction Motor Drive', *ibid*, p. 629
M. YAMAGUCHI 'Analysis of Commutatorless Motors', *Elec. Eng. Japan*, **89,** No. 9 (1969), p. 35
C. G. HELMICK 'Motor Fusing for Adjustable Frequency Inverter Systems', *IEEE Trans.*, **IGA-5,** No. 1, Jan./Feb. (1969), p. 40
E. W. DAVIS 'Applications of Inverter Drives', *ibid*, p. 45
R. W. JOHNSTON 'Modulating Inverter System for Variable-Speed Induction Motor Drive', *IEEE Trans.*, **PAS-88,** No. 2, Feb. (1969), p. 81
P. C. KRAUSE 'Analysis and Simplified Representations of a Rectifier-Inverter Induction Motor Drive', *IEEE Trans.*, **PAS-88,** No. 5, May (1969), p. 588
E. H. DINGER 'Digital Applications of a.c. Motor Controls', *IEEE Trans.*, **IGA-5,** No. 3, May/June (1969), p. 291
E. E. WARD 'Preliminary Investigation of an Inverter-Fed 5-Phase Induction Motor', *Proc. IEE*, **116,** June (1969), p. 980
K. Y. G. LI 'Analysis and Operation of an Inverter-Fed Variable-Speed Induction Motor', *Proc. IEE*, **116,** Sept. (1969), p. 1571
B. H. SMITH 'Variable Speed Twin Stator Induction Motor', *Electrical Times*, 14 Aug. (1969), p. 53
A. J. HUMPHREY 'Constant Horsepower Operation of Induction Motors', *IEEE Trans.*, **IGA-5,** No. 5, Sept./Oct. (1969), p. 552
W. SHEPHERD 'Slip Energy Recovery Using a Thyristor Inverter', *Electrical Review*, 12 Sept. (1969)
W. SHEPHERD 'Slip Power Recovery in an Induction Motor by the Use of a Thyristor Inverter', *IEEE Trans.*, **IGA-5,** No. 1, Jan./Feb. (1969), p. 74
R. F. FLACK 'Applications for Eddy Current Couplings', *Electrical Times*, 17 April (1969), p. 52
A. H. HOFFMAN 'Brushless Synchronous Motors for Large Industrial Drives', *IEEE Trans.*, **IGA-5,** No. 2, March/April (1969), p. 158
B. J. KABRIEL 'Simple Formulas for Voltage-Regulator Gains for Best Alternator Stability', *Proc. IEE*, **116,** No. 11, Nov. (1969), p. 1907
J. F. TONG 'Brushless Synchronous Motors', *Electrical Review*, 19 Sept. (1969), p. 413

W. FAIRNEY 'Thyristor Excitation-Large Set Pilot Scheme', *Electrical Times*, 13 March (1969), p. 52
J. V. WOODS 'New d.c. Motors for Thyristor Supplies', *Electrical Review*, 1 Aug. (1969), p. 163
E. E. RABEK 'Characteristics of a.c. Powered d.c. Motor Controls Using Three and Six Controllable Elements', *IEEE Trans.*, **IGA-5,** No. 2, March/April (1969), p. 187
J. GREEN 'The Use of Thyristors in the Excitation of Alternators', *IEE Conf. Pub.*, No. 53, Pt. 1, May (1969), p. 433

B. M. EMUNSON 'New NEMA Standards for d.c. Motors for Use on Rectified Power', IEEE 22nd Annual Conf. of Electrical Eng. Problems in the Rubber & Plastics Industries Akron, Ohio, 6–7 April (1970), p. 16
J. K. HAGGERTY 'Application Factors for Thyristor Converter d.c. Motor Drives', *ibid*, p. 73
T. WATANABE 'Transistor Motor, a Brushless d.c. Motor', *Toshiba Review*, Jan./Feb. (1970), p. 50
P. G. BHAGWAT 'Thyristor Commutation for Automatic Speed Control', *Electrical Times*, Feb. (1970), p. 68
H. OTT 'A New Electronic Motor for Tape Recorders', *Siemens Rev.*, **37,** No. 7 (1970), p. 420
R. A. FINK 'The Brushless d.c. Motor', *Control Engng.*, Pt. 1, June (1970), p. 75, Pt. 2, Aug. (1970), p. 42
A. DUNWORTH 'A Precise Digital Control System for a Commutatorless Electric Motor', *IEEE Trans.*, **IECI**, Feb. (1970)
'Thyristors for Induction-Motor Variable-Speed Lifting-Gear Drives', *Electrical Engineer,* Jan. (1970), p. 18
S. J. SHEHEEN 'Shaded Pole Motor Speed Control', *Instruments and Control Systems*, June (1970), p. 118
H. LARGIADER 'Design Aspects of Induction Motors for Traction Applications with Supply Through Static Frequency Changers', *Brown Boveri Review*, (1970), p. 152
L. J. JOCOVIDES 'Analysis of Induction Motor Drives with a Non Sinusoidal Supply Voltage Using Fourier Analysis', IEEE 5th Annual IGA Group Meeting, 5–8 Oct. (1970), p. 467
R. W. JOHNSTON 'Variable Speed Induction Motor Drive System for Industrial Applications', *ibid.*, p. 581
A. C. BETKE 'Special Purpose a.c. Converter Systems for Constant Horse-power Applications', *ibid*, p. 695
P. J. TSIVITSE 'Optimum Voltage and Frequency for Polyphase Induction Motors Operating with Variable Frequency Power Supplies,' *ibid*, p. 815
F. F. MAZDA 'Control of Induction Motors Using Thyristors', *Electronic Components,* July (1970), p. 777
A. JAGER 'Thyristorized Control System for a Two-Frequency Shunting Locomotive of Swiss Federal Railways', *Brown Boveri Review*, **10** (1970) p. 410
R. MEDERER 'Variable Speed a.c. Drives—Wide Choice Available', *Electrical Times*, 10 Sept. (1970), p. 69
S. DENYER 'Thyristors and Their Application to Motor Control Systems', *IEETE Journal*, Sept. (1970), p. 5

R. P. VERES 'New Inverter Supplies for High Horsepower Drives', *IEEE Trans.*, **IGA-6,** No. 2, March/April (1970), p. 121

'Static Excitation System with Thyristor Converter', *ASEA Journal*, **43,** No. 2 (1970), p. 41

I. A ZBOROVSKII 'Automatic Voltage Regulation of Large Synchronous Motors', *Elec. Technol. USSR (GB)*, **2** (1970), p. 142

R. T. SMITH 'Excitation Requirements for Rectified Output a.c. Generators', *IEEE Trans.*, **PAS-89,** No. 8, Nov./Dec. (1970), p. 1910

H. ZANDER 'Power Electronics in the Excitation of Large Synchronous Generators', Conf. on Power Electronics, Budapest, Hungary, 25–27 Nov. (1970) (Budapest Magyar Elektrotechnikai Egyesulet)

R. TUSCHAK 'Development of Semiconductor Excitation System for Synchronous Generators', *ibid*

J. G. BOYD 'Reliability of Electronic Generator Excitation Equipment', Proc. of the American Power Conf., Chicago, Ill., USA, 21–23 April (1970), p. 982

D. J. FEYRER 'Static Adjustable Speed Drives '68, '70, '72', IEEE 22nd Annual Conf. of Electrical Engng. Problems in the Rubber and Plastics Industries, Akron, Ohio, USA, 6–7 April (1970), p. 6

J. N. STRETTON-DOWNES 'Thyristor & Motor Control', *LSE Eng. Bull. (GB)*, **11,** No. 3, Dec. (1970), p. 18

A. M. MARTIN 'Thyristors for Motor Control—Changing Economics', *Electrical Times*, 8 Jan. (1970), p. 62

J. M. LAVALLEE 'The Basis of a Design for a Reversible Drive with Regenerative Braking', *Mullard Tech. Comm.*, No. 105, May (1970), p. 113

J. A. WILKINSON 'Design of Variable Speed Thyristor Controlled d.c. Multi-Motor Drives', *Electrical Review*, 7 Aug. (1970), p. 195

J. S. MICHAEL 'D.C. Motors Need Matching to Thyristor Supplies', *Electrical Times*, 5 March (1970), p. F11

P. G. MESNIAEFF 'Solid State Sets New Trends in Motor Drives', *Control Engineering*, Oct. (1970), p. 73

J. J. BATES 'Using Thyristors and Diodes to Improve Commutation', *IEEE Spectrum*, **8,** No. 1, Jan. (1971), p. 38

M. RAMAMOORTY 'The Transient Response of a Thyristor-Controlled Series Motor', *IEEE Trans.*, **PAS-90,** No. 1, Jan./Feb. (1971), p. 189

D. D. STEPHEN 'Increasing Use of Brushless Synchronous Motors', *Electrical Times*, 23 April (1971), p. 46

B. M. BIRD 'Induction-Motor Speed-Control Schemes Employing Stator-Voltage Variation and Pole-Change Windings', *Proc. IEE*, **118,** No. 3/4, March/April (1971), p. 537

J. FORNASIERI 'Frequency Controlled Drives with Squirrel-Cage Motors', *Brown Boveri Rev.*, **9** (1971), p. 407

E. C. BARNES 'Performance and Characteristics of Induction Motors for Solid-State Variable Frequency Drives', *IEEE Trans*, **IGA-7,** No. 2, March/April (1971), p. 212

'Thyristor Speed Control of a.c. Motors', *Electrical Equipment*, Oct. (1971), p. 24

M. F. ARNOLD 'Thyristor Drives for Motors', *Students Quarterly Journal (IEE)*, June (1971), p. 271

M. SYMES 'A Low Cost Variable Speed Drive', *Automation*, July/Aug. (1971), p. 9

F. F. MAZDA 'Electronic Commutators for Small Motors', *Electrical Times*, 17 Sept. (1971), p. 45

F. F. MAZDA 'What's New in Motors', *Electrical Review*, 12/19 Feb. (1971), p. 227

S. F. WRIGHT 'Automatic Voltage Regulation', *Electrical Equipment*, Jan. (1971), p. 20

W. FAIRNEY 'Developments in Thyristor Excitation of Large Generators', *Electrical Times*, 15 Oct. (1971), p. 37

P. WARTO 'Thyristorized Excitation for Large Turbo-Generators', *Brown Boveri Rev.*, **58,** No. 1, Jan. (1971), p. 41

H. K. LAUW 'Simulation of a Brushless Synchronous Machine with an Analogue Computer', *Ann. Ass. Int. Calcul. Analogique (Belgium)*, **13,** No. 2, April (1971), p. 60

F. W. KEAY 'A Solid-State Regulator for Electric Utility Applications' *IEEE Trans.* **PAS-90,** No. 4, July/Aug. (1971), p. 1537

F. F. MAZDA 'Machine Control', *Electronic Components*, 6 Aug. (1971), p. 853

R. BELL 'Design of Stepping Motor Feed Drives for NC Machine Tools', *Mach. & Prod. Eng. (GB)*, **118,** No. 3035, 13 Jan. (1971), p. 54

E. MEYER 'What are Railways Expecting from the Use of Thyristors in their Electric Locomotives?', *Bull. Ass. Suisse Elect. (Switzerland)*, **62,** No. 1, 9 Jan. (1971), p. 1

W. M. GREER 'Speed Control for Stepper Motor', *IBM Tech. Disclosure Bull. (USA)*, **13,** No. 10, March (1971), p. 3044

D. COOPER 'State of the Art in Power Conversion', *Electron. Equip. News*, Pt. 1, March (1971), p. 38; Pt. 2 , April (1971), p. 54; Pt. 3, May (1971)

T. E. BELING 'Drive System for Small or Large Angle PM Stepping Motors', *Comput. Des. (USA)*, **10,** No. 3, March (1971), p. 77

H. A. SCHEIB 'Application of Solid-State Electronics and Stepping Motors to Optical Printer Drive Systems', *J. Soc. Motion Picture & Televis. Eng. (USA)*, **80,** No. 3, March (1971), p. 193

M. KAPOOR 'SCR Supply for Continuous Rod and Merchant Mills', *Brown Boveri Rev.*, **6,** (1971), p. 242

P. C. SEN 'Single Silicon-Controlled Rectifier Reversible Drive Motor Control', *Int. J. Control*, **13,** No. 1 (1971), p. 65

'Solid State Controls in Domestic Electrical Appliances', *Electrical Times*, **159,** No. 10, 12 March (1971), p. 42

R. REIHS 'Thyristor-Controlled Main Drives for Machine Tools', *Siemens Review*, **38,** No. 3, March (1971), p. 119

L. J. SPALLA 'Current Feedback Drives for d.c. Torque Motors', *Instrum. & Contr. Syst.*, **44,** No. 5, May (1971), p. 106

'Thyristor Motor Control', *Electrical Times*, **160,** No. 7, 13 Aug. (1971), p. 29

H. J. GUTT 'The Design of a.c.—Controller-Fed Induction Machines for Brushless Variable-Speed Drives', *Siemens Review*, **38,** No. 1, Jan. (1971), p. 40

T. A. LIPO 'The Analysis of Induction Motors with Voltage Control by Symmetrically Triggered Thyristors', *IEEE Trans.*, **PAS-90,** No. 2, March–April (1971), p. 515

J. L. WATTS 'Speed Control of Squirrel-Cage and Synchronous Motors', *Mach. & Prod. Eng. (GB)*, **118,** No. 3039, 10 Feb. (1971), p. 217

G. J. BERG 'Speed Change of Induction Motors with Variable Frequency Supply', *IEEE Trans.*, **PAS-90,** No. 2, March/April (1971), p. 495

P. R. BASU 'A Variable Speed Induction Motor Using Thyristors in the Secondary Circuit', *ibid*, p. 509

S. K. DATTA 'A Novel Three-Phase Oscillator for the Speed Control of a.c. Motors', *IEEE Trans.*, **IGA-7,** No. 1, Jan./Feb. (1971), p. 61

S. D. T. ROBERTSON 'Torque Pulsations in Induction Motors with Inverter Drives', *IEEE Trans.*, **IGA-7,** No. 2, March/April (1971), p. 318

E. C. BARNES 'Performance and Characteristics of Induction Motors for Solid-State Variable Frequency Drives', *ibid*, p. 212

B. J. HODGSKISS 'Modern Developments in Electronic Speed Control', *Eng. Dig. (GB)*, **32,** No. 6, June (1971), p. 61

G. T. MOORE 'Selecting Speed Control for Electric Lift Trucks', *Allis-Chalmers Eng. Rev. (USA)*, **36,** No. 1 (1971), p. 12

W. FAIRNEY 'Developments in Thyristor Excitation of Large Generators', *Elec. Times*, **160,** No. 16, 15 Oct. (1971), p. 37

K. NITTA 'Dynamic Response of Separately Excited d.c. Motor Driven by a Thyristor Pulsating Power Supply', *Elec. Eng. Jap.*, **90,** No. 4, July–Aug. (1971)

P. PENNATI 'Frequency Control of Squirrel-Cage Induction Motors', *Ingenieur*, **83,** No. 38, Sept. (1971), p. E97

E. A. KLINGSHIRN 'Optimum Voltage and Frequency for Polyphase Induction Motors Operating with Variable Frequency Power Supplies', *IEEE Trans. Ind. & Gen. Appl.*, **IGA-7,** No. 4, July–Aug. (1971), p.[480

N. SATO 'Adjustable Speed Drive with a Brushless d.c. Motor', *ibid*, p. 539

R. C. LACH 'Paralleling Hot Strip Mill M-G sets with SCR's for Additional Power', *Iron & Steel Eng.*, **48,** No. 8, Aug. (1971), p. 89

J. F. BRESKA 'Adjustable a.c. Variable Speed Drives', IEEE Cement Industry Tech. Conf. Seattle, Wash., USA, 10–13 May (1971)

G. MELLGREN 'Thyristor Converters in Locomotives', *Direct Curr. & Power Electron.*, **2,** No. 2, May–June (1971)

W. B. GISH 'A Thyristor Replacement for Rheostatic Voltage Regulator', *IEEE Trans. Power App & Syst.*, **PAS–90,** No. 5, Sept.–Oct. (1971), p. 2079

M. GOTO 'Transient Behaviour of Synchronous Machine with Shunt-Connected Thyristor Exciter Under System Faults', *ibid*, p. 2218

M. ROSSION 'Some Considerations on Electric Crane Control', *ACEC Revue*, No. 2 (1971), p. 3

J. FORNASIERI 'Frequency-Controlled Drives with Squirrel-Cage Motors', *Brown Boveri Rev.*, **58,** No. 9, Sept. (1971), p. 407

C. BEATSON 'Thyristors Move in to Control the Motors of Today', *Engineer*, **233,** No. 6038, 25 Nov. (1971), p. 45

V. V. SASTRY 'Improved Two-Speed Single-Winding Single-Phase Induction Motors', *IEEE Trans. Power App. & Syst.*, **PAS-90,** No. 3, May (1971), p. 1355

F. KAPPIUS 'Electronic Motors for Industrial Applications', *Siemens Rev.*, **39,** No. 10, Oct. (1971), p. 453

W. T. GRANT 'Electrical Variable-Speed Drives', *Eng. Dig.,* **32,** No. 11, Nov. (1971), p. 59

A. BELLINI 'Digital Simulation of Inverter Fed a.c. Drives', Report n 1–08, Univ. Rome, Italy, (1971)

W. TRAUTNITZ 'D.C. Drives with Thyristor Converters in Shipyard Crane Installations', *Siemens Rev.*, **38,** No. 11, Nov. (1971), p. 508

P. BOWLER 'Power Electronics for the Control of Small Electrical Machines', Advance No. 11, Dec. (1971), p. 47

T. SHIMIZU 'Thyristor-Scherbius System Efficient for Motor Speed Control', *Toshiba Rev.*, No. 63, Nov. (1971), p. 13

P. C. SEN 'On Solid State d.c. Motor Control', IEEE 6th IGA Group Meeting, Cleveland, Ohio, USA, 18–21 Oct (1971), p. 417

T. MAENO 'A.C. Commutatorless and Brushless Motor', *ibid*, p. 25

R. O'TOOLE 'Adjustable a.c. Variable Speed Drives', *ibid,* p. 263

N. SATO 'A Brushless d.c. Motor with Armature Induced Voltage Commutation', *ibid*, p. 35

G. SIDDALL 'Trends in High Power Thyristor Applications', *Elec. Rev.,* **189,** No. 24, Dec. (1971), p. 847

P. G. MESNIAEFF 'Solid-State Adjustable Frequency a.c. Drives', *Contr. Eng.*, **18,** No. 11, Nov. (1971), p. 57

S. KAZI 'Variable Speed Drive by Eddy Current Couplings', *Elec. India*, **11,** No. 10, Oct. (1971), p. 8

B. L. GOSS 'Developments in a.c. Adjustable-Speed Drive Systems', Proc. of the American Power Conference, Chicago, Ill., USA, 20–22 April (1971) (Chicago, Ill., USA: Ill. Inst. Technol.), p. 369

D. SIMMS 'Thyristor Control to Advance through Technical Ability', *Contr. & Instrum.*, **4,** No. 1, Jan. (1971), p. 37

R. G. DANCY 'Selecting the Correct Motor Speed Control System', *Contr. & Instrum.*, **4,** No. 1, Jan. (1972), p. 33

J. M. D. MURPHY 'Solid-State Supplies for a.c. Variable Speed Drives', *Focus*, **161,** No. 3, 20 Jan. (1972), p. 19

T. PETERSON 'Starting of Large Synchronous Motor Using Static Frequency Converter', *IEEE Trans. Power App. & Syst.*, **PAS-91,** No. 1, Jan.–Feb. (1972), p. 172

P. W. FRANKLIN 'Theory of d.c. Motor Controlled by Power Pulses', *ibid*, p. 249

A. C. BETKA 'Special-Purpose a.c. Converter Systems for Constant Horsepower Applications', *IEEE Trans. Ind. Appl.*, **IA-8,** No. 2, Mar.–Apr. (1972), p. 126

F. F. MAZDA 'Motor Control Using Thyristors', *Instrumentation & Control Engineering*, Part 1, Feb. (1972), p. 14, Part 2, Mar. (1972), p. 17

W. C. BEATTIE 'Speed Control of d.c. Shunt Machine with Digital Circuitry', *Control & Instrum.*, May (1972), p. 47

CHAPTER 12

J. F. YOUNG 'Timers Using Unijunction Transistors', *Electronic Engng.*, Oct. (1967), p. 621

R. E. LITTLE 'How to Prevent Fatigue Failure', *Machine Design*, **39,** July (1967), p. 137

W. STEFFE 'Avoiding Second Breakdowns in Power Transistors', *Electronic Engineer*, Dec. (1967), p. 65

R. MINTON 'Design Trade-Offs for r-f Transistor Power Amplifiers', *Electronic Engineer*, 26 March (1967), p. 68

S. L. MARSHALL 'Second Breakdown in Transistors', *Semiconductor Products and Solid-State Tech.*, **10,** March (1967), p. 25

W. C. STEFFE 'Secondary Breakdowns in Power Transistors and Circuits', *IEEE Internat. Conv. Rec.*, pt. 6, **15,** (1967)

C. TURNER 'Carl Turner of RCA Explores Selection of Second-Breakdown-Resistant Transistors', *EEE—Circuit Design Engng.*, **14,** July (1967), p. 8L

R. J. BUCZYNSKI 'Commutating dv/dt and its Relationship to Bidirectional Triode Thyristor Operation in Full-Wave a.c. Power Control Circuits', IEEE Conf. Rec. of IGA Group, Pittsburg, (1967), p. 45

U. FLACHMANN 'A Large Water-Cooled Silicon Rectifier Installation of the Compact Type for an Aluminium Electroysis Plant', *Siemens Review*, (1968), p. 97

W. R. SPOFFORD 'Unlocking the Gates for UJT's', *Electronics*, April (1968), p. 56

R. M. MARSTON '20 Unijunction Transistor Application', *Radio-Electronics*, Pt. 1, June (1968), Pt. 2, July (1968), p. 38

D. E. LANCASTER 'Plastic Power Transistors', *Electronics World*, Feb. (1968), p. 50

D. M. BAUGHER 'Cut Down on Power Transistor Failures', *Electronic Design*, 9 May (1968), p. 72

D. COOPER 'Triacs Bid for a.c. Power Control', *Electro-Technology*, March (1968), p. 53

S. Wm. FIELDS 'The Triac—From Trickle to Triumph in Three Years', *Electronics*, 19 Feb. (1968), p. 169

D. COOPER 'Bi-Directional Thyristors for a.c. Power Control', *Electrical Review*, 15 Nov. (1968), p. 718

D. COOPER 'SCR's and Triacs—The Revolution Continues', *Electronics World*, Aug. (1968), p. 25

D. COOPER 'A.C. Power Control-Triac or SCR's?', *Control Engng.*, Aug. (1968), p. 80

R. GUNDLACH 'Rx for rf Power Transistors', *Electronics*, May (1969), p. 84

N. R. HOWARD 'p-n-p-n Switches', *Electronic Components*, Jan. (1969), p. 57

D. COOPER 'Power Logic Triac', *Electronic Components*, Pt. 1, April (1969), p. 401. Pt. 2, May (1969), p. 585

J. GAULT '200 Ampere Power Logic Triac', *Solid State Technology*, Feb. (1969), p. 52
F. W. PARRISH 'Applying Power Logic-Triacs', *Machine Design*, April (1969), p. 149
'Triacs in Control', *Instrument and Control Engineering*, Feb. (1969), p. 14
R. FASEL 'Superconducting Rectifiers', Low Temperature & Electric Power Conf. London, 24–28 March (1969) Pergamon, p. 303

H. A. HORST 'Water-Cooled Silicon Rectifiers of the Compact Type for Electrolytic Plants', *IEEE Trans.*, **IGA-6,** No. 4, July/Aug. (1970), p. 361
W. H. SCHROEN 'Characteristics of a High-Current High-Voltage Shockley Diode', *IEEE Trans.*, **ED-17,** No. 9, Sept. (1970), p. 694
Y. C. KAO 'The Design of High-Voltage High-Power Silicon Junction Rectifiers', *ibid*, p. 657
P. POLGAR 'A High-Current Metal-Semiconductor Rectifier', *ibid*, p. 725
R. DENNING 'Epitaxial $\pi - \nu$ n-p-n High Voltage Power Transistors', *ibid*, p. 711
G. A. LANG 'Thermal Fatigue in Silicon Power Transistors', *ibid*, p. 787
T. C. NEW 'High Power Gate Controlled Switch', *ibid*, p. 706
'Mounting Techniques for Press-Fit Rectifiers', *Electronic Components*, 20 Nov. (1970), p. 1313
L. L. LEHNER 'The Ubiquitous Power Transistor', *Electronics & Power*, Jan. (1970), p. 19
L. LEHNER 'Silicon Power Transistors—Design and Application', *Electro-Technology*, Feb. (1970), p. 29
L. MAGILL 'Watt-Megahertz Ratings Run Second to High Reliability in Foreign rf Power Transistors', *Electronics*, April (1970), p. 80
P. FRANSON 'Power Transistors, a Status Report', *Electronics World*, May (1970), p. 25
O. GREITER 'PUT—The Programmable Unijunction', *Wireless World*, Sept. (1970), p. 430
'Diac Triggering of Thyristors & Triacs', *Miniwatt Digest*, March/April (1970), p. 25
J. BUDEK 'Solid-State Switching Using Triacs and Thyristors', *Design Electronics*, May (1970), p. 59
J. HADAS 'Design for Reliability of Large Semiconductor Rectifier Apparatus', Conf. on Power Electronics Budapest, Hungary, 25–27 Nov. (1970) (Budapest: Magyar Elektrotechnikai Egyesulet)
L. FEJES 'Structure of Silicon Planar Transistors with High Breakdown Voltage', *Hiki (Hungary)*, **10,** No. 4 (1970), p. 13
C. W. GWYN 'Modeling of the Saturation Characteristics of High Voltage Transistors', *IEEE Trans.*, **NS-17,** No. 6, Dec. (1970), p. 63
D. NIEHAUS 'Neutron Hardness Assurance for Power Transistors', *ibid*, p. 154
G. ROSENBERG 'Statistical Analysis of Neutron-Induced Gain Degradation of Power Transistors', *ibid*, p. 160

'Novel Unijunction Transistor Circuit for High Accuracy Time Control', *Automation*, July/Aug. (1971), p. 13

G. D. GRIMSDELL 'Class D Power Transistor Circuits for Motor Control', *Electrical Times*, 28 May (1971), p. 37

J. H. JOHNSON 'You Can Depend on Today's rf Power Transistors', *Electronics*, Sept. (1971), p. 90

N. GRI 'Microwave Transistors—From Small Signal to High Power', *Microwave Journal*, **14,** No. 2, Feb. (1971), p. 45

M. J. SHILLING 'Thermal Cycling Rating System for Silicon Power Transistors', *New Electronics*, Dec. (1971), p. 36

C. D. MOTCHENBACHER 'Protect Your Transistors Against Turn-On or Testing Transient Damage', *Electronics*, Dec. (1971), p. 92

'Programmable Unijunction Transistors (PUT's), *Electronic Components*, Dec. (1971), p. 1275

D. R. GRAFHAM 'Thyristor Control of a.c. Power', *Electrical Review*, Pt. 1, 14 May (1971), p. 663. Pt. 2, 21 May (1971), p. 694

J. M. DIAMOND 'Triac Phase Control with a Line Frequency Control Signal', *Proc. IEE*, **118,** No. 1, Jan. (1971), p. 135

'Two Thyristors Make a Triac—and Bring Costs Under Control', *The Engineer*, 20 May (1971), p. 35

B. A. BLAND 'High Voltage Silicon Rectifier Stacks and their Applications', *Mullard Tech. Commun.*, **11,** No. 110, March (1971), p. 210

S. RUBIN 'Thermal Hysteresis and its Possible Effect in Restricting the Hotspot Free Operating Range of Some Power Transistors', *IEEE Trans.*, **ED-18,** No. 6, June (1971), p. 393

J. OLMSTEAD 'High-Level Current gain in Bipolar Power Transistors', *RCA Rev.*, **32,** No. 2, June (1971), p. 221

'Development of Power Semiconductor Switching', *Electrical Times*, **159,** No. 20, May (1971), p. 34

H. Q. TSERNG 'Performance Degradation of Bidirectional Triode Thyristors Due to di/dt Stress', *Solid-State Electronics (GB)*, **14,** No. 2, Feb. (1971), p. 111

H. K. GOSS 'Operation of Power Triacs', *Electron. Equip. News,* **13,** No. 2, May (1971), p. 20

A. D. COX 'Pedestal Control in a Diac-Triac Circuit', *Electron. Lett.*, **7,** No. 25, 16 Dec. (1971), p. 756

'Hotspots in Transistors', *Electron. Aust.*, **33,** No. 7, Oct. (1971), p. 18

J. De WARGA 'The Design, Application and Projection of a New 48 mm 2200 amp. Power Semiconductor', IEEE Internat. Electrical and Electronics Conf. Toronto, Ontario, Canada, 4–6 Oct. (1971), p. 82

D. J. FRANKLIN 'Power-Transistor Design Considerations for Better Forward-Secondbreakdown ($I_{s/b}$) Characteristics', IEEE 6th Annual IGA Group Meeting, Cleveland, Ohio, USA, 18–21 Oct. (1971), p. 851

P. P. WANG 'Thermal Instability of Multiemitter Power Transistors', *ibid*, p. 861

P. CHOWDHURI 'Transient-Voltage Characteristics of Medium-Power Silicon Rectifiers', *ibid*, p. 781

P. CHOWDHURI 'A Study on the Impulse Breakdown Phenomena of Silicon Power Rectifiers', *ibid*, p. 791

N. P. MOLIBOV 'The Turn-On Process for a p-n-p-n Structure with a Field

Region in the Emitter', *Radio Eng. & Electron. Phys.*, **16,** No. 4, Apr. (1971), p. 657

S. W. KESSLER Jr. 'Transcalent Silicon Power Rectifier', *IEEE Trans. Aerosp. & Electron. Syst.*, **AES-7,** No. 6, Nov. (1971), p. 1151

B. SEXTON 'Applications of Solid-State Power in Britain', *New Zealand Elec. J.*, **44,** No. 12, 25 Dec. (1971), p. 239

J. LANGE 'An Application of Device Modelling to Microwave Power Transistors', *IEEEJ. Solid State Circuits*, **SC-7,** No. 1, Feb. (1972), p. 71

W. HOEFT 'A Monolithic Power Driver for Complementary Power Transistors', *IEEE Trans. Broadcast & Telev. Receivers*, **BTR-18,** No. 1, Feb. (1972), p. 59

F. F. MAZDA 'Basic Semiconductor Power Devices', *Electrical Review*, Part 1, 14 Jan. (1972), p. 47, Part 2, 1 Jan. (1972), p. 83

C. R. WRONSKI 'GaAs Vapour-Grown Shockley Diodes and Semiconductor-Controlled Rectifiers', *IEEE Trans. Electron Devices*, **ED-19,** No. 5, May (1972), p. 691

V. J. LUKACH 'Thermal Cycling Ratings of Power Transistors', IEEE Proc. of 1972 Annual Reliability and Maintainability Symposium, San Francisco, Calif., USA, 25–27 Jan. (1972), p. 492

Index